Principles of Engineering Manufacture

Principles of Engineering Manufacture

THIRD EDITION

Stewart C. Black   BSc, MSc, CEng, FIEE, FIMechE
Principal Lecturer in Manufacturing
Department of Mechanical Engineering and Manufacturing Systems
University of Northumbria, Newcastle upon Tyne

Vic Chiles   CEng, PhD, BSc, MIEE
Senior Lecturer in Manufacturing
Department of Mechanical Engineering and Manufacturing Systems
University of Northumbria, Newcastle upon Tyne

A. J. Lissaman   CEng, MIMechE, FIProdE
Formerly Head of Department of Production Engineering
North Gloucestershire College of Technology, Cheltenham

S. J. Martin   CEng, FIMech, FIProdE
Formerly Principal Lecturer in Production Engineering
North Gloucestershire College of Technology, Cheltenham

A member of the Hodder Headline Group
LONDON • SYDNEY • AUCKLAND

Copublished in the Americas by Halsted Press
an imprint of John Wiley & Sons Inc.
New York – Toronto

First published as *Principles of Engineering Production* in Great Britain 1964
Second edition 1982
Third edition published 1996 by Arnold,
a member of the Hodder Headline Group,
338 Euston Road, London NW1 3BH

British Library Cataloguing in Publication Data
A catalogue record for this book is available for the British Library

Library of Congress Cataloging-in-Publication Data
Principles of engineering manufacturing / Vic Chiles . . . [et al.]. –
3rd ed.
p.   cm.
Includes bibliographical references and index.
ISBN 0-470-23558-6 (pbk.)
1. Production engineering.   I. Chiles, Vic.
TS176.P717   1995
670.42–dc20   95-21367

ISBN 0 340 63195 3
ISBN 0 470 23558 6 (in the Americas)

Typeset in 10/13 Times by MCS Ltd, Salisbury
Printed and bound in Great Britain by
J W Arrowsmith Ltd, Bristol

Contents

x   Contents

Preface to the Third Edition

Principles of Engineering Production has been a standard textbook for manufacturing engineers since its first edition in 1964. Messrs Lissaman and Martin revised, updated and enlarged the text in the second edition in 1982 to reflect the major developments, in machine tools, and the move to metrication.

The stated aim of the book was to help students obtain a first appreciation of some important aspects of engineering manufacture, and in this it succeeded. Thirty years have elapsed since the first edition, and obviously the nature of manufacturing has changed considerably during this time. Automation systems have gone through a major revolution, and machining techniques have continued to develop.

This third edition does not focus on manufacturing systems, JIT, etc., but introduces the reader to a wide range of manufacturing processes. JIT principles necessitate the selection of appropriate processes for varying manufacturing situations, and we have tried to bring out the technological aspects of this throughout this text. Changes in the industrial scene, with the decline in engineering and the increased mechanisation of other areas of manufacturing, have made it necessary to consider the wider field, not just engineering production, but engineering principles applied to manufacturing in general, whatever the product.

The changing nature of the engineering industry and the curriculum in schools has also had its effect on the type of student entering higher education. Whereas the production or engineering student used to arrive at university with a basic knowledge of engineering machinery, this knowledge now seems to have been channelled into the field of computing, and it has been necessary to introduce a larger content of explanation of basic techniques than previously.

In creating this edition we have been faced with severe problems as to which subject areas to expand; we have tried to make the best use of the space available and provided details of the fundamental principles behind each topic.

Many companies have helped us with this new edition, and we are grateful to them all. We would mention in particular Black & Decker, Renishaw, Rhodes and Traub, and special mention must be made of Sandvik for their liberal assistance, including permission to draw on their recent book *Modern Metal Cutting*.

It is our hope that this reworking of an established textbook will make it valuable to a new generation of students, and will help them to apply established techniques and principles to a wide range of manufactured product.

S. Black
V. Chiles
Newcastle, November 1995

Preface to the Second Edition

This book treats technical aspects of manufacturing with respect to metal machining and press-forming. Starting from a consideration of specification and standardisation, it goes on to deal analytically with the main aspects of the manufacturing processes giving due attention to the crucial matters of quality and cost.

The new edition, in SI units, is an enlarged revised version of the original book which first appeared in 1964. It incorporates the many changes necessitated by the metrication and revision of British Standards; all the relevant standards up to 1980 have been consulted.

Since the book first appeared there have been major developments in machine tools. This edition incorporates a new chapter, 'Control of Machine Tools' which gives a substantial introduction to numerical control and programming. A further new chapter deals with electro-discharge and electrochemical methods of machining, and the chapter on 'Statistical Methods of Process Control' has been extended to cover control by attributes. A Bibliography is added at the end of the book, listing further reading likely to be of interest to students.

Eight printings of the original work show that it met a real need. While courses leading to the higher engineering qualifications have changed considerably since 1964, there is now a growing awareness that such courses ought to include some consideration of manufacturing technology in order better to meet the needs of industry. Since this is a diverse subject involving considerable practical detail students may have difficulty in gaining a useful knowledge of the basic principles within the limited time available.

This book is designed to help 'A' level entrants to higher diploma and degree courses obtain a first appreciation of some important aspect of engineering manufacture. It should also be of service during their periods of industrial experience.

It is hoped that the extensive updating of this edition with respect to British Standards will again make the book useful as a reference for mature engineers.

The authors and publishers would like to express their thanks to firms which have supplied data and illustrations. They are particularly indebted to the British Standards

Institution, 2 Park Street, London W1A 2BS, for permission to reproduce extracts from their publications. Copies of Standards may be obtained on application to the Institution.

The specially drawn diagrams featured in the book have been prepared by Mrs E. M. Harris and the authors are extremely grateful for her valuable assistance. They also thank the Principal and Librarian of the North Gloucestershire College of Technology for allowing access to British Standards and other reference material held in the College library.

A. J. Lissaman
S. J. Martin

Extract from the Preface to the First Edition

Engineering manufacture is a diverse economic activity embracing all the work lying between a design and its execution. It calls for decisions which, if they are to be wisely made, ought to have a rational basis.

The *Principles of Engineering Production* ought to satisfy two criteria:

1. They should be developed logically from the elements of manufacturing activities.
2. Their application should tend to improve the quality of the work produced, or to lower its cost.

This book aims to develop and illustrate some important principles underlying engineering manufacture, principles which the authors believe come near to satisfying the above criteria. They are principles of wide application and apply equally to batch work and to large quantity production involving automatic machinery.

The text has been developed mathematically wherever appropriate, and it is hoped that the treatment will stimulate the teaching of the subject, as well as capture the interest of students. Mature engineers engaged in manufacture should find in this book much to guide and assist them in the analysis and solution of their day-to-day problems.

In a work deliberately planned to introduce greater rigour into the treatment of its subject, two difficulties of presentation have confronted the authors. A reasonably consistent set of symbols has had to be adopted for use throughout the book, and this has led to certain topics, e.g. Merchant's Theory of Cutting, appearing in symbols which differ from those of the original research papers. In order to illustrate certain points by means of worked examples, it has sometimes been necessary to over simplify practical detail in the interest of conciseness. It is hoped that readers will accept that the authors have pondered a great deal over both difficulties and that their decisions, however imperfect, solve in some degree both problems of presentation.

A. J. Lissaman
S. J. Martin

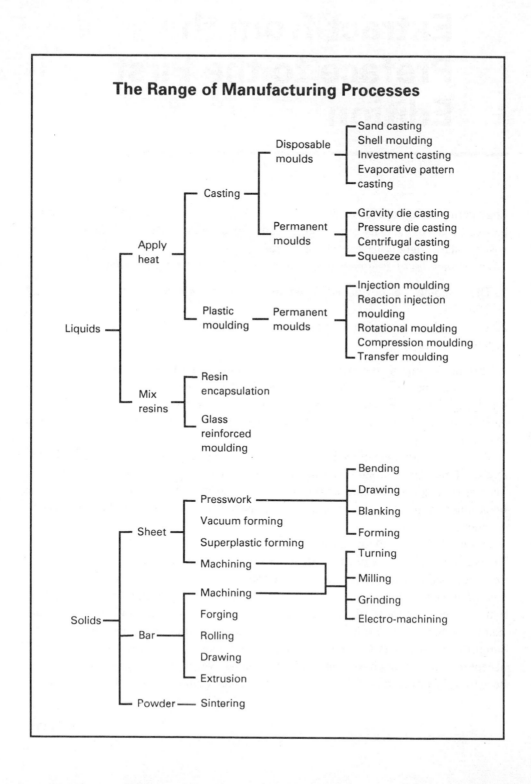

The Range of Manufacturing Processes

- **Liquids**
  - **Apply heat**
    - **Casting**
      - **Disposable moulds**
        - Sand casting
        - Shell moulding
        - Investment casting
        - Evaporative pattern casting
      - **Permanent moulds**
        - Gravity die casting
        - Pressure die casting
        - Centrifugal casting
        - Squeeze casting
    - **Plastic moulding**
      - **Permanent moulds**
        - Injection moulding
        - Reaction injection moulding
        - Rotational moulding
        - Compression moulding
        - Transfer moulding
  - **Mix resins**
    - Resin encapsulation
    - Glass reinforced moulding

- **Solids**
  - **Sheet**
    - Presswork
      - Bending
      - Drawing
      - Blanking
      - Forming
    - Vacuum forming
    - Superplastic forming
    - Machining
      - Turning
      - Milling
      - Grinding
      - Electro-machining
  - **Bar**
    - Machining
      - Turning
      - Milling
      - Grinding
      - Electro-machining
    - Forging
    - Rolling
    - Drawing
    - Extrusion
  - **Powder**
    - Sintering

1 Manufacturing

1.1 Introduction

The use of tools and skills to make items to improve the environment is one characteristic which separates man from animal. Another is the ability to organise oneself and others to obtain a better standard of life for everyone. These two aspects combine together in the area we term 'Manufacturing'.

In this book we are looking at the science of making things. We are not studying a particular type of product, or the handling of particular materials, but are looking at the art of making, and the processes we look at can be applied to many materials, and many situations. We are looking also at the way in which we organise the utilisation of these processes, and the basic structures we create to facilitate this organisation.

No matter what material we start from, the operations to make our product will fall into one of two basic categories. Either we will be deforming our raw material, or we will be cutting it, i.e. removing material from it, but it is important to remember that the finished product will be made in a series of operations, and will incorporate techniques from each area.

Modern manufacturing encompasses an ever-increasing variety of processes (see the diagram opposite this page), and the engineer's challenge is to select the most economic combination of processes to make a product of high quality at the right price. Products range from high-priced aerospace items, such as jet engines, which are made in low volumes, to everyday items such as razor blades which are made by the million, in an extremely competitive market place. To meet this challenge the manufacturing engineer needs to have a broad knowledge of the ways in which materials can be processed, and the shapes that can be formed by these processes.

In order to introduce this concept of process variety, we will examine a common consumer product – in this case a Black & Decker hedge trimmer (see Figure 1.1). This is a typical mass-produced consumer product using a surprisingly wide range of processes in its manufacture. The requirements are wide-ranging, and are satisfied by a variety of metal and plastic parts, metal for strength, conductivity, and wear resistance, plastic for weather protection and insulation, and overall there is a

Figure 1.1 Hedge trimmer: a typical consumer product.

consideration of the aesthetic appearance and ergonomic design. There is not space to examine each part in detail, so only some of the major parts (Figure 1.2) are considered, as listed in Table 1.1. Only a few parts are listed and a rough description of the materials and operations. The reader should examine everyday products to see how varied the requirements are.

It is obvious that a great deal of know-how and experience is required for the successful design of such a product, and to continuously improve it and the methods used in its manufacture, to maintain a successful presence in a competitive market.

In the succeeding chapters we will introduce the reader to the variety of processes available, and their basic principles, and then will study the main processes in greater depth.

Table 1.1 Component parts of a hedge trimmer.

Component	Requirement	Material	Process
1. Body shell	Insulation and weatherproof	Plastic	Injection moulded
2. Handle	Good grip, strong and rigid	Plastic	Injection moulded
3. Finger guard	Safety, semi-rigid	Plastic	Injection moulded
4. Motor	Provides power	Steel	Punched laminations
(This is a sub-	Conductor, strength,	Copper	Insulated wound coils
assembly produced	support/shape, insulation	Steel	Machined shaft
on automatic		Plastic	Moulded frames
machines)		Paper	Cut and folded
5. Gearbox	Provides accurate location	Zinc alloy	Die cast
	and bearings for gears	Copper	Sintered
6. Switch	Insulation, contacts	Plastic	Moulded
		Copper	Pressed
7. Gears	Strength and wear resistance	Steel	Machined and cut
8. Blades	Sharpness and wear resistance	Steel	Stamped and ground
9. Gearbox cover	Keep out dirt and keep in grease	Steel	Pressed
10. Screws	Tensile strength	Steel	Headed and rolled
11. Blade guides	Rigidity, wear resistance	Steel	Pressed

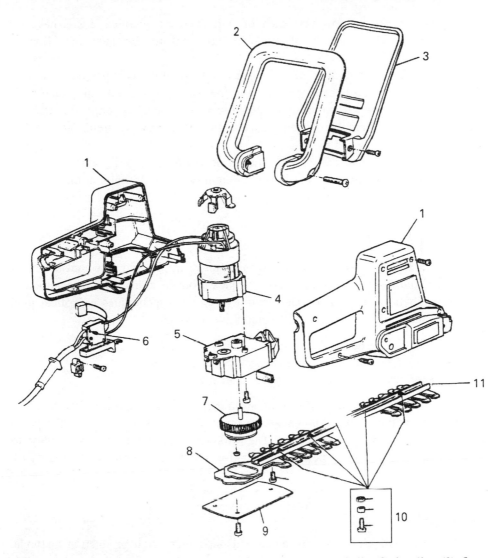

Figure 1.2 Component parts of a hedge trimmer: (1) body shell, (2) handle, (3) finger guard, (4) motor, (5) gearbox, (6) switch, (7) gears, (8) blades, (9) gearbox cover, (10) screws, (11) blade guides.

1.2 Types of production

There are three basic types of production, characterised by the way in which the operations, and the control of the operations, take place.

In *job production* the product is built by an operator (or group of operators) working on and completing an item before proceeding to the next job, which may be different or may be another of the same type.

In *batch production*, however, each of a number of identical items have an operation completed on each of them in turn before the next operation is commenced.

In *flow production* each item of a number of identical items is passed to another operator for the second operation as soon as the first operator has completed their operation on it, and so on through a number of operators until the item is complete, the first item being completed while the following items are still in process.

It should be noted that the type of production is not necessarily associated with any particular volume of production, and different types of production may be used at different stages of manufacture of a product.

Job production

Where technology involved is low, its organisation is simple, and may consist simply of a 'make complete' instruction, but on complex products a project control structure may be required, with planning and control using such techniques as critical path analysis. Whatever type of control is required, there must be:

1. Clear definition of objectives.
2. Agreement on quantifiable results at specified times.
3. Supervision which is empowered to take decisions.

The working group may be made up of a number of skills. Generally the level of operator skill required in job production is higher, since operators must be capable of working with minimum instruction. The worker may have a dual responsibility: to the job supervisor for work performance and to the skill supervisor for quality of workmanship.

Batch production

As quantities increase, work may be carried out under batch production methods. This requires that the work be broken down into a series of operations, suited to skills or equipment, and that each operation be completed on all parts before the next operation starts. This is the most common technique used in manufacturing, typically a machine being set for an operation and all parts scheduled for that operation have that operation completed before the lot are moved on for the next operation, either on the same machine after resetting, or on another machine. Thus only one part is in work at any time, all the others of the batch being at rest.

Batch production is inherently flexible. Deliberate intervals may be built into the schedule between operations to allow for the queueing of the work prior to key operations, this stock of work ensuring that the machine always has work waiting for it, and idle time is eliminated. It also makes it possible for priorities to be set such that late or rush work can 'jump the queue' while low priority work can be used as a filler.

Batch operation will inevitably result in a longer time elapsing between origination and completion of the first item than would have been the case with jobbing production, due to the rest periods for each unit while work was carried out on other units within the batch, and by the time spent between operations. This is, however, offset by reduced setting time, particularly if the set-up is complex, as one lot of setting serves for the whole batch, and by increased operator efficiency as each operator gains experience through the batch. The effect of this time increase will be to increase the capital tied up in work in progress. The presence of the buffer areas, however, allows the production area to absorb uneven loading and increase utilisation.

It is normal in batch production to have plant laid out by function, and for the parts to pass from area to area as the operations demand. This incurs delay in transporting the parts from operation to operation, with lost time when transport is not available, and even loss of materials if locations are not adequately recorded and movements properly controlled.

Flow production

If all rest periods are eliminated giving continuous value addition, the result is flow production. As the work on each unit is completed it is passed to the next stage immediately, and the next stage of work commences without any waiting. This requires that the machines or work stations must be arranged adjacent to those of the previous operation, and in accordance with the sequence of operations to be performed. To ensure flow, all operations must be of equal length, and all operations must be carried out within the flow, including any inspection required.

To make flow production economic there must be substantially continuous demand, either in fact or by arranging for storage of output during low demand periods. The latter can result in a financial penalty in the form of storage costs. The product itself must be standardized, as a flow line by its nature cannot tolerate significant variations. Materials used must be to specification and be available at the right time. In a flow line there is no time to fit a part that is difficult: all components must assemble easily. There is no time for rectification and all parts must be on hand as required, since any delay reverberates along the whole line.

Stages must be balanced accurately. The output of a line is governed by the throughput of its slowest stage, and if the stages are not matched some stages will have idle time. This applies as much to actual as to allowed time: a line may in theory be perfectly balanced but a slow operator will slow the whole line, not only curtailing output but causing frustration to other operators.

Operations must be closely defined. Method analysis is essential in the setting up of a flow system and operators must be trained to work rigidly to the prescribed methods. Work must conform to quality standards. If an operation is incomplete or substandard it cannot be rectified, as can be done in batch conditions, since subsequent operations require its immediate presence. It is a requirement therefore that tooling is maintained to a high standard, and operators work accurately and consistently.

Equipment must be provided for each operation. It cannot be shared, since work must move forward continuously. This may well result in underutilisation of equipment, and if the equipment is extremely costly, and the work flow cannot be organised so that all operations requiring it are carried out at the one stage, the cost of duplication may be so great as to render flow production uneconomic.

Maintenance must be of a high standard, and carried out outside working hours, since breakdown or stoppage of a machine will bring the line totally to a halt. Preventitive maintenance must be practiced, and possibly standby equipment and extensive spares stock will be required.

The advantages of flow production are several:

1. The total labour content will be reduced by virtue of increased operator efficiency brought about by the high degree of planning and tooling involved.
2. Product reproducibility will be improved, as will quality, for the same reasons.
3. WIP will be at a minimum, reducing the working capital required. The need for close scheduling will similarly reduce the raw material stocks.
4. Handling costs are reduced, and potential damage through mishandling is eliminated.
5. Shop loading and shop controls are simplified. Weaknesses in plant, method, materials, or personnel are quickly highlighted for action.

Continuous production

This type of concept may be extended to give continuous production, with the line running 24 hours a day, 7 days a week, generally in process industries, e.g. oil refining, although the use of automation to replace human operators is extending this into more usual product areas. Under such conditions the capital investment in plant is very high. Maintenance must be carried out 'live' or at least with the minimum of down time, or delayed to planned shutdown periods.

'Jobbing production' and 'mass production'

It is important that these terms should not be confused with the terms 'job production' and 'flow production' used above.

'Jobbing production' is a term for production carried out solely against non-recurring, or potentially non-recurring, customer order.

'Mass production' refers only to the scale of production, and makes possible large investment in jigs and fixtures, and the dedication of machine tools, thus saving both operation and setting times, or may even justify the building of special purpose machines, or the complete automation of the processes.

Either may use any or all of the types of production outlined above. Typically component parts of a product would be produced under batch control, while the final assembly may be made under flow conditions.

Group technology

This modern development, combining some of the characteristics of flow production with those of batch production, is sometimes considered as a production technique in itself.

Products are grouped into families having similar characteristics, and facilities are allocated to perform the operations relating to these characteristics. These are laid out in an area specific to the group, like a flow line, and a team of operators are allocated. Groups are arranged to function independently, and work is processed from group to group using batch control techniques. Operations which do not fit into a group are carried out separately under batch control.

Group technology enables the advantages of the flow line to be obtained within a batch controlled environment and over a variety of products.

1.3 Manufacturing economics: time and cost estimates

Manufacturing must be carried out at the lowest cost consistent with the quality and functionality of the product. The cost of any manufactured item is made up of three factors:

1. Material cost.
2. Cost of direct labour.
3. Overheads.

The first of these is easily determined for each product going through the factory, since the Bill of Materials should list all materials required, and the purchasing department should be able to provide current prices for these.

In manufacturing parlance, material is everything which is purchased for incorporation into the product which leaves the factory. Thus within the material cost area we can have:

1. *Raw materials*, i.e. basic materials on which work will be carried out in order to make the components which will become parts of the finished product.
2. *Purchased components*, i.e. parts which will be incorporated in the product, but which are not manufactured in-house. These may be further divided into:
   (a) *Proprietary parts* i.e. parts which are product of another manufacturer as part of their product range.
   (b) *Sub-contracted parts* i.e. parts which are manufactured to the company specification and requirements by an external supplier. Parts may be sub-contracted as a matter of policy, or may be sub-contracted only on occasion when the internal work load is excessive.
3. *Packaging*. It is often forgotten that the packaging of a product forms, for cost purposes, part of the product itself.

The cost of direct labour should be equally available, as the work content and class of labour required should be available from the planning sheets defining methods and allowed times. By direct labour is meant that labour which is directly concerned with the manufacture of the product, i.e. the cost of the people who are *actually performing work* on the product. This cost is not merely the hourly rate paid to the employee factored by the amount of time he is involved with the product, but is the *real* cost to the company of that employee, i.e. an hourly rate which incorporate the employee's holiday payments, National Insurance contributions, sick leave entitlement, and other welfare provisions.

It is in the third area, that of allocation of overheads to a specific product, that difficulty occurs. In order to function, the company must have some auxiliary functions and facilities which are not directly concerned with the actual manufacture of the product. These facilities may be shared by a number of products, and their total cost will be divided among the product ranges and then passed down as a charge against each item produced. Typical of these costs are the cost of providing the buildings and their services, the cost of providing supervisors and managers, the cost of providing a purchasing department, and all the other service departments. These provisions are ongoing, in that they are expenses generated during each period.

The method of allocating these overheads varies from company to company. The traditional technique is to express these as a percentage of the direct labour bill, and then use this percentage to allocate the overheads in proportion to the labour cost for the product. However, if one product is heavily mechanised, and another is labour intensive, allocation on this basis will penalise the labour-intensive product, so other means of allocation of overheads must be found.

Some items of expenditure related to a specific product will occur only at the inception of the product, or at spot instances during the product life. Examples of this are the design costs, the tooling costs, and the purchase of machines specifically for that product. Logic requires that these costs are recovered within the cost of the product by making a charge on each item produced. This is referred to as *amortisation*. The technique is simple: the total cost of the design and tooling are divided by the number of items the manufacturer expects to produce, and the result forms an overhead on the product.

Departmental overhead

Typically, in a small company overheads will be charged on an overall basis, but in a bigger company it is necessary to determine more accurately the distribution of costs, and overheads will be divided into those which are specific to each department, and those which are company-wide.

From the above it will be seen that the cost of manufacture is made up of (material + labour + departmental overheads + general overheads). The objective must be to arrive at a method of overhead allocation which fairly reflects the utilisation of these facilities in the manufacture of the product concerned. In a labour intensive department it is logical to allocate the cost to product on the basis of manhours used, but in an automation intensive department it is more logical to allocate on the basis of machine hours used. In some operations it may be of benefit to vary the technique from department to department, with a view to arriving at a product costing which reflects the true cost of producing the product.

Budgetary control

The ability to establish the historical cost of product is of interest to the accountant in determining the past performance of a manufacturing operation, but is of little use to the manager, who requires a continuous feedback of performance of the unit. The techniques of budgetary control provide the tools for this.

Forecasts are required of the performance of the operation, and these are then used as a yardstick against which the actual performance can be measured on a month-to-month or week-to-week basis.

The basis of the budget or business plan is the sales forecast. From this the manufacturing personnel obtain details of what sales are expected, and when, and can build up their own plan showing the labour and material requirements, and the plant and equipment required to support this. This plan can in its turn be translated into a budget for the operation by the application of forecast labour rates and forecast material costs, and applying the expected departmental and administrative overheads. The difference between these costs and the value of the forecast product will then indicate the expected profitability of the operation, and this can be set against the actual expenditure to give an immediate measure of performance. In reality there will be fluctuations from period to period, particularly if the period is short, and the normal technique is to measure not only the performance during the period, but also the year-to-date performance.

Because in most manufacturing the costs of materials will fluctuate in the course of a year, and there will be wage increases, and more importantly there will be efficiency variations in the manufacturing processes, it is normal to use set up standards of time and cost valid for the period of the budget, and to use these in the budget calculations. This has the value of establishing notional costs to the product which can be of help in determining selling prices, but in addition it provides earlier warning of potential problems, and enables early corrective action, since the variations from the standards are immediately obvious.

1.4 Safety in manufacturing

Since safety regulations vary from country to country, and from time to time, only the principles of safe working will be considered.

Safety in the workshop can be divided into three categories, the enclosing of dangerous machinery, the provision of safety equipment, and the promotion of safe working practices.

Enclosure of machinery

Moving parts of machinery should be enclosed to prevent the accidental entrapment of hands or other parts of the body in them, or wounding by them. Thus cutters will be enclosed except for the aperture through which cutter accesses the workpiece, while gear trains and drive belts will be fully enclosed. Where it is necessary for a worker to reach into a danger area, provision will be made to ensure that the machine cannot be actuated while there is danger of injury, either by the interlocking of safety guards or by arranging the controls so that the hands must be withdrawn to operate the controls, and must remain on the controls while the machine is in operation. (See Figure 1.3.)

For automatic machines the separation of worker and machine is ensured by the provision of fencing with interlocked gates, or photoelectric interlock devices.

Safety equipment

The risk of injury can be minimised by the provision of safety equipment. In any area where there is heavy equipment in use or heavy items are being handled, there is a risk of injury to the feet. In some areas safety footwear is a legal requirement. Stylish modern safety footwear offering protection to the toe area is not costly, and the wearing of this generally within the factory environment should be encouraged. The wearing of soft footwear, such as trainers, should be actively discouraged.

Approved safety glasses should be worn for all machining operations, and anywhere that there is any danger of flying particles.

In areas where work is being carried out overhead, or where stacking or high storage takes place, a danger exists from falling objects. Dependent on the nature of the processes, safety canopies should be provided (e.g. on lift trucks) or 'hard hats' should be worn.

Steps should be taken to minimise noise levels in the factory environment, either by eliminating the cause or by enclosing the offending equipment. Where this is not possible, ear plugs or ear defenders should be used.

It is not only in from machinery that danger exists in the manufacturing environment. Many industrial injuries occur when moving or lifting product or equipment. Where heavy objects must be handled, suitable lifting equipment should be provided. European law now stipulates that no person should be required to lift a greater weight than 20 kg unaided. Even for lesser lifts, employees should be instructed in the correct lifting techniques.

Hazards that cannot be eliminated should be clearly identified, and made clearly visible by suitable painting and/or the fitting of warning lights. This applies

Figure 1.3 Guarding of a mechanical press.

particularly to works transport. It is good practice with automatic machinery to fit the equipment with a flashing beacon which operates when the machinery is live.

Most importantly, every piece of powered equipment should be provided with safety stop provisions accessible both to the operator and to those around, providing a means of immediately stopping the equipment in an emergency.

Safe working practices

Safety in the workshop depends not only on the provision of guarding and safety equipment, but on the development of safe working practices.

Every worker should be adequately trained for the tasks undertaken, and should be aware of the areas of inherent danger. Safety equipment must be worn where necessary, and guards must not be removed from machinery. All equipment must be kept in good order.

Sensible dress should be worn, giving protection to the body. Properly designed overalls are to be preferred, avoiding loose items which might lead to entrapment. Long hair should be enclosed in a hair net or cap.

Where it is necessary to operate equipment without the guards in operation, e.g. during maintenance or setting, such work must be limited to trained and certificated personnel. In many areas this is a legal requirement. Special care must be taken when such personnel are working on the machinery that the machinery cannot be accidentally started by other persons.

Most industrial accidents are caused by carelessness, either on the part of the victim or a third party.

SAFETY IS EVERYONE'S RESPONSIBILITY!

2 Primary Forming Processes

2.1 Introduction

The primary manufacturing processes for a product may be considered as the translation of the basic raw materials into a form suitable for further utilisation. The basic raw material will have been formulated or blended to a suitable composition, and now requires to be converted into billet, blank, rod, sheet, wire, or whatever form is most suited for the task in hand. In this the material may be processed in either liquid or plastic form, and it is useful to consider this as a categorisation of the processes.

Manufacturing materials may be of a metallic or non-metallic nature. It is useful at this initial stage to briefly review these.

Metals

Most metals in their pure states lack some of the characteristics which are required in the final product. Copper, for example, in its pure state is highly electrically conductive, but has little mechanical strength. Metals, however, may be alloyed together to achieve characteristics not present in the parent materials. Certain other materials, although not truly metals, can also be alloyed with metals to give enhanced characteristics to the alloy, notably carbon, which has a major effect on the hardness of steels. Hardness or softness of metals can also be affected by heat treatment, the rate of heating or cooling affecting the grain structure and thus the mechanical characteristics of the metal. In the course of this book the terms *annealing* and *hardening* will occur: annealing is the raising of the metal to a temperature at which the grain restructuring takes place and then cooling under controlled conditions to control the grain size to give maximum ductility, whereas hardening is a similar process incorporating rapid cooling to give controlled grain size for maximum hardness.

Distortion of the grain structure by mechanical working will also result in a hardening of the material, and in many manufacturing operations inter-stage annealing is required to restore the ductility prior to further working.

Synthetic materials

A *plastic* is usually defined as a *synthetic organic material* that is solid in its final form but is fluid at some stage in the processing and is shaped by heat and pressure. Sometimes the term *polymer* is used referring to any substance in which several or many thousand molecules or units are joined into larger and more complex molecules. Plastics are broadly classified as thermosetting or thermoplastic.

Thermosetting plastics are formed to shape with heat, with or without pressure. The heat first softens the material, but as additional heat or chemicals are added the material is hardened by a chemical change called polymerisation and cannot be resoftened.

Thermoplastic materials undergo no chemical change in moulding. They may be remelted by heat, and are hardened by cooling.

Natural non-metalic materials

Manufacturing industry also uses a large number of non-metallic materials of natural origin. The manufacturing techniques outlined in this book are equally applicable to such materials.

Wood and the various processed boards and papers based on natural fibre have a large place in manufactured product. The techniques used for working these do not basically differ from those used for metals of similar form.

Ceramics also play a substantial part in manufactured product, and the manufacturing techniques are not significantly different from those used for more conventional materials.

Even *textile materials* must be considered as coming within the scope of the manufacturing engineer, since the processes once exclusive to the textile industries are now being used extensively in other parts of industry, while machining techniques from the engineering industry are being applied to textile manufacture.

2.2 Casting

Casting is the process of pouring a material in a liquid form into a mould and allowing it to solidify to produce the desired object. This liquid–solid transition may be achieved by heating and cooling, by dissolving and precipitating, or by chemical reaction, according to the characteristics of the material being worked.

The moulds used may be made of a variety of materials, dependent on the materials to be cast and the surface finish required. The oldest form of casting still in current use is sand casting, used not only in its original form but in a number of modern derivatives. Other materials used for moulds include plaster, metal, and even rubber. The mould may be made from a pattern, or the shape of the object to be produced may be cut into the material of the mould itself.

Sand castings

Two methods are available for the production of sand castings, classified according to the type of pattern used. These are *removable pattern* and *disposable pattern.*

With a removable pattern sand is packed around the pattern in such a way that the cavity can be split and the pattern removed. Disposable patterns, on the other hand, are similarly packed in sand, but remain in position and are destroyed by the hot metal when it is poured in. Normally made from a material such as polystyrene, which is vaporised by the molten metal.

The moulds themselves are classified according to the materials used to make them.

Green sand moulds

This term may seem misleading at first, since the sand used is normally dark brown or black. The term 'green' refers to the fact that the sand is uncured. The mould is made in a flask that has two parts, the *cope* (top part) and the *drag* (lower part). Had the part been more complex and required a three part mould the centre part would have been termed the *cheek.*

Skin dried moulds

A binding agent is introduced to the sand around the pattern, and the mould is then dried with warm air or flame to harden the surface and drive out moisture. Two different techniques are used to achieve this hardened surface, either the sand around the pattern is mixed with the binder, and the rest of the box filled with green sand, or the mould is made of green sand and the surface is then sprayed with the binder in liquid form. Binders used include linseed oil, gelatinised starch, and other materials in solution.

Dry-sand moulds

An extension of the skin-dried mould process, in this case all the sand in the mould contains a binding agent. The moulds are oven baked before use to harden them and dry them out. Dry-sand moulds hold shape better when pouring and give less problems with gassing due to presence of moisture.

CO_2 moulds

In this process the binder mixed with the sand is sodium silicate. After forming the mould, CO_2 is fed into the mould under pressure. This causes the mixture to harden.

Casting techniques

Each of these mould techniques requires some way of constructing the recess in the sand that will eventually give the required part.

For simple symmetrical parts this may be done by forming the sand using a shaped board drawn across or turned in the surface of the cope and/or drag to scoop out a recess corresponding to the part. This technique is termed a 'swept' mould, the board used, shaped to the profile required, being termed the 'sweep'.

Most castings, however, make use of a removable pattern for the part required. This pattern may be made of any stable material and will differ from the finished part in a number of important respects. It will be slightly larger, since the material will shrink as it cools, and if the pattern were to be the size of the part required, the cast part would be undersize. Allowance will also have been made for the provision of extra material on surfaces which will require machining to achieve accuracy or finish. Then since it is necessary to extract the pattern from the mould easily, the pattern will be free from undercuts and will have those surfaces perpendicular to the mould separation given an inclination, termed 'draw', to help them come free from the mould readily and without damage to the mould.

Since the mould will be in two parts to allow for extraction of the pattern, the pattern will itself be split along the mould line, to allow it to sit flat down on the moulding board when filling the drag. The parts will be dowelled together to give accurate location of the second half relative to the first half when filling the cope.

Holes in the casting will not appear in the pattern unless these are vertical to the part line and are large in diameter, and can afford to have a large draw. Instead separate cores will be used, and bosses will be added to the pattern which will form locations in the mould for these cores. These bosses will be of such a length as will ensure that the cores are adequately located.

The cores themselves will be made of a bonded sand, to make them handleable, and will be moulded in dies of wood or metal. As the cores are disposable items, and are broken up to remove them, cores may have undercuts and recesses.

Runners and gating may also be incorporated in the pattern, but more usually these are put in separately by hand, or using separate sprue pins.

Removable patterns are commonly made of wood or aluminium, since they should be light to handle and easily worked. The surface of a wood pattern is sanded and varnished to give a smooth surface which will withdraw easily from the sand. Some form of detachable grip is required on the split face to facilitate extraction. For wooden patterns this is normally a spike driven in, for metal patterns a tapped hole or holes.

The procedure in making a mould from a pattern is to set the half pattern on the moulding board, place the drag over it, and sieve sand over it, pressing the sand down by hand then ramming it to compact it onto the pattern either by hand or machine. The top surface is then levelled off with a straight bar termed a 'strike rod' and a 'bottom board' placed on top (Figure 2.1). The lot is then turned over and the moulding board removed. The second half of the pattern is positioned, the surface

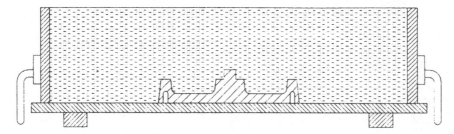

Figure 2.1 Half pattern.

of the sand sprinkled lightly with a dry parting sand, and the cope is placed on the drag (Figure 2.2). A sprue pin is inserted about 25 mm from the pattern to provide an entry for the molten material, and the cope filled with sand and rammed, and struck. The sprue pin is then withdrawn, and a funnel-shaped entrance to the sprue scooped out to make pouring easy. Further similar holes, termed risers, may be made by inserting further pins. These will aid in the rapid filling of the mould. Vents, small holes pierced in the top sand but not contacting the cavity, may be made in the top of the mould to further aid the escape of gases when pouring (Figure 2.3).

The mould is then carefully separated, and the pattern withdrawn. A gate is cut linking the sprue to the cavity, and gates linking any risers to the cavity. The surfaces of the mould may be sprayed with a coating to improve the surface finish. Any cores are positioned, then the cope and drag are united for pouring. Weights are placed on top of the cope to hold the top down to prevent the incoming metal floating the cope.

Disposable patterns are normally made of polystyrene, and are made complete with gate. They are machined or carved from blocks of polystyrene, and may be fabricated as several parts joined with adhesive. Since the pattern does not have to be removed from the mould, undercuts are possible. Holes may be cored to prevent break-up of the mould during casting, the cores being set into the polystyrene before moulding (Figure 2.4).

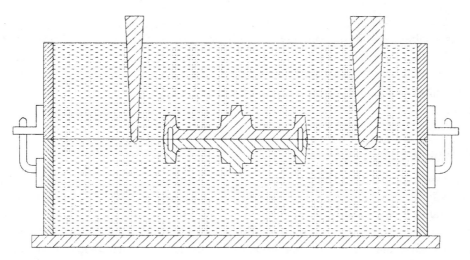

Figure 2.2 Sprue pins.

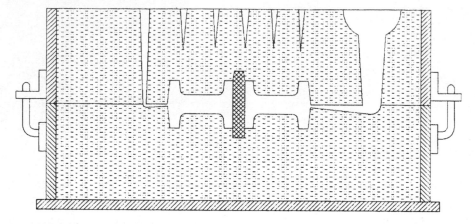

Figure 2.3 Risers and vents.

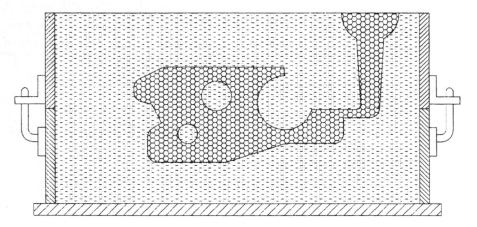

Figure 2.4 Disposable pattern.

Disposable patterns are normally placed on a follow board, i.e. a board profiled to take the one half of the casting, and the drag rammed in the usual way. The drag is then inverted, the follow board removed, and the cope fitted, filled, and rammed, then vented. No parting sand is used. After casting, the mould is broken from the casting.

Shell moulding

If a resin and an accelerator are added to the dry sharp sand and well mixed in, the material will harden over a period of time to give a hard mould. This curing may be accelerated by the application of heat. The resin–sand mixture is expensive, so it is normally used to form a shell around the pattern, and this shell is then supported with dry sand, although it will occasionally be used as a complete moulding material if the parts are small (Figure 2.5).

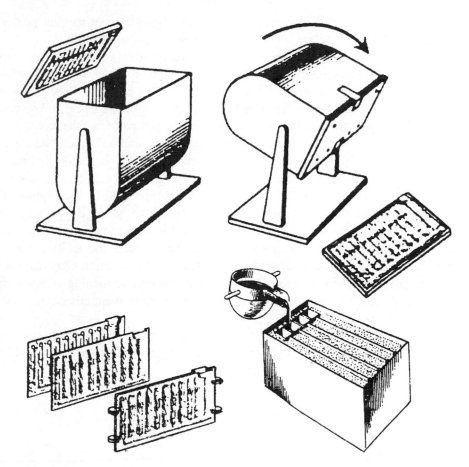

Figure 2.5 Shell moulding.

This technique has been developed into a mechanised process. The pattern in this case is made of metal, the two halves making up the part being mounted on separate metal sheets. The pattern half is preheated to around 230 °C, sprayed with a silicon release agent, then has a sand–resin mix either blown onto it, or is placed on a dump box containing the sand–resin mixture, whereupon the dump box is inverted, covering the pattern with the mixture, then after a short delay turned back up again, when the mixture not in contact with the hot pattern falls back into the dump box, leaving a coating of part-cured mix adhering to the pattern. In either case, the pattern with the shell on it is cured in an oven, then the shell re-moved. Top and bottom halves are assembled with clips and resin adhesive, and are then ready for pouring.

Die casting

To obtain accurate castings of good finish, cutting down on the need for future machining, metal moulds may be used, sometimes with the material to be cast fed in under pressure. Such techniques are limited to materials with lower melting points, otherwise fusion may take place between the metal being cast and the die.

The technique of casting into metal moulds is referred to as 'die casting'. No pattern is used, the dies being cut by machining the desired profile into the die blocks. Dies are basically in two parts, which separate to release the cast part. Core pins and loose parts, termed 'draws' can be added to provide complex shapes, holes, and undercuts. Ejector pins are used to ensure removal of the cast part from the mould.

There are several processes for die casting. The most commonly used process is *pressure die casting*. Here the molten metal is forced into the mould under pressure. Because the metal is held under pressure while cooling the resultant casting conforms very closely to the die dimensions, and surface finish. Two types of machine are available, the *hot chamber* (Figure 2.6) machine, where the melting pot is incorporated in the machine and the injection ram is immersed in molten metal all the time, and the *cold chamber* (Figure 2.7) machine, where the melting pot is separate, and the molten metal is transferred by ladle to the injection chamber. Use of the hot chamber technique is restricted to lower temperature melting alloys such as the range of zinc-based alloys, since higher temperature materials cause rapid corrosion of the immersed ram.

In the *low pressure casting* (Figure 2.8) process the metal mould is located over an induction furnace. An inert gas under pressure is fed to the furnace pot, and this

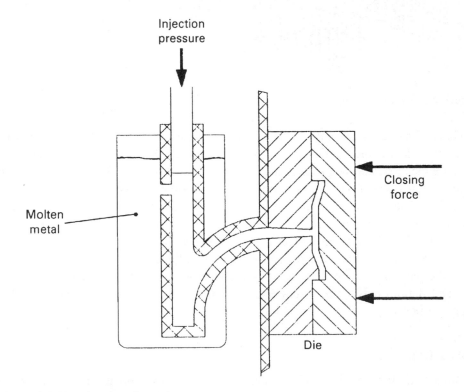

Figure 2.6 Hot chamber die casting.

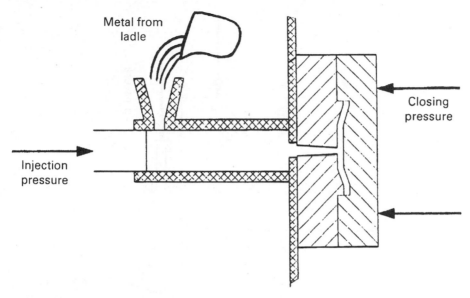

Figure 2.7 Cold chamber die casting.

forces the molten metal up through a heated riser stalk into the mould where it is cooled. Vacuum may be applied to the mould to aid filling and remove entrapped air. The resultant casting is dense, free from inclusions and oxidation, and of good dimensional accuracy. The cycle time is reasonably fast, and scrap and remelt is low.

In *gravity die-casting* (Figure 2.9) the metal mould is clamped and heated, then the metal is poured directly into the mould. No pressure is used, although the mould may be provided with a long neck to give a head of metal which will provide both a reservoir of hot metal and a small pressure head to help eliminate shrinkage.

Slush casting is a variant of the gravity process. Molten metal is poured into the mould, which is then inverted to allow any metal which has not chilled on to the die to run out, resulting in a hollow casting, the thickness of the wall depending on the rate of chilling of the mould.

Pressed casting is a further variant. Here a predetermined amount of metal is poured into the die, then a core plunger is forced into the cavity, forcing the metal against the cavity walls. When the metal has chilled the plunger is withdrawn and the mould opened, giving a cored casting.

Another variant of the slush casting technique is *centrifugal casting*, used to cast such items as pipes, and similar items. Here a measured quantity of molten metal is poured into a spinning mould, and centrifugal force spreads it evenly around the mould walls.

Plaster or ceramic moulds

Processes have been developed using moulds of plaster or ceramic, the mould material being chosen to suit the temperature requirements of the material being

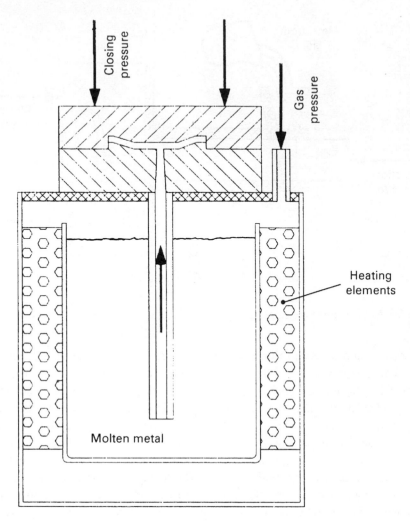

Figure 2.8 Low pressure die casting.

Figure 2.9 Gravity die casting.

cast, and to give a mould which will retain dimensional stability and resist the erosion of the liquid metal during pouring.

Plaster moulds may be made utilising a removable pattern by first supporting the pattern in a tray, and pouring plaster into the tray to immerse the pattern to half its depth. The lot may then be placed in a vacuum chamber to remove entrapped air. When this plaster has set, the surface is coated with a release agent, then further plaster poured in to completely immerse the pattern. The lot is again vacuumed. When this has set, the two halves are separated and the pattern removed. Sprues are cut in, then the two halves are clamped together and the metal poured into the mould. The resultant casting has a good surface finish. The need to withdraw the pattern from the plaster mould means that the casting must be free from undercuts and have generous draw on the faces, and coring of holes is difficult.

Investment casting

Sometimes called 'lost wax' casting, investment casting overcomes these problems. The pattern used to prepare the mould is made of wax, or sometimes a plastic. The pattern or patterns are joined to a 'stalk' or sprue also of wax to form a 'tree' of patterns. A metal cylinder is placed over this tree and sealed to the wax base, then the cylinder is filled with the 'invest', a specially formulated plaster, and allowed to set under vacuum, the vacuum chamber is vibrated to aid the expulsion of trapped air. An alternative technique used is to build up a coating of the invest by successive dipping of the wax into the liquid invest, obviating the need for the metal container.

The prepared moulds are then placed in an oven and heated gently to dry off the invest and melt out the bulk of the wax, then the temperature is raised to bake the invest and burn off any residue of wax. The temperature is then reduced to give a suitable mould temperature for pouring.

The moulds may be poured under gravity, but it is more usual to place the moulds on a vacuum box and pull a vacuum on them while pouring. Air is drawn out through the slightly porous mould, giving good filling and resulting in high quality accurately dimensioned castings. Alternatively they may be poured under centrifugal action, the flask being spun rapidly and centrifugal force drawing the metal in and completely filling the mould.

After cooling invest is removed from the castings by pressure-jetting or vibratory cleaning.

This process has been developed to a high degree, allowing the casting of parts of great complexity or close tolerance which cannot be produced by any other process.

The wax patterns may be produced by carving, by low pressure injection moulding, or by conventional injection moulding. The moulds used for these may be in metal or rubber, or a combination. Rubber moulds, made by vulcanising rubber over a replica of the desired part made from steel or brass, or by using liquid synthetic rubbers poured over a replica made from plastic, are commonly used with low pressure wax injectors. Metal moulds, either directly machined into aluminium blocks or made by pouring lead or bismuth alloy over a steel or brass replica, are used in modified injection moulding machines. Polystyrene resin is sometimes used instead of wax.

Complex structures can be cast in one piece by assembling multiple wax parts together by melt fusion. Complex cores may be cast in plaster and then have the wax cast around them, to give thin-wall parts of high complexity. An alternative technique for complex interior forms is to make a mould of the desired interior in a soluble wax, then place this in the cavity of the mould for the wax and inject normal wax around it, then finally remove the soluble core wax by dissolving it out. The resulting wax is then processed in the normal way.

Rubber moulds

These are used for specialised applications, where the materials being cast are liquid at low temperatures.

The patterns used are normally brass. These are placed between disks of unvulcanised rubber, the surfaces of which have been chalked to prevent top and bottom vulcanising together, and placed under a heated platen press. The rubber flows around the patterns, and cures to form two halves of a mould.

The disks are separated and the patterns removed, after which sprues are cut in to allow the flow of material from the centre of the disk to the cavities. In use the disks are clamped between steel plates and rotated at high speed, and the metal poured into the centre, from where it is thrown into the cavities by centrifugal force.

2.3 Moulding

Whereas in casting the material was liquid when constrained in the mould, and was then solidified, in moulding the material is handled in a plastic state. This plastic state may be achieved by heat, pressure, or chemical means, or by a combination of these.

Compression moulding

The oldest of the processes, this is mainly used for the moulding of thermoset materials, most commonly phenolics and aminos (ureas and melamine) in combination with a range of inert fillers. Alkyds and polyesters may also be moulded this way.

A matched set of male and female dies are needed. Compression moulding (Figure 2.10) requires that the exact amount of material be placed directly in the mould. This material may be in powder form or in a preform or pellet form, normally preheated to the region of 135 °C before being placed in the mould. The press is then closed, usually hydraulically, causing the material to flow to fill the mould cavity. Heat is conducted from the mould walls (up to 200 °C) and under

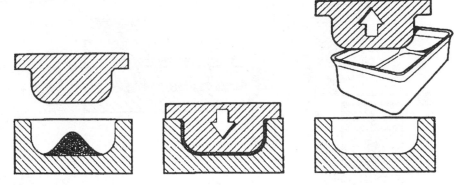

Figure 2.10 Compression moulding.

action of heat and pressure (30–50 N/mm^2) the moulding material polymerises into a solid mass which cannot be remoulded. This cure time varies between 0.5 and 3 minutes dependent on the configuration of the part. When curing is complete the die is opened and the part ejected.

Because of the difficulty of measuring an absolutely accurate charge, tools are designed with clearances which will allow a small amount of excess material to escape as flash, either horizontally or vertically. This must be removed by a secondary operation. Parts and moulds are designed to position this flashing in as unobtrusive a location as possible.

Another version of the compression mould is the transfer mould. In this case the charge is not placed directly into the mould, but into a separate chamber attached to the mould proper, and the material in its plastic state is forced from this chamber into the mould proper, either by the closure of the mould itself (pot transfer) or by the action of a separate plunger after the closure of the mould proper (plunger transfer). Multi-cavity layouts are possible with these types of mould, and parts can be moulded to tighter tolerances than for simple compression moulding. Cycle times are shorter, but against this must be set the waste material of sprue and runner systems.

Injection moulding

This is probably the most used technique for the quantity production of plastic parts. It is a very versatile process capable of producing intricate parts at fast production rates with good dimensional accuracy. Moulds are costly, however, and this limits the process to high production volumes.

The basic injection moulding process (Figure 2.11) involves the injection under high pressure of a predetermined quantity of heated and plasticised material into a relatively cool mould, where the material is allowed to solidify before the mould is opened and the part ejected.

The injection moulding machine (Figure 2.12) is made up of two basic sections, the clamp unit and the injection unit. The clamp unit resembles a press. It positions

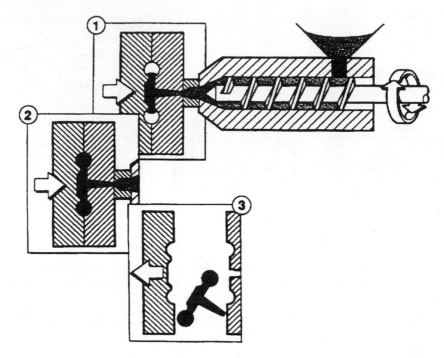

Figure 2.11 Injection moulding process.

and closes the mould halves, and it provides the large forces necessary to prevent the separation of the mould during material injection, when forces in the region of -40 N mm^2 of projected mould area may be encountered. Clamp units of injection moulding machines are usually horizontal, and either of mechanical toggle type or fully hydraulic type. The function of the injection unit is to melt and feed plasticated

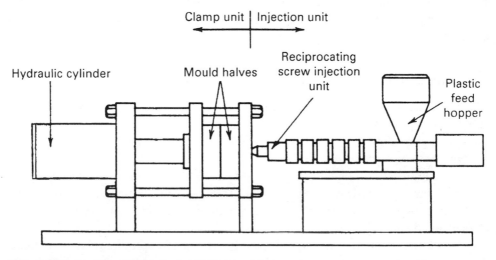

Figure 2.12 Layout of injection moulding machine.

material into the mould. Two variants of feeder are used, the reciprocating screw and the plunger type. In the latter a plunger loads material into the heater chamber, from which it is ejected by another plunger, while in the former the screw serves both as loader and injector.

Operation of the screw type, the most common in industry, is as follows. In the barrel of the unit is a helical screw capable of both rotational and longitudinal motion. The feed hopper feeds the granulated material to the rear of this screw, from where it is carried into the heated extruder body. As the material advances down the barrel, it is heated from the barrel walls and work is done on it by the action of the screw, changing the material into a molten state. As the material is fed forward, the screw is moved backwards while still rotating, until the chamber in front of the screw contains the requisite amount of material. The screw rotation then stops, and the stationary screw moves forward, acting as a plunger to eject the molten plastic through the injection nozzle into the mould. After a few seconds, to allow the gate sections to the mould to solidify, the screw begins to rotate again, starting a new injection cycle (Figure 2.13).

Moulds are cooled by channeled water flow, and are opened as soon as the parts are sufficiently rigid to be ejected. Some materials permit the use of cool moulds, and give short cycle times, while others, particularly the more crystalline materials, require warmer mould temperatures and slower cooling times, and thus longer cycles. Thermosetting materials may also be injection moulded, and require the use of steam heated, not cooled, moulds to give proper material curing. Moulds may vary from simple single cavity to multi-cavity moulds with multiple cores and multiple separations. Moulds may also have heated runners, keeping the material in the runners fluid and avoiding the need for separation and handling and recycling of waste.

Design considerations

Broadly similar considerations apply to parts whether injection moulded, compression moulded, or die cast. Since the process is one of flowing material into a die cavity, it is essential that anything which might obstruct or otherwise interfere with the material flow should be avoided.

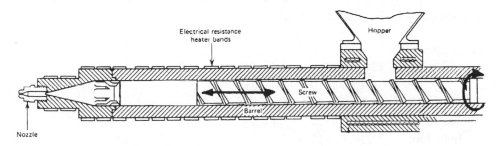

Figure 2.13 Injection moulding using reciprocating screw.

As far as possible, wall thickness should be constant. This wall thickness will depend on the structural strength and dimensional stability of the part, and also the material and process to be used. Proper distribution of stress and effective use of material can best be achieved by strategic adjustment of slope, contour, and shape of the part. Every attention should be given to such adjustments before relying on increases of wall thickness, which may prejudice the economic manufacture of the part. Any variation in wall thickness should be gradual, and in the direction of the material flow, and transitions should be smooth, with steps avoided whenever possible. Irregularities in thickness will cause irregularities in setting and contracting, with internal parts stresses, and warping or sink marks.

Where possible, sharp corners, both internal and external, should be avoided. Conner internal radii should normally be at least one-third of the wall thickness. From the point of view of die costs, undercuts should be avoided as these make ejection difficult, if not impossible. Some materials can stand flexing when being ejected, but the ejection arrangements are liable to increase the cost of the tool significantly.

To ease ejection, a taper or draft is normally applied to all surfaces normal to the parting plane of the mould. Walls, ribs, internal cavities, bosses, etc. should have a minimum taper of 1° for easy ejection. Although smaller draft angles can be used if absolutely necessary, this will complicate the ejection, perhaps requiring separate draws.

Stiffening may be carried out by provision of ribs and carefully chosen sections. Such ribs and sections should be designed so that the rules above for wall thickness are observed.

Interior bosses should be located in corners, as an integral part of the wall wherever possible, and the height should not exceed twice the boss diameter. Exterior or free-standing bosses should be ribbed with gussets to provide support and aid material flow. Bosses should be cored to provide uniform wall thickness, and should be provided with sufficient draft to ensure easy mould removal (Figure 2.14).

Lugs and projections should be eliminated wherever possible. Where they are necessary, they should be adequately reinforced with ribs, again taking care to ensure even wall thickness.

Moulded holes are classed as blind, through, or step holes. Through holes are preferred since the mould pins can be supported in both halves of the mould. The pin for a blind hole is more susceptible to damage. Recommended minimum dimensions for holes (see Figure 2.15 on p. 30) are:

Through holes
　　Compression: $D/W = 1/3$ for $3 < D < 6$ mm
　　　　　　　　　　　 $= 1/1$ for $D < 3$ mm
　　　　　　　　　　　 $= 1/4$ for $D > 6$ mm
　　Transfer:　　 $D/W = 1/15$
　　Injection:　　 $D/W = 1/15$
　　Die casting:　 $D/W = 1/15$

'Exterior' (free-standing)
boss

'Interior' (integral-wall)
boss

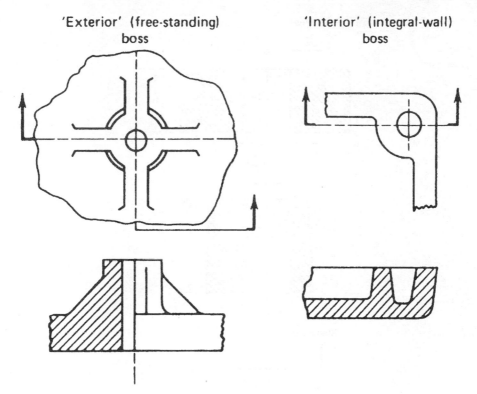

Figure 2.14 Bosses in castings.

Step holes

Compression: $(D + d)/W < 1/2$
Transfer: $(D + d)/W < 1/3$
Injection: $(D + d)/W < 1/3$
Die casting: $(D + d)/W < 1/3$

Blind holes

$$E_{MIN} = D/6$$

Compression: $D/W = 1/1$ for $D > 3$ mm
             $D/W = 1/2$ for $D < 6$ mm
Transfer: $D/W = 1/3$
Injection: $D/W = 1/3$
Die casting: $D/W = 1/3$

If fixed pins are used, a draft angle of at least 1° is recommended. Where parallel holes are required, retractable pins are used, the pins being retracted before ejection of the part from the die.

Care must be taken in positioning holes, particularly those close to edges, or proper flow will not be obtained around the hole.

Where parts will be subject to numerous assembly and dissembly operations during usage, or where additional strength is required, inserts may be fitted. These

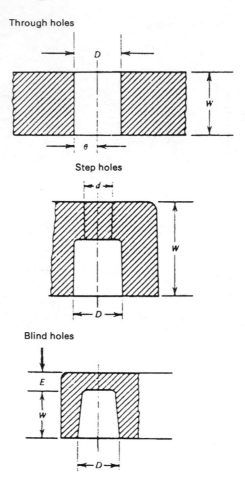

Figure 2.15

may be moulded in or press fitted, and are shaped and knurled to give positive grip. Where assembly is 'once and for all' self tapping screws are normally used in cored holes.

2.4 Forging

The equivalent plastic deformation process for metals is the process of forging. This may be carried out cold, but for large deformations heating is required. Within their elastic range metals behave very closely in accordance with the well known mathematical laws. Once the yield point is exceeded the behaviour is less exactly predictable, especially for conditions of combined stresses. There is a growing volume of work on plasticity in which mathematical treatments are developed (see Bibliography) but for most of the common metal working processes semi-empirical methods still seem to be the most useful. The object of such methods is to enable the results of practical tests to be used as a means of predicting behaviour under processing conditions somewhat similar to those of the tests. Standard tensile and compressive tests are the simplest tests providing information about the behaviour of metals within the plastic range.

Figure 2.16 shows the load–extension graph of a tensile test of a ductile metal which has no distinct yield point. Up to the load represented by point A, the material is elastic. Consider the load to be raised slowly to point B, and then to be removed. During the unloading the load–extension graph will follow line BC parallel with AO; distance OC represents the permanent extension caused by the load, distance CD represents the elastic contraction (recovery) which occurs as the load is removed. The shaded area OABD represents the amount of work necessary to cause the deformation OC.

If the 'overstrained' material is again loaded a new load–extension graph will have its origin at C, its yield approximately at B and its breaking point at approximately the same point as would have resulted in carrying the first test through to failure. The results of working the material to point B within the plastic range are:

1. The yield point, and hence the safe working stress for the material, is raised.
2. The ductility, measured by the amount of elongation occurring before fracture, is lowered.

Two important deductions relevant to metal deforming at room temperature can now be made:

1. The elastic recovery which occurs on release of the deforming stresses is inevitable, and there will be minor changes of the dimensions when cold-deformed parts are ejected from the tools. 'Spring-back' in sheet-metal presswork typifies this.

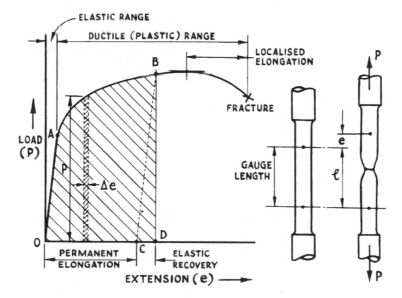

Figure 2.16 Tensile test of a ductile material.

2. During the plastic deformation each increment of elongation, Δe, will require a specific amount of work, as represented in Figure 2.16 by $P \times \Delta e$. From the shape of the graph between points A and B it can be seen that each successive increment of deformation will require a slightly larger amount of work to be done on the material; the resistance to 'working' rises steadily during the deformation. An alternative way of looking at this is to regard the yield point as rising steadily from A to B as the permanent extension is increased from O to C.

If tests of material properties at elevated temperatures are conducted information relative to the hot working of metals may be similarly examined. Figure 2.17 illustrates, for a 0·09 per cent mild steel, changes in the initial yield stress and changes in the ductility as revealed by a percentage elongation test, caused by heating the steel to various temperatures. Figure 2.18 illustrates successive changes

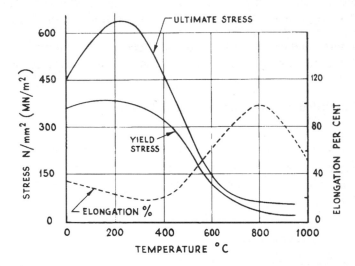

Figure 2.17 Influence of temperature on the properties of a 0.09% carbon steel (after Lea).

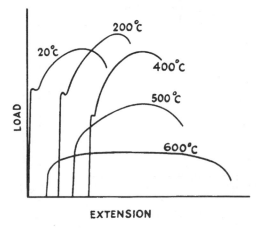

Figure 2.18 Influence of temperature on the load–extension diagrams for a low carbon steel.

in the load–extension diagrams of a similar steel, caused by heating. From Figure 2.17 it is obvious that by working the steel at temperatures around 800 °C the following advantages accrue.

1. The amount of work required, and the magnitude of the forces necessary to work the metal, are both greatly reduced.
2. The working (ductile) range of the metal is greatly extended.
3. The elastic recovery consequent upon the removal of the deforming stress is very small because of the low yield point.

Hot working is easier to perform than cold working, but as a secondary manufacturing process it is subject to the following disadvantages.

1. The finished components contract on cooling. Since the finishing temperature may not be uniform throughout the component, and may not be accurately known, exact dimensional allowances cannot be made to counteract this. An accuracy better than IT 12, see Table 16.3 (p. 511) is difficult to attain by hot working of steels.
2. Suitable working temperatures for steels give rise to oxidation and scaling. This results in rough component surfaces, particularly when the surface finish of the tooling begins to deteriorate.
3. Dimensional changes occur in the tools due to heating; abrasive wear of the tools steadily reduces their accuracy.

For these reasons hot working cannot compete with cold working in regard to dimensional accuracy and to surface finish of the resulting product.

True stress–strain curves

In Figure 2.16 the load is seen to fall as the breaking point is approached, a condition which is the reverse of that occurring between points A and B. The reason is to be found in the reducing cross-sectional area (necking) which precedes fracture. If the reduced area is taken into account the actual stress in the yielding material can be shown to rise continuously. Figure 2.19 shows a true stress–strain curve developed from the load–extension curve of Figure 2.16. Notice that the yield point rises progressively, and in approximate linear relationship to the strain between points A and B. In the region of point B the ductility of the material peters out; there is a further sharp rise of stress up to the breaking point, accompanied by very little strain.

A much more complete picture of the plastic working properties of a metal is gained by extending the tests into the compression range. Figure 2.20 shows the compression curves for load–compression and for true stress–strain. The curves are drawn in the third quadrant because compressive stress and strain are regarded as of opposite sign to tensile stress and strain. The nature of the ultimate failure of the material differs as between tension and compression, a fact which has some bearing upon the success of certain cold-deforming techniques.

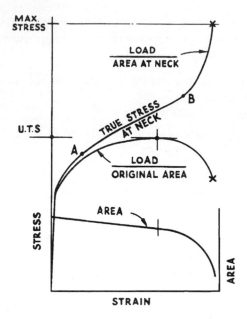

Figure 2.19 True stress–strain curve for a tensile test on ductile metal (after Salmon).

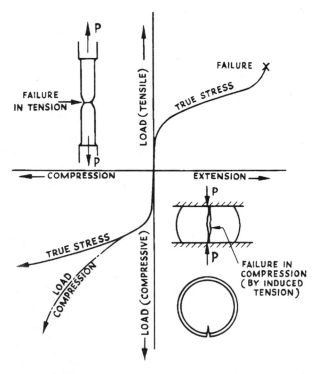

Figure 2.20 Behaviour of a ductile material under tensile and compressive loads.

Ductile metals deformed under compressive loads admit of a much greater amount of deformation before failure than if they are deformed under tensile loads. When such failures do occur they tend to be caused by induced tensile and shear stresses. The stress at the expanding periphery in the example of the cylindrical compression test specimen illustrated in Figure 2.20 is a typical example.

Work done in compressing a perfectly plastic material

A first approximation to the work done in deforming a metal can be made, subject to two important basic assumptions:

1. That the material is perfectly plastic, represented by the fact that the yield stress σ_y is a constant.
2. That there is no friction between the deformed material and the anvils which exert the deforming load and against which deforming displacements occur.

It follows from the second assumption that the deformed specimen will remain a parallel cylinder throughout the test and not exhibit the barrelling effect, caused by friction at the anvils, shown in Figure 2.20.

Figure 2.21 shows the deformation assumed to occur, and the resulting load–compression curve for the idealised compression test. Any change in volume during compression will be extremely small. If A = cross-sectional area of the specimen,

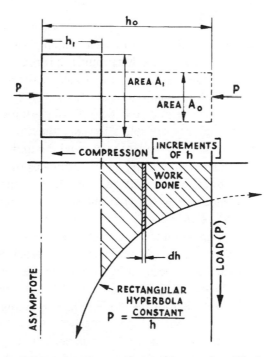

Figure 2.21 Work done in deforming a given volume of perfectly plastic material, no frictional resistance occurring.

h = length of specimen, V = volume of specimen and P = deforming load, then:

Original volume $(A_0 h_0)$ = final volume $(A_1 h_1)$

The deforming load at any position is given by:

$$P = \sigma_y A; \text{ but } A = \frac{V}{h}$$

hence

$$P = \frac{\sigma_y V}{h} = \frac{\text{Constant}}{h}$$

and the graph is a rectangular hyperbola. The work done in deforming the material is represented by the shaded area under the graph.

$$\text{Work done} = \int_{h_1}^{h_0} P \, dh = \sigma_y V \int_{h_1}^{h_0} \frac{dh}{h}$$

$$= \sigma_y V \ln \frac{h_o}{h_1} \tag{2.1}$$

$(\ln = \log_e)$

Note on units: given σ_y in N mm^3, V in mm^3, the work done will be in N mm. The energy equivalent of 1 N mm of work is 10^{-3} J.

This is an important result. Much of the experimental work done in order to obtain empirical data for the determination of the forces acting in plastic forming processes is based upon the concept of the principal logarithmic deformation represented by ln (h_0/h_1). If equation 2.1 is divided by the deformed volume the result, $\sigma_y \ln (h_0/h_1)$, represents the work done to deform a unit volume of material.

In order to relate equation 2.1 to true conditions for deforming metals, a realistic value must be given to σ_y and a reasonable allowance must be made for the work done in overcoming friction between the workpiece and the tools. Figure 2.22 shows the type of test data available, from which realistic estimates of σ_y can be made. If the mean value of σ_y for the range of deformation which occurs is obtained from test results, a reasonable estimate of the frictionless work of deformation can be made. (See Feldmann, H. D., *Cold Forging of Steel*.)

The effect in a simple compression test, of friction at the anvils, is to give rise to an increase in the load P necessary to cause yielding. Siebel has shown that the approximate increase in yield stress caused by friction at the anvils is given by σ_y $\mu d/3h$, where μ is the coefficient of friction and d and h the diameter and height of the specimen respectively. A method of developing this approximate expression is as follows.

Let k = resistance to deformation when friction is neglected, i.e. $P_0 = kA$, and let k_f be the additional resistance to deformation caused by friction such that the true deforming load $P_1 = (k + k_f) A$. Suppose the load P_1 to be just sufficient to cause a

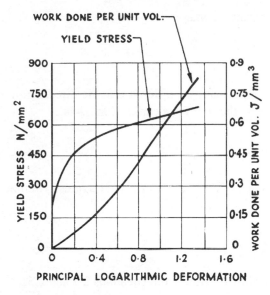

Figure 2.22 Plastic deforming properties of a 0.07% carbon steel (after Feldman).

small deformation Δh, during which the elemental ring shown in Figure 2.23 moves outwards by an amount $x/r\,\Delta r$.

Work done at the anvil in overcoming friction,

$$(P_1 - P_0)\Delta h = Pf\,\Delta h$$
$$= k_f \pi r^2\,\Delta h \qquad (2.2)$$

Frictional resistance to sliding of elemental ring (2 anvils)

$$= 4\pi k\mu x\,dx \qquad (2.3)$$

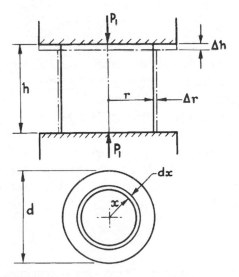

Figure 2.23 Elemental ring in compression.

By property of constant volume

$$\Delta r = \frac{-r}{2h}\,\Delta h$$

the difference in sign representing a difference in the direction of change, hence the outward displacement of the ring for a contraction Δh is

$$\frac{x}{r}\,\frac{r}{2h}\,\Delta h = \frac{x}{2h}\,\Delta h \qquad (2.4)$$

By equations (2.3) and (2.4), the work done against friction in displacing the elemental ring

$$= 4\pi k\mu x\,dx \times \frac{x}{2h}\,\Delta h \qquad (2.5)$$

Total work done against friction, from equations 2.2 and 2.5,

$$k_f \pi r^2(\Delta h) = \frac{4\pi k\mu(\Delta h)}{2h} \int_0^r x^2\,dx$$

$$= \frac{2\pi k\mu(\Delta h)r^3}{3h}$$

$$k_f = \frac{2\mu kr}{3h} = k\left[\frac{\mu d}{3h}\right]$$

It follows that for a resistance to deformation k, the total load on the anvils must be

$$P = Ak\left[1 + \frac{\mu d}{3h}\right] \qquad (2.6)$$

The solution assumes that the cylinder remains parallel during deformation which cannot occur unless $\mu = 0$, hence the proof is not rigorous.

It follows from equation 2.6 that the ratio d/h is significant in relation to the amounts of work to be done against friction. If h becomes very small in relation to d, much more work will be needed to overcome friction than to cause the frictionless deformation. For this reason it is impossible to continue the forward extrusion process to eject the entire billet. In sheet-metal presswork the drawing of relatively thin sheet is more difficult than the drawing of thicker sheet partly because of the relatively large frictional forces involved.

Example 2.1

A billet of aluminium 25 mm diameter × 38 mm long is compressed between flat parallel steel anvils to a length of 19 mm. The initial yield stress is 65·4 N/mm² and after a 50 per cent reduction the yield stress is 82·7 N/mm². Find the frictionless work done in deforming the material and the mean force which would produce this amount of work. If $\mu = 0·14$, find the maximum load to be exerted.

Solution

$$\text{Mean yield stress} = \frac{65.4 + 82.7}{2} = 74.05 \text{ N/mm}^2$$

$$\text{Volume} = \frac{\pi}{4} \times 25^2 \times 38 = 18\,653 \text{ mm}^3$$

Frictionless work of deformation, by equation 2.1,

$$\text{Work done} = 74.05 \times 18\,653 \times \ln \frac{38}{19} \times 10^{-3} \text{ J}$$

$$= 958 \text{ J}$$

Mean deforming force (neglecting friction)

$$P_{\text{max}} = \frac{958}{19} \times 10^3 = 50\,420 \text{ N}(50.42 \text{ kN})$$

Final diameter of work, assuming frictionless deformation,

$$d^2 \times 19 = 25^2 \times 38$$
$$d = 35·36$$

Maximum deforming force occurring at the end of the compression,

$$P_{\text{max}} = 82.7 \times \frac{18\,653}{19}\left[1 + \frac{0.14 \times 35.36}{3 \times 19}\right] \text{ by equation 2.6}$$

$$= 82.7 \times 981.7(1 + 0.087) = 88\,250 \text{ N}(88.25 \text{ kN})$$

The operation is within the capacity of a 10-tonne (force) press.

□   □   □

It would be wrong to expect high accuracy from such calculations; the theory employed is not rigorous, the yield stress (or more accurately the resistance to deformation) is not exactly known. However, such estimates as can be made are of considerable practical value; they enable presses of adequate capacity to be selected and provide data from which stresses in the tools can be reasonably apportioned.

Some metallurgical aspects of hot and cold working

The effects of plastic deformation upon the structure of metals and alloys has been the subject of considerable metallurgical research. There is a trend towards explanation of the plastic behaviour of metals in terms of dislocation theory, and this may be profitably studied by students whose courses include metallurgy as a separate subject. It seems desirable to include within this chapter a brief outline of some important metallurgical aspects of plasticity.

Commercial metals are polycrystalline. The crystals have an ordered arrangement of atoms which may be represented by a 'lattice', Figure 2.24(a). Single crystals have considerable ductility and distort by 'slip' along favourably oriented crystallographic planes when subjected to a sufficient stress, Figure 2.24(b). The crystals have a random orientation among themselves, and much of the strength of a metal is due to the fact that the slip planes lie in different directions in the different crystals and that at the crystal boundaries atoms are shared, i.e. have a position in the lattice of two or more crystals. Dislocations are 'faults' in the atomic lattice, i.e. line discontinuities. Their density in an annealed material (total length of the lines of misfit per unit volume) is about 10^3 mm/mm^3.

Commercial metals are never 'pure', and due to *minor* segregation at the ingot casting stage, a proportion of the impurities are deposited along the grain boundaries. It is one of the functions of hot working to break down the large crystals which form due to the slow cooling of the ingot, and to break up and disperse segregations. Figure 2.25 illustrates the effect of mechanical working and of the reheating of a metal. Large equi-axed crystals are distorted under the

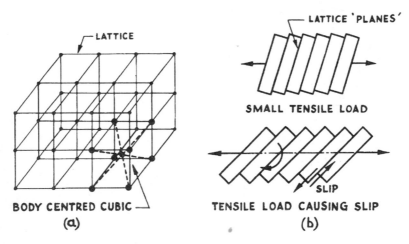

Figure 2.24 (a) Lattice structure of a crystal; (b) deformation by 'slip'.

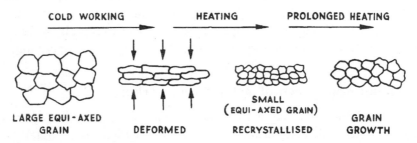

Figure 2.25 Effects of mechanical working and heating on the grain structure of a metal.

deforming load (by 'slip'). The deformed crystals are unstable above a certain temperature, and on heating will recrystallise to form small equi-axed crystals. If the heating is prolonged, larger equi-axed crystals will 'grow' from these small crystals.

In hot working the sequence is a continuous one, recrystallisation immediately follows severe distortion of the crystals. In cold working the temperature of the material remains below the recrystallisation temperature. Changes of mechanical properties which arise during cold working are mainly the result of the deformation of the crystals by 'slip', so that ultimately the number of 'slip planes' remaining unused is small. An annealing process subsequent to cold working will change both the structure and mechanical properties of the material. Figure 2.26 shows the relationship between recrystallisation temperature and the degree of cold work; Figure 2.27 shows the relationship between the grain size resulting from an annealing process and the amount of cold work to which the material has been subjected. The very coarse grain is that caused by the *critical* amount of cold work. In sheet-metal presswork it is necessary to avoid the critical amount of cold work where subsequent heat treatments occur, otherwise a large-grained 'orange peel' surface defect will result.

One result of the dispersion of segregations during hot working is that commercial forms of metal tend to have directional properties. Sheet and strip metals used in presswork will bend across the direction of rolling more successfully than along the direction of rolling; sharp bends made along the direction of rolling may cause cracks to appear on the tension side of the bend. Deep-drawn cups frequently have localised

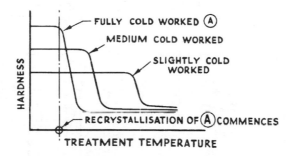

Figure 2.26 Influence of the amount of cold work (as revealed by hardness) upon recrystallisation temperature.

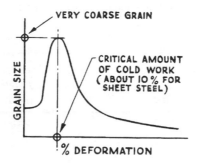

Figure 2.27 'Critical' amount of cold work.

elongations round their rim (ears) caused by differing properties of ductility along and across the strip. One of the advantages of cold forging, as compared with conventional machining, is that the flow pattern of the material may relate more satisfactorily to the stresses which the component has to carry.

The general metallurgical concepts stated above need to be modified when considering certain alloys. The solid–solution alloys behave very much as pure metals. More complex alloys do not conform to the simple pattern of the pure metals. Figure 2.28 shows the changes in the strength and ductility which occur in brasses as the zinc content is increased.

The α brasses are both hot and cold working.

The $\alpha + \beta$ brasses are hot working because at the working temperature of about 750 °C the α phase is entirely absorbed into the β phase. Any brass containing the γ phase is brittle and of little use except for low-grade casting.

Table 2.1 draws attention to the remarkable change in mechanical properties of a 70/30 brass subjected to cold work.

Hammer forging

The billet of heated metal is hammered either with hand tools or between flat dies in a steam hammer. Only simple shapes can be made using the basic process, and considerable operator skill is required.

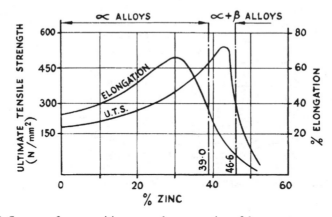

Figure 2.28 Influence of composition upon the properties of brasses.

Table 2.1

Condition of 70/30 brass	0.1% proof stress N/mm^2	Tensile strength N/mm^2	% Elongation
Cold-rolled – annealed	92	325	70
50% reduction by cold working	510	570	8

Rotary swaging

Rotary swaging (Figure 2.29) is a specialised form of forging. Here the dies are set within a ring of rollers. The dies themselves rotate about the workpiece and open and close rapidly. The shape of the die faces is such that they provide a reducing entry followed by a parallel portion, and material fed into the dies is tapered down to a reduced diameter by a series of forging blows. This technique is much used to reduce the ends of tubing to allow one tube to socket into the end of another. It is the standard technique for reducing the diameter of the end of a rod or bar.

Drop forging

Drop forging (Figure 2.30) is a development of simple forging, using closed-impression dies instead of hammers. The billet is placed between matched dies, and the impacting of the drop forging press forces the hot metal to conform to the die shape. A number of blows will be struck, each changing the shape of the billet progressively. For complex forgings more than one die set may be used.

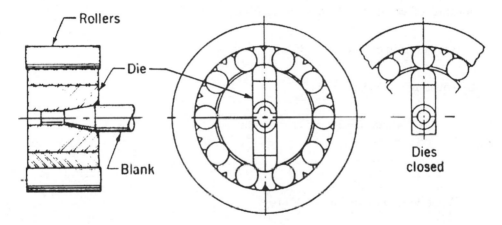

Figure 2.29 Rotary swaging.

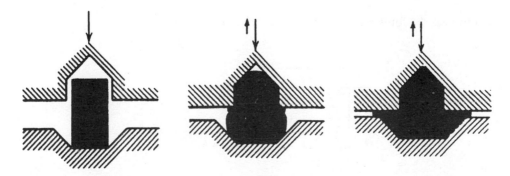

Figure 2.30 Drop forging.

Press forging

Instead of using progressive impact, press forging employs a slow squeezing action in a single operation. Very high pressures are required, but the accuracy of the resultant part is high. This work may be carried out hot or cold. The action may be one of pure forging, or may involve extrusion. Accurate calculation of the forces is difficult because of the involved mechanics of the plastic yielding of metals. However, reasonable estimates are required in order to select press capacities and to design suitable tools. Since cold extrusion (forging) of steel makes the most severe demands on equipment, this topic will be considered further.

The following semi-empirical approach is widely used. Equation 2.7 shows that the frictionless work of deformation can be calculated.

Work done $= \sigma_y V \ln(h_0/h_1)$, from which it is seen that the principal logarithmic deformation $\ln(h_0/h_1)$ gives a proportional indication of the work to be done.

Figure 2.31 shows a cylindrical billet and the extrusion which could be produced from it. If the frictionless work done on a perfectly plastic material is considered it can be shown that the work done in extrusion to reduce the diameter is equal to the work required in compression to return the metal back to the billet dimensions. By application of equation 2.7, substituting h_1/h_0 for h_0/h_1,

$$\text{Work done in extrusion} = \sigma_y V \ln \frac{h_1}{h_0}$$

By equating volumes,

$$\frac{A_0}{A_1} = \frac{h_1}{h_0}, \text{ where } A = \text{area of cross-section;}$$

hence

$$\text{Work done in extrusion} = \sigma_y V \ln \frac{A_0}{A_1}$$

(2.7)

$$\text{Mean force on punch } P_E = \sigma_y A_0 \ln \frac{A_0}{A_1}$$

where P_E is the force required to extrude, friction neglected.

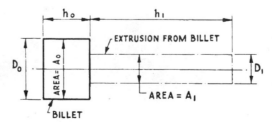

Figure 2.31 Cylindrical extrusion from a cylindrical billet.

It appears from equation 2.7 that the load on the punch (or ram) required for extrusion is related to three main variables:

1. σ_y, which will rise steadily according to the amount of deformation involved;
2. A_0, the area of billet to which the ram load is transferred;
3. In (A_0/A_1), which is a measure of the extent of the deformation required.

Research work by Pugh and Watkins (see Bibliography) has enabled the following empirical formulae for the cold extrusion of a number of non-ferrous metals to be established. The basic formula is:

$$p = a \ln \frac{A_0}{A} + b \tag{2.8}$$

where p is the extrusion pressure at the punch in N/mm^2 units:

$\quad A_0$ = cross-sectional area of the billet;
$\quad A$ = cross-sectional area of the extrusion.

The constants a and b for various materials are shown in Table 2.2.

Somewhat higher values of p generally occur as the speed of extrusion is raised.

The values of a and b automatically allow for the friction which occurs during the extrusions.

Figure 2.32 shows the experimental results in their graphical form for both forward and backward cold extrusion of a steel, En 2A, now 040A12. The empirical laws given above were formed by fitting equations to similar graphical results. Note that the backward extrusion of canisters does not give a graph quite the same as for forward extrusion. The reason for this lies partly in the fact that the ratio In (A_0/A) is not the exact principal logarithmic deformation for producing canisters.

Researches at PERA on the cold extrusion of low carbon steels have enabled Tilsley and Howard to construct a similar empirical equation for these steels,

$$p = \sigma_y\left(3.45 \ln \frac{A_0}{A} + 1.15\right) \tag{2.9}$$

Table 2.2 Constants for the pressure of extrusion, ram speed 140 mm/min.

Material	a	b
99.5% pure aluminium	223	106
ERHC copper	722	80
70/30 brass	1377	11
Tin	92	46
Zinc	325	177
Lead	42	39

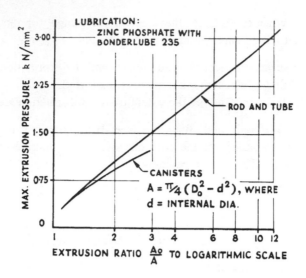

Figure 2.32 Extrusion pressures for En 2A steel, current equivalent 040 A12 (after Pugh and Watkins).

where σ_y is the yield stress of the particular steel after annealing. Pugh and Watkins, and Feldmann, also give information from which the deforming pressures for steels can be determined. For any particular composition of steel other than low plain carbon steel (0.1 – 0.3 per cent carbon), reference should be made to the available test date for that steel because the extrusion pressure rises with increase in carbon content and with alloying additions of nickel and chromium.

Example 2.2
Determine a procedure, and determine the extrusion pressures and loads. for producing a cold extrusion in 040A12 steel to the drawing Figure 2.33.

Solution
Before proceeding with the detail work, a few guiding principles need to be stated.
1. The slenderness ratio l/d of a backwards extrusion punch required to form a canister, should not exceed 3 when cold extruding steels, or the punch may fail by buckling.
2. Certain extrusion ratios reduce p to a minimum value, and for the backwards extrusion of canisters, $A_0/A = 2$ is a desirable ratio for this reason.
3. Generally, for canisters, a greater reduction can be effected by a forward extrusion than by a backward extrusion.
4. A maximum punch stress of 3700 N/mm² should not be exceeded.

From the drawing, Figure 2.33, it can be seen that the outside diameter of the finished cup is 25.6 mm. It might be possible to start with 24 mm diameter bar, coin this out to 25.6 mm diameter, and then backward extrude the canister. The required extrusion ratio would be

$$\frac{A_0}{A} = \frac{25.6^2}{25.6^2 - 18^2} = 1.98 \text{ (near to the ideal value)}$$

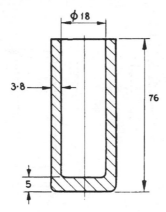

Figure 2.33 Example 2.2, extrusion.

However, the required length of punch would be 71 mm, making the punch slenderness ratio $71/18 (= 3.94)$, which is too high.

For this reason it will be necessary to start by making a backward extrusion as a preform, and to increase the depth of this by a forward extrusion. Possible billet diameters are 28 mm or 32 mm. If, as seems probable, the 32 mm diameter will give a solution, calculations must be made to confirm this.

Final extrusion ratio,

$$\frac{A_0}{A} = \frac{32^2 - 18^2}{25.6^2 - 18^2} = 2.11$$

which seems suitable.

Volume of component

$$= \frac{\pi}{4} [76(25.6^2 - 18^2) + (18^2 \times 5)]$$

$$- 21\,054 \text{ mm}^2$$

Required length of 32 mm diameter billet

$$= \frac{21\,054}{\frac{\pi}{4} \times 32^2} = 26.18 \text{ mm}$$

It will not be possible to forward extrude the entire volume of material for reasons given on p. 57, so an allowance must be made for a discard and trimming of the top of the canister. A 10 per cent allowance on the length of the billet is adequate.

The billet required is 32 mm diameter × 28.8 mm long and will be cut from bar stock and coined as shown in Figure 2.34(a). The coining operation assists in getting a symmetrical backward extrusion. The coined slug will be annealed, pickled, given a phosphate treatment (Bonderised) and then oiled. The phosphate coating, which itself acts as a lubricant, absorbs some of the oil and so further improves the lubrication.

The slug will then be backward extruded. Figure 2.34(b) shows the dimensions and tooling, dimension h must be calculated from the remaining sizes.

By equating volumes:

$$h(32^1 - 18^2) = 32^2(28.8 - 5)$$
$$h = 34.82 \text{ mm}$$

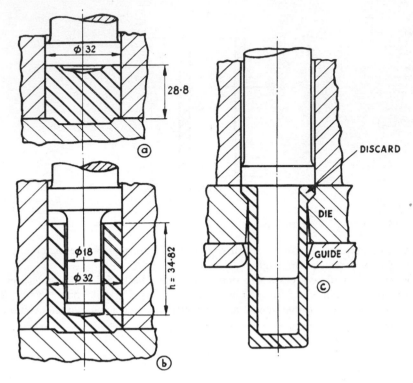

Figure 2.34 Tooling for a cold extrusion in steel.

The slenderness ratio of the required punch is

$$\frac{34.82 + 2.5}{18} = 2.07$$

which is quite satisfactory.

The extrusion ratio is $\dfrac{32^2}{32^2 - 18^2} = 1.463$

This is an acceptable ratio.

For 040A12 steel in the annealed condition, σ_y is about 310 N/mm^2
By equation 2.9, extrusion pressure, $p_1 = 310 \, (3.45 \ln 1.463 + 1.15)$

$$= 764 \text{ N/mm}^2$$

Punch load $= \dfrac{\pi}{4} \times 18^2 \times 764 \times 10^{-3} = 195 \text{ kN}$

The canister so produced will now be forward extruded. The yield stress σ_y will be much higher than the 310 N/mm^2 used above unless the canister is annealed prior to the forward extrusion, however, the total amount of deformation required is quite possible without interstage annealing. Reference to load compression graphs show that after cold extrusion to a ratio of 1.463, σ_y will be about 560 N/mm^2.

Note

The overall extrusion ratio,

$$\frac{32^2}{25.6^2 - 18^2} = 3.09$$

is also the product of the separate extrusion ratios, i.e. $1.463 \times 2.11 = 3.09$. Extrusion in steel may be taken to overall ratios of between 3 and 6, depending upon composition, without interstage annealing.

$$\text{Extrusion pressure} \quad p_2 = 560(3.45 \ln 2.11 + 1.15)$$

$$= 2087 \text{ N/mm}^2$$

$$\text{Ram load} = \frac{\pi}{4}(32^2 - 18^2) \times 2087 \times 10^{-3} = 1147 \text{ kN}$$

(i.e. about 115 t force)

Figure 2.34(c) shows the arrangement for forward extrusion.

□   □   □

Extrusion speeds

Experimental work reveals that there is generally some increase in the extrusion load accompanying an increase in extrusion speed, although the reverse may occur if the rise of temperature caused by the increased rate of working is sufficient to lower the yield stress of the material being worked.

Brasses and other non-ferrous alloys are extruded at speeds between 12 and 65 mm/s, the rate depending on the capacity and power of the equipment employed. Small extrusions carried out on crank-type presses may be done at relatively high speed, 150 mm/s.

Cold forging of steel

Extrusion techniques have been combined with hydraulic press techniques of squeeze-forging to give the process now generally described as cold forging of steel. The object is to form a component from a suitable billet without having to machine away surplus material; a cold-forged and thread-rolled socket screw is a typical product of the technique. The development is dependent upon special tool steels capable of withstanding compressive stresses up to 3700 N/mm^2, and upon presses such as the one installed by PERA, a 300 t double-action hydraulic extrusion press designed specially for this type of work. Figure 2.35 shows how backward and forward extrusion may be combined in a cold-forging operation, while Figure 2.36 shows a typical component produced by the cold forging technique.

The economic advantages may be summarised as follows:

1. Material saving: this can be as high as 70 per cent over conventional machining methods.

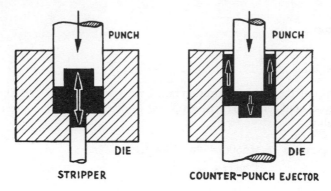

Figure 2.35 Cold-forging operations (after Feldmann).

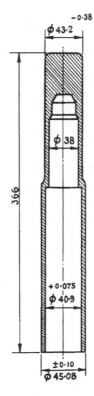

Figure 2.36 Typical cold extrusion forging in steel.

2. Reduction of process time: machining processes may be completely eliminated and the extrusion forging processes can be performed quite rapidly.
3. Dimensional accuracy: a tolerance of IT 11 is normal, but IT 9 can be achieved by introducing a 'sizing' operation. By special coining or calibration operations, IT 8 or IT 7 can be achieved. The resulting surface finish, 0.6–0.8 μm R_a, value, is superior to most single-point machining operations and equal to much commercially ground work.

4. Enhanced material properties: a 300 N/mm^2 UTS steel may cold work to 600 N/mm^2 UTS in the product, the hardness increasing from about 90–100 HB up to 200–250 HB. Figure 2.37 shows the changes in the mechanical properties of the material resulting from the process. There is, of course, some fall in ductility.

Economics of cold forging

Exploitation of the process depends upon adequate technical information at the design stage, and upon expensive specialised tool design and manufacture. For these reasons the quantities required in order to compete with conventional machining methods are rather high, but the quantities required may fall as the process becomes more highly developed and more widely used.

For multi-purpose installations capable of handling a reasonable range of work, the economic minimum quantity is about 3000–10 000 components, depending upon the complexity of the part. Where highly specialised single-purpose installations are required (as in automobile manufacture) the economic minimum quantity will be 100 000–500 000 parts, depending upon complexity.

The maximum advantages cannot be obtained without complete reconsideration of component design, because a component originally conceived as a machined drop stamping may not be ideally suited to cold forging. Satisfactory exploitation of new manufacturing processes must start at the component design stage, hence the need for designers and draughtsmen to study the principles involved.

Upset forging or heading

Upset forging or heading (Figure 2.38) is a form of forging where only a portion of the material is deformed in a die, the remainder being constrained to its original shape. The normal application is to the end of a bar to produce a head on the bar, hence the name. The maximum length of stock headed cannot be greater

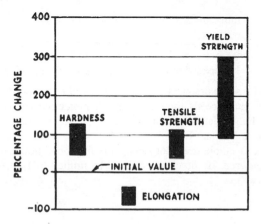

Figure 2.37 Changes in the mechanical properties of steel produced by cold extrusion (after Galloway).

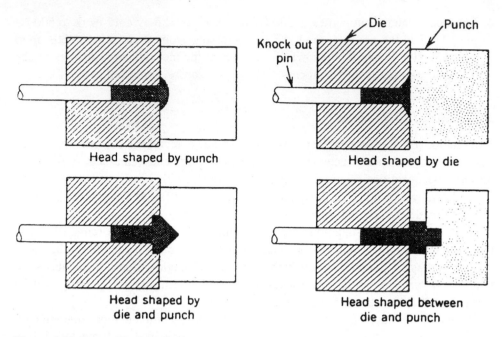

Figure 2.38 Upset forging principles.

than three times the stock diameter, or the stock will crumple instead of forging into a head.

Bolts and nails are made by cold forming the heads on to the end of the bar at very high speeds then cutting off to length, but components larger than about 12 mm diameter usually have the portion to be forged heated to a high temperature.

Example 2.3

Consider a drive shaft 25 mm diameter as sketched in Figure 2.39.

Volume of head = $\pi/4 \times 75^2 \times 10 = 44\,178.6$ mm^2
The bar cross-sectional area is $\pi/4 \times 25^2 = 490.9$ mm^2
Hence the length of bar needed to make the head is $44\,178.6/490.9 = 90$ mm

We now have to consider some practical constraints derived from shopfloor experience. If the unsupported length of bar exceeds three times its diameter the bar will buckle instead of forming a head, *but* longer lengths than 3d can be upset in stages provided the bar is supported in dies of not more than 1.5d.

It is obvious that the example does not satisfy the first condition, and so an intermediate forging operation is required. If the clearance is to be limited to 1.5d, then the diameter of the intermediate blank will be

$1.5 \times 25 = 37.5$ mm

and its head length will be

$44\,178.6 \div (\pi/4 \times 37.5^2) = 40$ mm

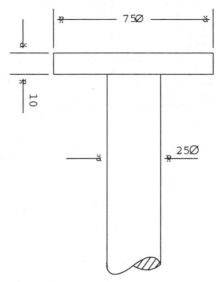

Figure 2.39 Example 2.3.

The bar must now be forged to the final shape. The ratio of length to diameter is now 25/ 37.5, i.e. 2 and, as this is less than 3, the part formed blank can be put straight in to the final die to provide the finished part.

□   □   □

Coining

Coining is a specialist application of forging. Punch and die carry different forms, and the exterior of the punch fits into the upper part of the die cavity. The blank is thus constrained from lateral flow. A single impact forms the relief design on each side of the piece.

Hobbing

Hobbing (Figure 2.40) is a plastic flow technique used to produce mould cavities for plastic moulding and die casting. A hardened steel form or hob is forced into a block of soft steel to form the cavity. Several pressings, carried out under a large hydraulic press, with intermediate annealing, may be necessary before the cavity is satisfactory. During the hobbing operation the flow of metal in the blank is restrained from lateral movement by a heavy retaining ring around the block.

The advantage of hobbing is that several identical cavities of good surface finish can be produced from the one hob.

Impact extrusion

Impact extrusion (Figure 2.41) is a technique for producing thin-wall containers with formed bases. Much used for toothpaste-type tubes, and food cans, the process has

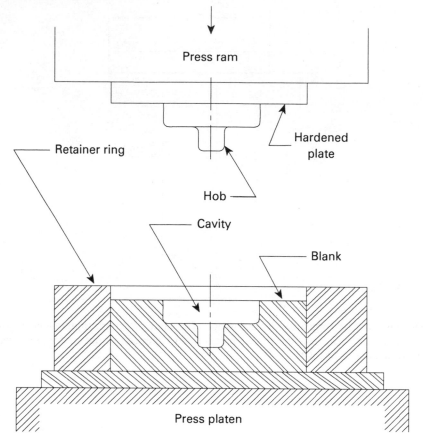

Figure 2.40 Hobbing.

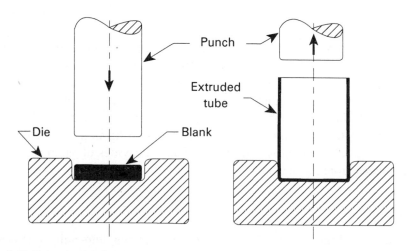

Figure 2.41 Impact extrusion.

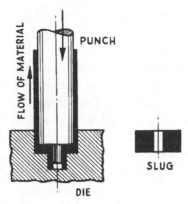

Figure 2.42 Cold impact extrusion.

also been applied to engineering parts. While normally a cold process, some materials require the use of heated billets.

The billet is placed in a die cavity, and is struck with a punch a single blow of considerable force. The punch is not close fitting in the die, but has clearance equal to the wall thickness desired. Under pressure the metal becomes plastic and is extruded through this gap to form the body of the container, while the bottom of the die and the punch end together form the remaining material into the bottom of the part in what is essentially a concurrent forging operation.

The technique differs from other forms to extrusion mainly in the extrusion speed (100–350 mm/s) and in details of the tooling. Figure 2.42 shows a typical cold impact extrusion as carried out on a crank-type press, and the comparatively simple tooling involved. It is much simpler to produce a small deep canister in aluminium by impact extrusion than by deep drawing.

2.5 Continuous extrusion

Extrusion is a process used to form continuous shapes by forcing material through a die. The growth of the use of aluminium structural sections, and a variety of complex plastic sections, in long lengths has made it an important process for the provision of complex section profiles to close tolerances. A centre piece can be positioned in the die aperture, provided means can be devised of hold it against the forces involved, to give a shaped hole up the centre of the extrusion. It is also used to produce film, sheets, filaments, tubes, and for the coating of wires and other metal shapes.

While the principles are the same, the techniques used for metal extrusion differ from those used for plastics due to the nature of the raw material and the forces involved.

Extrusion of metals

Hot extrusion of copper alloy bar in a variety of sections dates from 1894, when George Alexander Dick took out a patent for his horizontal extrusion press. This worked on the principle shown in Figure 2.43 (a) where a ram acts on the back of a heated billet located in a substantial chamber, forcing the material through the die located at the front of the chamber. This is referred to as *forward extrusion*. It was noted that the pressure required to extrude the material fell off as the ram advanced, due to the decreasing force requirement to overcome friction of the billet moving forward in the chamber. To reduce this, the *indirect or backward extrusion* process was developed, where the extrusion die is mounted on the face of the ram, thus cutting out the movement of the billet, and therefore the friction (Figure 2.43(b)).

The flow pattern of extrusion has been studied by means of the slip-line field theory and also by practical tests. Slugs of soft material are divided axially on a diameter, and a square mesh network is marked on the axial sections. After extrusion the halves of the slug are separated and the resulting distortion of the network examined. Figure 2.44(a) shows the flow pattern of a typical forward (or direct) extrusion. Note the 'dead' metal zone at the corner of the billet chamber and the shearing of the billet resulting from this. Figure 2.44(b) shows the flow pattern of a typical backward (or indirect) extrusion to form a canister, where taper on the punch end, and the more limited extent of the deformation, have practically eliminated any dead metal at A.

Ram forces in extrusion

Experimental studies have been made by fitting pressure-recording equipment to hydraulic presses and plotting the pressure against the ram travel. Figure 2.45 shows the curve for backward and for forward extrusion of aluminium at a temperature of 45 °C. The curves from O to A and O to B show the initial build up of the load, during which there is elastic flexure of the press and a small yield of the billet necessary to

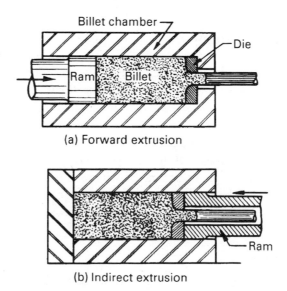

(a) Forward extrusion

(b) Indirect extrusion

Figure 2.43 Continuous extrusion processes.

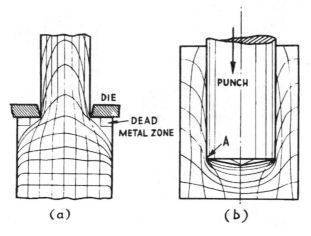

Figure 2.44 Flow patterns in extrusion: (a) forward extrusion (after Pearson); (b) backward extrusion (after Feldmann).

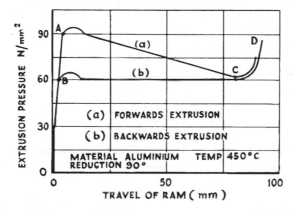

Figure 2.45 Extrusion pressures (after Pearson).

force it against the wall of the billet chamber. Pressure for forward extrusion is seen to fall as the work proceeds from A to C, while for backward extrusion the pressure remains approximately constant. This difference is due to the friction which arises in forward extrusion, during which the entire billet must slide forward in the billet chamber. Both curves show a sudden rise from C to D. This occurs as the billet-end gets nipped between the punch and the die, and is due to the rise of yield stress caused by deforming and by the fact that the friction force depends upon the ratio d/h. There is a small amount of the billet which cannot be extruded and which must be cut off as a discard from the end of the extrusion.

Figure 2.46 shows an indicator diagram for a cold extrusion of the type obtained by Feldmann. The diagram shows the initial spread of the billet up to point A, the elastic compression up to B and the fairly uniform extrusion force which is then required. The useful work done in extruding the part is represented by the area

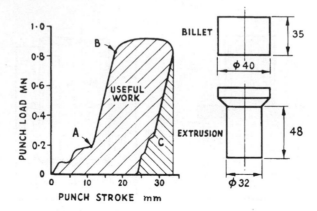

Figure 2.46 Load–displacement diagram for a cold extrusion in steel (after Feldmann).

enclosed by the graph, area C represents work lost as the elastic strains in the component and equipment are released. The very high punch load necessary for a cold extrusion in steel should be noted.

Typical hot extrusion temperatures are:

Magnesium and its alloys	280–320 °C
Aluminium and its alloys	450–490 °C
Brasses	700–750 °C
Copper	800–880 °C

Steels can be hot extruded in the 1000 to 1250 °C range but at these temperatures problems of scaling and die wear are considerable.

Extrusion of plastics

The plastic material in pellet or powder form is hopper fed onto a rotating helically-flighted screw which drives it down a heated barrel to the die. The molten plastic is forced through the shaped orifice of the die, assuming the required profile, and is then cooled rapidly by cold air blast, water spray, or immersion in cold water. When it is cool enough it is carried along by a take off mechanism and cut to length. There is usually some difference between the die profile and the finished section due to drawing down of the plastic after leaving the die (Figure 2.47).

An extension of the extrusion process provides for the continuous feed of a material, e.g. copper wire, through the centre of the die while extrusion is taking place to give a consistent concentric plastic coating.

Design of extruded parts

Simplicity of sectional profile is always desirable. Complex profiles, such as those

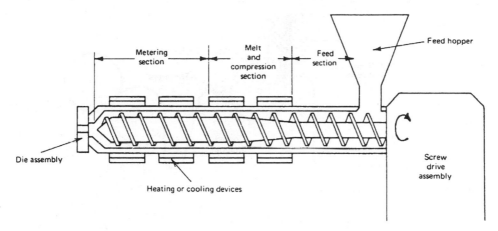

Figure 2.47 Extrusion machinery.

requiring a hollow within a hollow, or a projection within a hollow profile, should be avoided. As in moulding, uniform wall thickness is to be preferred, and balanced wall thicknesses are essential. Corners should be radiused wherever possible.

Thin sheet and film can be extruded by the use of a spreader chamber with a choker bar (Figure 2.48). This serves to spread the material to extrude it through a slot. The resultant sheet may be further thinned by passing it between calendering rollers prior to cooling and either spooling it on to a reel or shearing it into suitable lengths.

A similar technique is used in the making of plastic coated fabrics. Here the extrusion technique used is that for producing sheet or film, and the plastic is

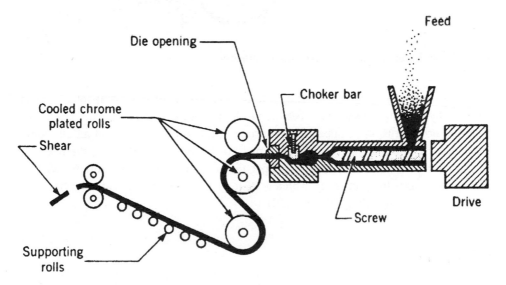

Figure 2.48 Extrusion of thin sheets and film.

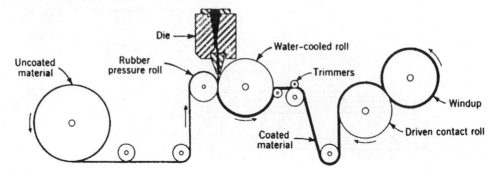

Figure 2.49 Extrusion coating process.

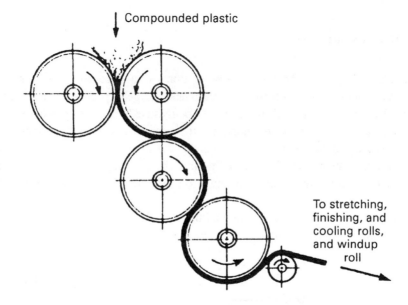

Figure 2.50 Forming film by the calendering process.

extruded directly on to the unreeling fabric, and while soft is compacted into the material by passing the sandwich between rollers (Figure 2.49).

While not extrusion, it appears appropriate at this point to mention an alternative technique for the manufacture of sheet or film. Instead of extruding, the plastic in its granular form is spread from a hopper to pass between heated rollers, where it softens and fuses to form sheet, which is then passed over further calendering rollers before cooling and reeling (Figure 2.50).

2.6 Rolling

This plastic deformation technique is widely used with metals to transform the material from billet to sheet or rod form (Figure 2.51). There are two categories of plastic

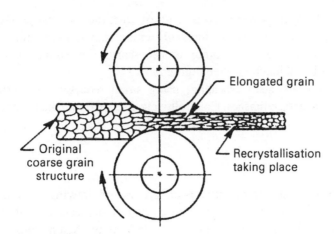

Figure 2.51 Rolling of sheet.

deformation, hot and cold. Hot working is defined as the working of materials at a temperature above the temperature of recrystallisation. When metal is hot worked, the forces for deformation are less, greater deformation can take place without fracture, and the mechanical properties are relatively unchanged. For cold working greater forces are required, and the strength and hardness of the metal is increased.

Hot rolling is used to produce sections and sheet material from billets, particularly for steel. The billet of cast steel is passed between rolls, either smooth rolls having progressively smaller gap in the case of sheet, or progressively changing form, if section rolling is required. The effect of rolling is to produce a refining of the grain

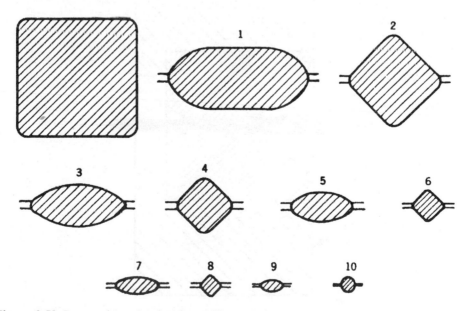

Figure 2.52 Progressive reduction from billet to rod.

structure and to eliminate porosity. As a result of this, the physical properties of the material are improved. Because of the high temperatures, oxidisation takes place rapidly at the surface, and descaling is necessary. For similar reasons, close tolerances cannot be maintained. (See Figure 2.52.)

Cold rolling is used, after descaling, using small reductions, to produce close tolerance and good surface finishes. Cold rolling is also used to induce work hardening.

2.7 Drawing

Another process of plastic deformation is the process of drawing (Figure 2.53). This can be considered as an inversion of the extrusion process, in that the material instead of being pushed through the die is pulled through. Again the process is used in both hot and cold forms, and may be used to make both solid and hollow sections.

Since there must be a portion of material through the die to get a grip on to commence drawing, the process of drawing is normally used as a secondary operation, working on rod which has been produced by rolling or extrusion, and working it down to smaller size and improved finish. Repeated drawing operations, with annealing in between, are used to produce fine wires and tube.

Drawing is done on a drawbench. One end of the material to be drawn is reduced by swaging to allow it to pass through the die, and this reduced portion is

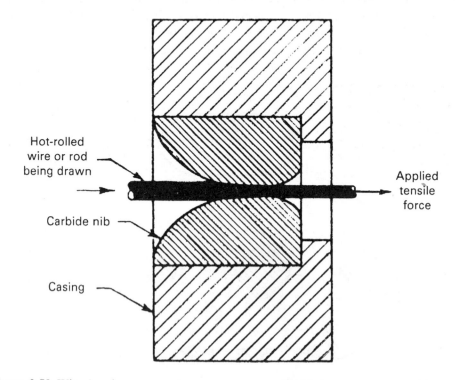

Figure 2.53 Wire drawing.

attached to the drawing mechanism, which may take the form of a carriage drawn by a chain drive, or in the case of a wire drawing bench a capstan round which the wire passes. The material is then pulled steadily through the die, reducing it in section.

When drawing tubular material it is common to control the inside diameter and finish of the tube by feeding the tube over a stationary mandrel. This is not essential, however, and tubing is often drawn through a round die without mandrel. Another technique is to draw the tube over a wire, which is later removed. Small bore hypodermic tubing is produced in this way. The same technique is used to draw composite wires, for example overhead conductor wires where the outer is of copper to give good conductivity, but a steel core is incorporated for strength.

2.8 Blow moulding

Blow moulding is a technique for making one-piece plastic hollow articles. The method consists of stretching a hot thermoplastic tube by air pressure, then cooling it against a relatively cool mould (Figure 2.54).

An extruder plasticises the material and forces it through a die to produce a parison (a hollow tube of plastic) of a suitable diameter for the product. The mould is positioned on either side of this parison, and closed over it, pinching one end and enclosing a blowing pin or compressed air entry pipe at the other. The enclosed parison is inflated by compressed air and expands onto the relatively cool wall of the mould where it hardens to the mould surface shape. When cool enough the mould is opened and the product removed.

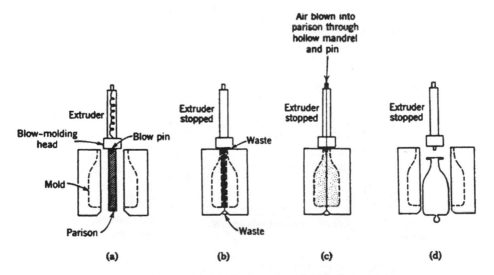

Figure 2.54 Blow moulding: (a) complete extrusion of parison; (b) close mould; (c) blow and cool; (d) eject.

As outlined above, this is an inefficient method, since the extruder is working only intermittently. In commercial operation the extruder is arranged to work almost continuously, with the moulds moving away for blowing, cooling, and ejection. Such a machine is shown here diagramatically (Figure 2.55). Three cavities are shown on a rotary wheel clamping around a continuously extruded parison. The moulds can be timed to clamp at the vertical, or slightly before or after, and the

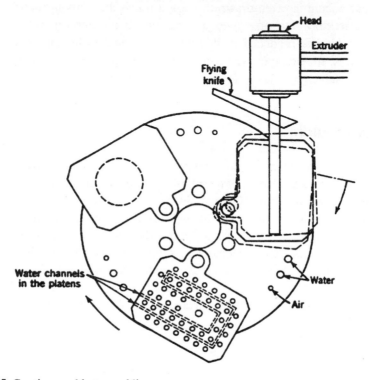

Figure 2.55 Continuous blow moulding.

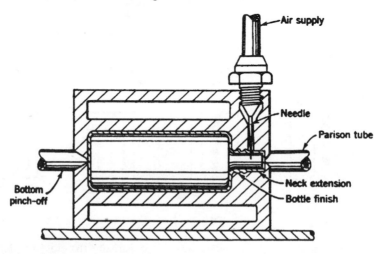

Figure 2.56 Bottle mould.

parison is severed by a flying knife positioned and timed to allow the severed end to feed into the next cavity as it arrives in position.

The mould is shown in more detail in Figure 2.56, illustrating how the air is fed into the neck of a moulded bottle through a small extension which is later removed. The moulded bottle is thus produced continuously and requires only a minimum of trimming.

The shape of the mould will determine the thickness of the wall at any point, and care must be taken with both mould design and parison positioning to ensure that excessive thinning does not take place. Corner radii should be generous, and deep grooves should be avoided.

2.9 Hydraulic forming

A similar process to the blow moulding process discussed above is used in the production of metal parts such as bellows, and T-pieces. The base material is annealed copper or brass tube, and this is placed within a mould corresponding to the finished object and is firmly clamped at its ends. Hydraulic fluid is then pumped in through one end, causing the pipe to expand to fill the cavity. This technique can be used to produce complex hollow shapes.

2.10 Rotational moulding

Rotational moulding is a technique for forming hollow plastic parts. A measured quantity of material is placed in a shell-like split female mould, which is then heated and rotated simultaneously about two orthogonal axes, causing the fluid material to disperse over the mould surface. The mould is cooled while still rotating, and the mould opened to give the finished article. The process is particularly suitable to larger parts.

Because the process is a centrifugal action, inward projections will have thinner walls and outward projections thicker walls. This can be utilised in the design of stiffening ribs. The rotational action also limits the minimum distance between adjacent walls of a rotationally moulded article. Wall thickness may also be controlled by heating or insulating sections of the mould to prevent material build-up on that portion of the mould.

2.11 Moulding of reinforced materials

The use of fibre materials to reinforce plastics is widespread. Mention has previously been made of the use of such materials in injection moulded parts. In such application the reinforcing material is normally in the form of short rovings, mixed with and injected with the carrier material. A considerable industry has also built up using woven materials as well as rovings, mainly with epoxy and polyester resins, to provide large reinforced plastic parts, or to provide strength to vacuum formed parts.

Techniques for making such large parts involve the use of liquid resin systems in combination with fibrous reinforcement. A male or female single surface mould is used, giving good surface finish on one side only. The moulds are inexpensive, such materials as wood or plaster being often used. The technique is to apply a gel-coat to the surface to ensure good surface finish, then to follow this up with layers of reinforcement and further resin, ensuring that the resin thoroughly impregnates the material.

Hand lay-up is the simplest process for this (Figure 2.57). Resin and mat are placed on the mould by hand, with brushes and squeegees used to distribute the resin and eliminate air bubbles. Since parts are normally large, and the process is relatively slow, slow room temperature curing resins are normally used, with

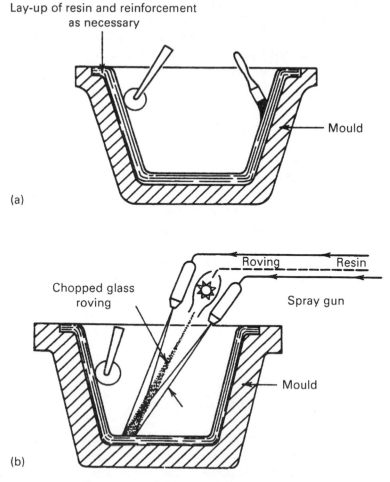

Figure 2.57 Lay up of reinforced material: (a) hand lay up; (b) mechanical lay up.

external heat, usually radiant, applied to speed up the curing as appropriate and when required.

The process is unpleasant, and many of the materials used give rise to allergic reactions. There is some concern for the health of the workers involved.

The spray-up process is essentially similar. Continuous glass fibre roving is fed through a fibre chopper and spray gun. Roving and fibre are sprayed together onto the mould surface building up the part wall thickness as desired. Since chopped roving is being used, as against woven mat or preform, the strength of the finished article is not equal to that of a good hand-laid product.

To achieve denser, higher quality products, hand or sprayed parts may, during the curing process, be subject to compacting forces by bag-moulding, i.e. by placing a flexible sheet over the lay-up and using vacuum or pressure to apply uniform force over the surface of the moulding. This may take place in an autoclave, providing heat as well as pressure.

Another technique used is filament winding. In this a continuous roving or tape is passed through a bath of resin then wound on a mandrel. When cured the resultant product is very strong, due to the high reinforcement content and its excellent disposition. The process is, of course, limited to axis-symmetric products.

Similar techniques are also utilised using preforms of reinforcing material, coating these preforms by immersion or applying resin by spray (or both) before and during positioning on the mould, to give optimum distribution of reinforcement.

3 Working of Sheet Materials

3.1 Vacuum forming

Vacuum forming, or to use the more correct terminology, thermoforming, is the process of forming a thermoplastic sheet into a three-dimensional object. The sheet is softened by heat, then forced against a mould by vacuum or air pressure, or a combination of these, then cooled.

There are a variety of techniques used:

1. *Vacuum forming* (Figure 3.1). The softened sheet is clamped around the perimeter of a female mould (a) and then is drawn down into the mould by extracting the air from the mould through small holes in the mould provided for this purpose. The softened sheet is drawn down into contact with the mould, and will take the shape and surface finish of the mould (b). After cooling the part is lifted from the mould, aided by application of air pressure in place of the vacuum. To assist the movement of the sheet, when forming deep shapes it is common to 'plug assist' the moulding process (Figure 3.2) by using a formed plug to push the softened sheet down into the mould before applying vacuum. The shape of this plug will control the stretch of the material and prevent unwanted thinning. Plugs may also be used to provide extra definition of local areas, for example the plug hole area of a moulded bath or shower.
2. *Pressure forming* (Figure 3.3). In this case air pressure is applied to the top of the sheet, usually through the assisting plug, to blow the material into contact with the mould. Air pressure may be used by itself, or in conjunction with a vacuum beneath the sheet.
3. *Drape forming* (Figure 3.4). Instead of a female mould, a male mould may be used, and the softened material lowered over the mould form, when it is pulled down by the application of vacuum through small holes in the mould.

The choice of method depends on the finish required on the product: drape forming will not provide as sharp an outer surface as will vacuum forming, but on the other hand drape forming will provide a more detailed inner surface.

Preferred materials for the process are those with good hot melt strength, i.e. become stretchable when heated to just below melting point. Generally used materials are polystyrene, ABS, PVC, Acrylic, cellulosics, polycarbonates, polyethylenes, and polypropylenes.

The process is suited to large thinwall parts with generous radii and flowing contours. Very large parts can be formed. Depth of draw is limited: the ratio of total mould surface to flat area should not exceed 3 : 1 and the ratio of draw length to part diameter should not exceed 0.5 : 1 for female drape forming, 2 : 1 for male drape forming, and 4 : 1 for snap back forming. The more generous the corner radii the better: these should if possible be of the order of four times the material thickness.

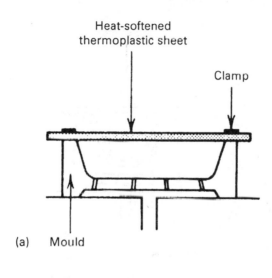

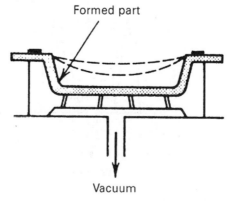

Figure 3.1 Vacuum forming.

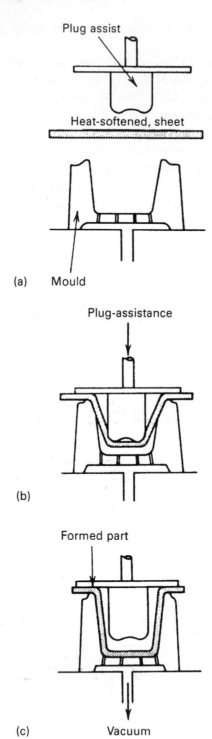

Figure 3.2 Plug-assist vacuum forming.

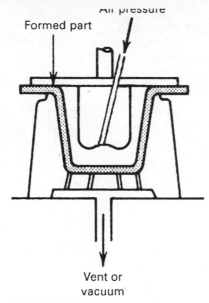

Figure 3.3 Plug-assist pressure forming.

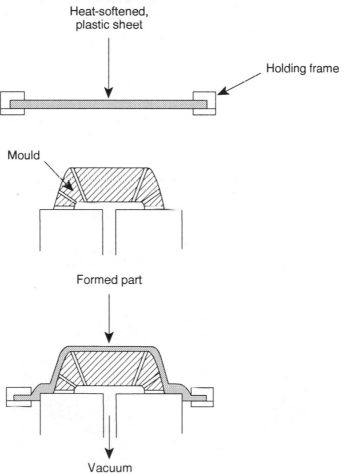

Figure 3.4 Drape forming.

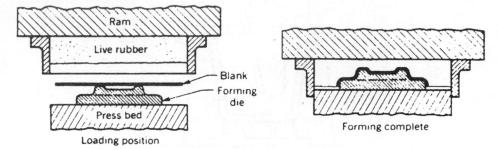

Figure 3.5 Rubber forming.

3.2 Rubber forming

The equivalent process for metal sheet is provided by rubber forming. The annealed sheet is placed over a former and a platen carrying a pad of hard rubber is pressed down on it, forming the sheet over or into the mould, according to whether a male or female former is used (Figure 3.5).

3.3 Superplastic forming

Certain extremely fine grained metals, notably zinc/aluminium and titanium/ aluminium alloys, exhibit superplastic characteristics under certain conditions of

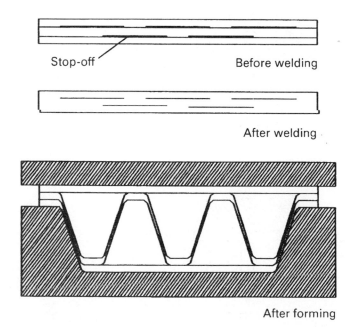

Figure 3.6 Superplastic forming.

temperature, allowing very large elongations to be produced by very low deforming forces, and yet exhibit substantial strength at normal temperatures. Under controlled conditions, therefore, these materials can be formed into complex shapes using processes normally confined to polymers, e.g. compression moulding, extrusion, vacuum forming, and blow moulding. This offers the advantage of low tooling costs, because of the low strength of the material at forming temperatures, ability to form complex shapes, weight and material savings, and a virtual absence of stress within the formed parts. However, the characteristics of the materials are that they must be formed at very low strain rates, and thus process times are long.

The importance of superplastic forming to the aircraft industry is found in the ability to combine diffusion-bonding with superplastic forming to provide light-weight panels with high stiffness to weight ratios. Sheets are first diffusion bonded along selected locations, then the unbonded sections between are expanded superplastically into a mould using air pressure (Figure 3.6).

3.4 Embossing

In embossing the punch and die are machined to mate with each other, with allowance for the thickness of the blank between them. The operation is a drawing or stretching operation, and one side is the reverse of the other. While normally used on metal sheet, plastic sheet may be embossed using heated platens.

3.5 Shearing

The following notes are based on the shearing of metals, but it will be appreciated that the findings are applicable to the shearing of other materials.

For simple shearing of a sheet of material along a straight line a guillotine is used, either hand or powered. The material is clamped on the bed of the guillotine by a clamp bar, with the line of cut along the edge of the bed, and the knife of the guillotine descends, creating a shearing action between itself and the edge of the bed, an action similar to the cutting of paper with a pair of scissors.

For shapes the blade is replaced by a shaped punch and the edge of the bed by a correspondingly shaped die plate. This is termed 'blanking'. A dieset incorporating guide pillars is provided to maintain alignment between punch and die. Such a blanking tool is shown diagramatically in Figure 3.7. Precise clearance is maintained between punch and die (5 per cent of the metal thickness when blanking steel) to give correct cutting conditions.

Punch and die are made from hardened alloy steel to resist wear. In operation the punch is pressed down on to the sheet material exerting a shear load which exceeds the shear strength of the material, and thus shears out the part, pushing it through the die. The stripper plate is provided to prevent retention of the material on the punch when the press is raised.

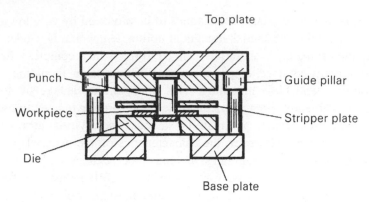

Figure 3.7 Blanking tool.

Blanking of this sort is carried out in a single action crank press, shown diagramatically in Figure 3.8. This press is driven by a motor which drives a flywheel. When the operator engages the clutch the rotating flywheel is coupled to the crankshaft, which rotates driving the ram of the press downwards. The energy stored in the flywheel provides the force necessary to push the punch into the die and then returns the punch to the starting position, whereupon the clutch disengages and the motor drives the flywheel up to speed ready for the next stroke, replacing the energy used in the blanking operation.

Figure 3.9 shows the progressive deformation and the development of a shear fracture during the shearing process, together with a typical load–penetration graph as obtained experimentally by Swift.

Figure 3.10 illustrates the effect of clearance on the piercing of a moderately ductile metal which work hardens and begins to develop cracks at an early stage of penetration. If the clearance is suitable the cracks run one into the other. The resulting hole is slightly tapered, the work done in shearing is somewhere near minimum, but the maximum load on the punch is almost independent of the clearance and is given by:

$$P_{max} = \text{metal thickness } (t) \times \text{perimeter} \times \text{ultimate shear stress of sheet}$$

Figure 3.11 shows the effects of varying the clearance upon the sheared edges of a blank. The amount of clearance required to give a reasonably clean edge varies with the thickness and hardness of the stock to be sheared. Swift suggested the values shown in Table 3.1.

Figure 3.12 illustrates a load–penetration curve for shearing where normal clearance is employed. It can be seen that a first estimate of the amount of work required for a shearing operation is given by,

$$\text{work done} = \text{max punch load} \times \% \text{ penetration} \times \text{thickness}$$
$$= P_{max}pt,$$

where the percentage penetration (p) represents the proportional depth to which the tools sink into the metal before cracks run one into the other. Table 3.2 gives some

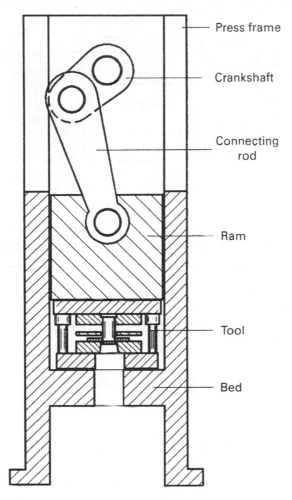

Figure 3.8 Single action crank press.

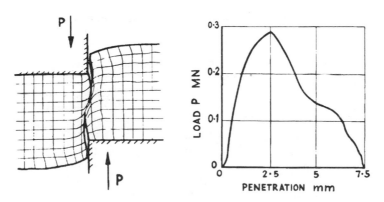

Figure 3.9 Shearing of mild steel, no clearance employed.

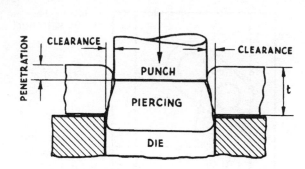

Figure 3.10 Influence of 'clearance' when piercing a hole.

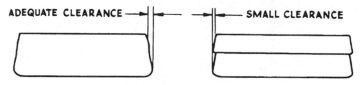

Figure 3.11 Effects of varying the amounts of clearance upon the edges sheared.

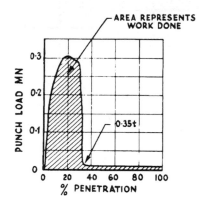

Figure 3.12 Load–penetration graph and work done in shearing, normal clearance.

Table 3.1 Clearance on tools for blanking and piercing.

Material	Hardness HV	Clearance as a % of thickness
Mild steel	94–144	5–10
70/30 brass	77–110	0–10
Copper	64–93	0–10
Zinc	61	0–5
Aluminium	21–28	0–5

Table 3.2 Percentage penetration and resistance to shearing.

Material	Annealed		Partially cold worked	
	σ_s N/mm^2	p%	σ_s N/mm^2	p%
Mild steel	240	50	297	38
Brass	220	50	360	20
Copper	150	55	195	30
Bronze	245	25	290	17
Aluminium	55	60	90	30

typical values for the resistance to shearing and percentage penetration which occurs under average conditions of clearance (σ_s is the ultimate shear stress).

In order to blank metal of substantial thickness or components of extensive contour, the work done in shearing may be spread over a larger length of press stroke by the introduction of *shear* on the tools. Figure 3.13 shows *shear* as applied to press tools.

Example 3.1

A hole, 100 mm diameter, is to be punched in steel plate 5.6 mm thick. The material is a cold-rolled 0.4 per cent carbon steel for which the ultimate shear stress is 550 N/mm^2. With normal clearance on the tools, cutting is complete at 40 per cent penetration of the punch. Give suitable diameters for the punch and die and a suitable shear angle for the punch in order to bring the work within the capacity of a 30 t press.

Solution

Suitable clearance on tools say 10 per cent of work thickness.
Punch diameter = 100 mm (determines smallest opening).
Die diameter = $100 + 2(0.1 \times 5.6) = 101.1$ mm.
Max load on punch (without shear)

$$= 550 \times 100 \ \pi \times 5.6 \times 10^{-3}$$
$$= 968 \text{ kN}$$

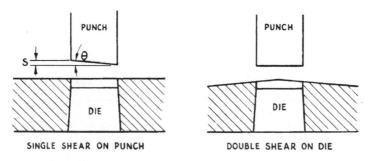

Figure 3.13 Application of shear to press tools, (s = depth of shear on punch).

Let s mm = the depth of shear required to reduce the punch load to 300 kN.
Work available during shearing stroke = (300 s)J.
Work required to shear hole = 968 × 5.6 × 0.4 = 2168 J.
Equating the amounts of work,

$$300\ s = 2168$$
$$s = 7.23 \text{ mm}$$

$$\text{Angle of shear, } \tan \theta = \frac{7.23}{100} = 0.0723 \qquad \theta = 4.2°$$

□   □   □

It must be remembered that shear distorts the resulting pressings, so a large flat pressing would need to be pressed with a flat punch, with the shear applied to the *die*. In this way the waste metal will be deformed to the shear angles on the die, but the pressing will be pushed by the flat punch and thus will remain flat.

The assumption has been made that shear on the punch will spread the load uniformly over the working portion of the stroke. This is not strictly true for piercing a circular hole. Also there is an additional load, required to bend the piercing to the punch face contour, which has not been included. Since the amount of shear angle suggested is already fairly large, it would be advisable to transfer the operation to a slightly larger press if available, say 35–40 tonne.

Finish blanking technique

Swift investigated blanking conditions in which a close-fitting punch having a sharp edge was used in conjunction with a die having a radiused edge, as shown in Figure 3.14. It was discovered that very smooth-edged blanks were produced, probably due to the ironing effect as the material is forced into the die, but that considerably more energy was required than for conventional blanking. The Production Engineers Research Association (PERA) further developed this process, now referred to as the 'finish blanking' technique.

Savings effected by this process in eliminating profile trimming of a copper blank. as reported by PERA, have been substantial.

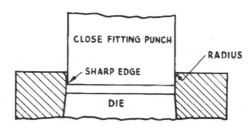

Figure 3.14 'Finished blanking' technique.

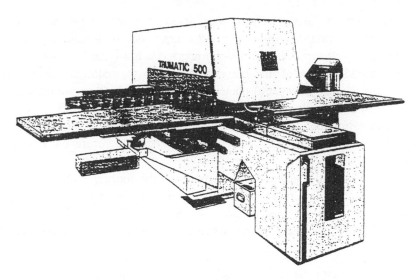

Figure 3.15 Turret press.

Turret presses

For a part having a number of holes punched in it the dieset will incorporate a number of punches and correspondingly located dies, allowing all the holes to be punched at one stroke of the press. Such a dieset is expensive to manufacture and maintain, and is limited to producing only the part for which it was designed.

To overcome these limitations, and to provide flexibility, the turret press (Figure 3.15) was developed. A number of punches and their corresponding dies are mounted in a turret rotating beneath the ram of the press. A number of strokes will be utilised, moving the workpiece to alternative positions, and rotating the turret to change punch and die as required, to produce the required part. Action is not limited to punching holes: the use of rectangular and shaped punches allows the production of profiles as a series of small steps (termed 'nibbling'). Movement of the part (held by an edge in clamps capable of movement in X and Y axes) beneath the press ram may be manual, controlled by a template, with punch selection being carried out manually, or both may be computer controlled. While the output rate is much slower than a multiple punch dieset, the flexibility offered by turret presses allows for low volume manufacture.

3.6 Bending

The simplest technique for producing a simple bend in sheet is the folder, where the body of the material is clamped to the bed of the machine with the fold line along the edge, and a hinged plate then folds the part overlapping the edge. Such an operation does not lend itself to high production rates, however, and most production bending work is done on a press or press-brake.

The bending of a metal strip or bar gives rise to plastic deformation in compression on one side, and in tension on the other side of a neutral plane. As the deformation proceeds the width on the compressed side of the bar increases, while that on the stretched side of the bar decreases and to maintain the moment of resistance for the area, the neutral plane displaces towards the compressed side of the metal. Figure 3.16 illustrates this. The shift of the plane must be taken into account when estimating lengths of material prior to bending; the principles are generally well known and are set out in engineering reference books under 'allowances for bends'.

Figure 3.17 shows the general arrangement of the tooling required to produce bends in strip metal. The load on the punch will vary during the operating cycle according to the work being done.

A punch load P_B is required to produce the bends. The conditions are as

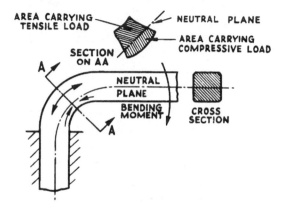

Figure 3.16 Displacement of neutral plane during bending of a bar.

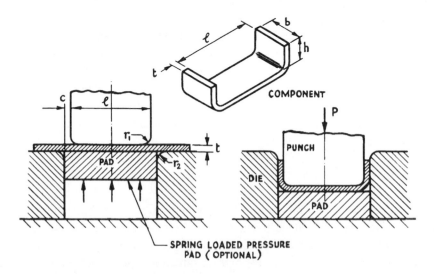

Figure 3.17 Press tool for a simple bending operation.

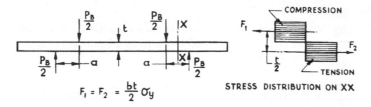

Figure 3.18 Approximate stress distribution in a bending operation.

represented in Figure 3.18. Once the material in the region of the bend is stressed to the plastic state the stress distribution can be represented with approximate accuracy.

$$\text{Moment of resistance to bending} = \frac{bt}{2}\sigma_y\frac{t}{2} = \frac{bt^2}{4}\sigma_y$$

The moment acting is obtained by assuming concentrated loads to occur at the points of tangency between the punch and die radii, and the strip, hence:

$$\text{Bending moment} = \frac{P_B}{2}a$$

Distance $a = r_1 + c + r_2$ at the start of bending but reduces as the punch descends and may be taken as equal to the metal thickness t for estimating P_B.

$$\text{Equating moments,}\ \frac{P_Bt}{2} = \frac{bt^2}{4}\sigma_y$$

$$P_B = \frac{bt}{2}\sigma_y \qquad\qquad (3.1)$$

The solution ignores the effects of friction.

As the metal strip is drawn into the die there will be friction between the component and the die walls, which will rise to a high value if dimension c is made smaller than the strip thickness t. Assuming the material is stressed up to the yield point, and that the coefficient of friction is given by μ, the maximum punch load required to overcome this friction is given by:

$$P_F = 2\mu bh\sigma_y \qquad\qquad (3.2)$$

If c is much smaller than t, ironing will occur and the forces will be considerably higher.

In order to 'set' the bends, and flatten (planish) the workpiece, it is sometimes required to 'bottom' the press. To do this effectively the punch load must be sufficient to take the material trapped beneath it just above the yield stress. The required force for planishing is given by,

$$P_p = bl\sigma_y \qquad\qquad (3.3)$$

Example 3.2

Figure 3.19 shows the dimensions of a bracket required to be bent from a flat strip of mild steel of 450 N/mm^2 yield stress. If the work is drawn to a depth of 18 mm below the top surface of the die, find the punch force required for:

1. Bending.
2. Overcoming friction, assuming the sides of the component are very slightly ironed (take $\mu = 0.15$).
3. Planishing at the end of the stroke.

Solution

By equation 3.1

$$P_B = \frac{20 \times 3 \times 450 \times 10^{-3}}{2} = 13.5 \text{ kN}$$

By equation 3.2

$$P_F = 2 \times 0.15 \times 20 \times (18 - 11) \times 450 \times 10^{-3}$$
$$= 18.9 \text{ kN}$$

By equation 3.3

$$P_P = 20 \times (32 - 14) \times 450 \times 10^{-3}$$
$$= 162 \text{ kN}$$

□   □   □

The results show quite clearly that for the planishing operation the press capacity must be much higher than is necessary to accomplish the bending operation. Because the force to overcome friction may exceed that required to bend the material, equation 3.1 does not satisfactorily indicate the press load capacity for bending.

Figure 3.17 and Figure 3.20 illustrate various types of bending tools. A feature of cold bending which frequently requires attention is the 'spring-back' which occurs due to elastic recovery of the strained material, as illustrated in Figure 3.21. Ways of countering this effect are:

1. Overbending, e.g. $\theta = 88°$ in Figure 3.20 (a) in order to produce 90° bend.
2. Overbending by means of an 'ironing' effect as illustrated in Figure 3.22.
3. Slight ironing and planishing to 'set' the bends.

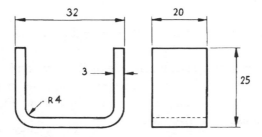

Figure 3.19 Bracket.

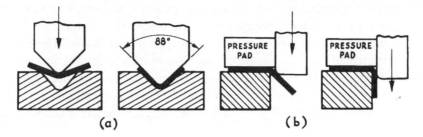

Figure 3.20 Various methods of bending by means of press tools.

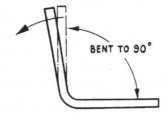

Figure 3.21 'Springback' in sheet metals bends.

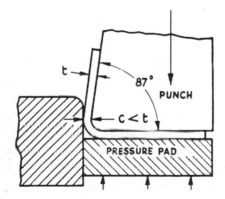

Figure 3.22 Overbending by ironing.

Again the above findings are primarily related to the working of sheet metal, but will be applicable to non-metals having suitable properties. Application to plastic materials may require that provision should be made for preheating the plastic material along the line of the bend.

3.7 Drawing

Drawing is a technique for producing three-dimensional parts from sheet material. The flat sheet is clamped between a pressure plate and the top face of a die, and then a punch having clearance between its flanks and the die aperture pushes the material through the die, stretching and deforming it to produce the part. The force on the pressure plate must be such that the material is allowed to slide into the die, but must be great enough to prevent wrinkling of the sheet as it is drawn in. The punch has not only to stretch and deform the metal, but also overcome the friction created by this pressure plate. Limitations on the shear strength of the material being drawn may make it necessary to have several successive drawing operations to produce a complex part without tearing. Figure 3.23 shows diagramatically a simple drawing operation.

While most drawing operations can be carried out on a single acting press as described for blanking, using spring action or die cushion to provide the pressure plate forces, the double action press (Figure 3.24) has been developed to give better control over complex drawing operations. The press has two concentric rams, the outer actuating the pressure plate while the inner drives the punch. These are normally operated by heavy duty hydraulic cylinders, and allow the speed of the draw to be controlled, as well as the pressure. Power requirements are greater than for a single acting press, since the motor must require the maximum power required at any instant.

Stress in deep drawing

Swift has shown that stress analysis applied to deep drawing is an involved matter. Rules for deep drawing remain largely empirical, and research has been confined almost exclusively to the drawing of cylindrical cups. For other shapes the theoretical discussion of the deforming mechanics is too complicated to lead to any useful general rules.

Figure 3.25 illustrates the deforming stresses when drawing a cylindrical cup, and Figure 3.26 shows the stresses acting upon a small element of material positioned within the flange, as depicted by Swift. It is clear that the element will tend to increase in thickness as it is drawn towards the wall of the cup. Figure 3.27 shows the changes of thickness caused by the drawing process.

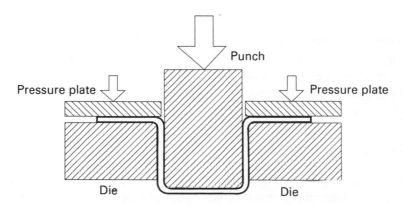

Figure 3.23 Deep drawing.

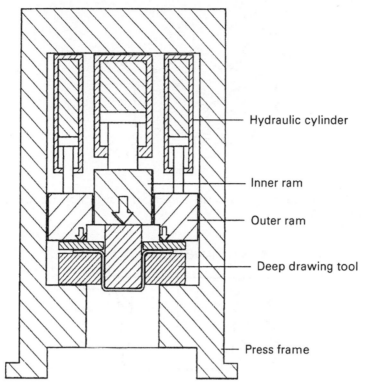

Figure 3.24 Double action press.

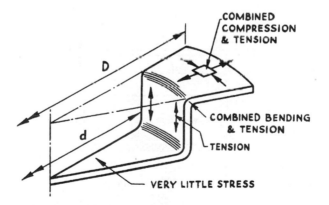

Figure 3.25 Deforming stresses in metal drawing.

The punch force in 'drawing' is limited to the maximum tensile load which can be carried by the wall of the cup, and this sets a limit to depth of flange which can be drawn.

Figure 3.28 shows the relationships between a blank and the resulting drawn cup. The ratio D/d is called the drawing ratio. The maximum value of the drawing ratio

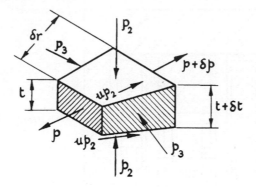

Figure 3.26 Stresses acting upon a small element of flange material during a drawing operation (after Swift).

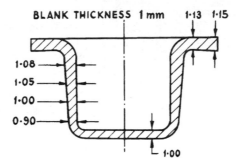

Figure 3.27 Plastic strain in a drawn cup.

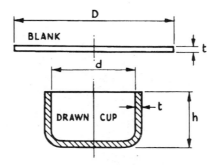

Figure 3.28 Relationship between blank and drawn cup.

depends upon the thickness ratio t/D. If it is assumed that no change in metal thickness occurs, the diameter of blank required to draw a given cup may be obtained approximately by equating surface areas.

$$\frac{\pi D^2}{4} = \frac{\pi d^2}{4} + \pi dh$$

$$D = \sqrt{(d^2 + 4dh)} \qquad (3.4)$$

From figures published by Crane (see Bibliography), and by reference to test results given in Swift's researches, Table 3.3 has been constructed to show the maximum drawing ratios for high-grade sheet steel.

Both Swift and Crane show that $D/d = 2$ represents about the maximum drawing ratio likely to be achieved under good conditions, and under general manufacturing conditions somewhat lower first reductions are advised. Some materials are relatively weak in tension, e.g. pure aluminium cannot be drawn to quite the high ratio of sheet steel.

The material properties which principally determine how well a metal may draw are:

1. Ratio of yield stress to ultimate stress (σ_y/σ_t), the lower the ratio the better.
2. Rate of increase of yield stress relative to progressive amounts of cold work; too rapid work hardening is usually an indication of a poor drawing property.

Professor Swift was able to show that the maximum drawing ratio was achieved by observing the following rules:

1. The drawing radius of the die to be $10t$.
2. The pressure pad force to be about one-third of the punch force, more rather than less.

An empirical formula for the punch load during drawing can be developed from two basic facts established by Swift's researches:

1. Cups tend to fail by tearing of the wall near the bottom when the ratio D/d exceeds 2.
2. The drawing load increases in an approximate straight line relationship with respect to the drawing ratio.

When $D/d = 2$ the punch force is given approximately by

$$P_{max} = \sigma_t \pi dt$$

where σ_t is the UTS of the sheet.

Table 3.3 Drawing ratios for sheet steel (first draw) where a pressure pad is employed.

Thickness ratio, $t/D \times 100$	Maximum drawing ratio, D/d
0.15	1.43
0.20	1.54
0.30	1.67
0.40	1.82
0.50	1.91
Above 0.50	2.00

For other drawing ratios,

$$P = \sigma_t \pi dt \left[\frac{D}{d} - 1 \right]$$ according to item 2 above.

Where P is required as a means of selecting press capacity, a degree of safety can be introduced into the expression and some allowance made for friction between the cup and the die walls, by adding 30 per cent to the above, i.e.

$$P = 1.3\sigma_t \pi dt \left[\frac{D - d}{d} \right]$$

$$= 1.3\sigma_t \pi t [D - d] \tag{3.5}$$

This expression does not make any allowance for ironing the walls of the cup and should be used without making such an allowance only where the clearance between the punch and die is about 5 per cent larger than the blank thickness.

Example 3.3

A cup, 105 mm inside diameter × 90 mm deep, is to be drawn from steel sheet of deep-drawing quality 1 mm thick. Determine the blank diameter and a suitable punch diameter for the first draw. Give the probable dimensions of the cup resulting from the first draw and estimate the press capacity required, assuming $\sigma_t = 415 \text{ N/mm}^2$.

Solution
By equation 3.4

$$D = [105^2 + (4 \times 105 \times 90)]^{1/2} = 221 \text{ mm}$$

Thickness ratio, $\dfrac{t}{D} \times 100 = \dfrac{1.00}{221} \times 100 = 0.453$

By Table 3.3, safe drawing ratio $= 1.82$

Diameter for first draw, $d = \dfrac{221}{1.82} = 121 \text{ mm}$

The cup cannot be produced by a single-stage draw, and the reduction must be divided between the operation here considered and a redrawing operation. If d is taken as 121 mm the amount of the reduction at the redrawing operation is rather small, and it would be better to have a somewhat lower reduction at the first draw. (See p. 90 for redrawing ratios.)

Let the diameter of first draw be $d = 130 \text{ mm}$

$\dfrac{D}{d}$ for first draw $= 1.7$

$\dfrac{D}{d}$ for second draw $= 1.24$, which is quite suitable.

Estimated depth of first draw when $d = 130$, by equation 3.4:

$$130^2 + (4 \times 130)h = 221^2$$
$$h = 61.43 \text{ mm}$$

Estimated capacity of press required, by equation 3.5:

$$P = 1.3 \times 415 \times \pi \times 1 (221 - 130) \times 10^{-3}$$
$$= 154.3 \text{ kN } (15.43 \text{ t force})$$

□   □   □

If a double-action press is used it might just be possible to do this job on a press of 15 t capacity. However, since the maximum punch force is only available near the end of the stroke, the maker's capacity charts should be consulted. If a single-action press is used the punch load will be increased by the pressure-pad load (because there is no separate part of the press mechanism to provide this), and an increase of one-third P must be made. It might then be just possible to do the job in a 20-ton single-action press, but again the maker's charts should be consulted.

Redrawing

Where it is impossible to obtain the required depth of cup at one draw, as in the above example, there are several ways of deepening the cup after the first draw. These are:

1. Redrawing, either by direct or by reverse redrawing methods.
2. Ironing, which will restore a more uniform wall thickness to the cup.
3. Pressure sinking, a method applicable to cups which are short relative to their diameter, and which increases the thickness of the cup walls. (See Willis in Bibliography.)

Figure 3.29 shows successive stages in producing a cup, and Figure 3.30 shows the normal form of a redrawing tool where a pressure pad is employed to prevent wrinkling of the work. Swift found that a throat angle of 10° gave the best results for most materials, but that 15° was superior for the redrawing of aluminium. Figure 3.31 illustrates reversed redrawing; the process does not permit of any greater reduction, but

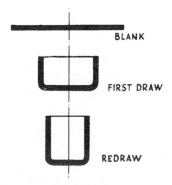

BLANK

FIRST DRAW

REDRAW

Figure 3.29 Successive stages of deep drawing.

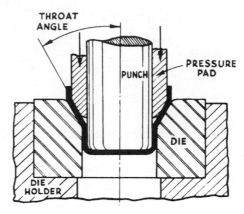

Figure 3.30 Press tool for redraw operation.

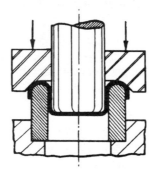

Figure 3.31 Reverse redrawing.

has the advantage of 'working' the material more uniformly than direct redrawing. For this reason it is sometimes used on brasses which have to be heat treated to avoid season cracking; there is then less likelihood of coarse grain developing due to critical amounts of cold work having occurred. Swift emphasised that the reverse redrawing die should have a single-radius curve over which the metal flows, not separate radii joined by a flat.

Redrawing is generally done in descending ratios for the subsequent draws given approximately by:

(D/d) 1.43, 1.33, 1.25, 1.19, 1.14 and 1.11.

For thin material a first redraw of 1.33 or 1.25 may be sufficiently high. Several redraws may be possible without inter-stage annealing, but the total reduction made must be within the ductile range of the material as revealed by its percentage elongation.

Ironing

This process is similar in principle to the cold drawing of tubes on a mandrel. There is a reduction in wall thickness and a consequent lengthening of the cup. Figure 3.32 shows typical ironing conditions.

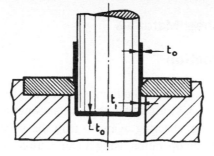

Figure 3.32 'Ironing' of a cup.

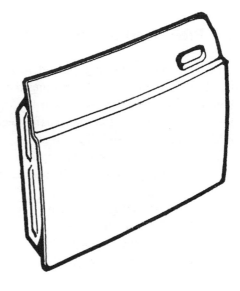

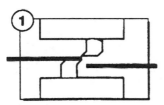

Blank die to cut out blanks from coiled material

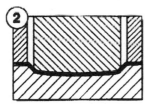

Draw die to shape blanks

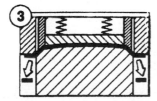

Trim die to cut off excess material

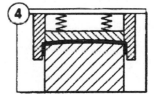

Flange die to make the initial bend for flanges

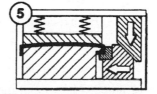

Cam flange die to bend the flanges further inside

Figure 3.33 Manufacture of car door panel showing complexity of press dies for panel processing.

3.8 Practical application

In practice a component may have a number of different press operations carried out in its manufacture. Figure 3.33 illustrates this; the production of the car door panel involves blanking, drawing, shearing again to trim the panel, bending to form the flange, and a further folding operation using a cam action to give horizontal folding to bend over the flanges.

3.9 Press load curves

Energy from the motor is stored in the flywheel. Engaging the clutch links the flywheel to the crankshaft, pushing the ram and the top tool downwards onto the bottom tool. The clutch is designed to transmit a certain torque, but the mechanical leverage exerted by the crankshaft mechanism varies throughout the stroke. This relationship is analysed as follows:

Let F_t = tangential force arising from flywheel energy

F_l = compressive force in connecting rod

R = crank radius

l = connecting rod length

P = force on ram

α = current crank angle

Considering the triangle a b c

$$\frac{F_t}{\sin(\alpha + \beta)} = \frac{F_l}{\sin 90}$$

Thus $F_l = \dfrac{F_t}{\sin(\alpha + \beta)}$

$$P = F_c \cos \beta$$

Thus $P = \dfrac{F_t \cos \beta}{\sin(\alpha + \beta)}$

And $F_t = \dfrac{P \sin(\alpha + \beta)}{\cos \beta}$

$$= \frac{P \sin \alpha \cos \beta + \cos \alpha \sin \beta}{\cos \beta}$$

$$= P \sin \alpha + P \cos \alpha \tan \beta$$

For most presses the second term in this expression is less than 10 per cent of the first term, since the connecting rod is long in comparison to the crankshaft throw, making angle β, and thus tan β, small. The second term may thus be ignored, giving the simplified relationships

$$F_t = P \sin \alpha$$

and

$$P = F_t / \sin \alpha$$

This can be used to plot the relationship between the crank angle and the available ram force. This curve is termed the *load curve* for the press.

Example

Construct the load curve for a single action press which is rated at 250 tonnes at a crank angle of 30° from bottom dead centre.

We have $F_t = P \sin \alpha = 250 \sin 30° = 125$ tonnes.

Now F_t is the force produced by the stored energy in the flywheel, the crankshaft and the press frame have been designed to withstand the force in the press at the *rated* angle of 30°. Thus we should take F_t as a fixed maximum value which should not be exceeded. We can use this value to calculate the resulting ram forces at various crank angles:

α	$F_t / \sin \alpha$	P tonnes
5°	125/0.087	1434
10°	125/0.174	720
20°	125/0.342	365
30°	125/0.500	250
40°	125/0.643	194
60°	125/0.866	144
80°	125/0.985	127

The graph of these shows two danger areas in the use of this mechanism (Figure 3.35).

1. At crank angles greater than 30° the mechanism would be overloaded by a force of 250 tonnes. Thus the point at which the press is rated is a very important feature of every press. The ISO standard rates at 30°, the figure used above, but unfortunately different countries and press manufacturers have adopted different standards. An American press will usually be rated at 1/2″ from BDC while a British press is probably rated at 5 mm from BDC. These are major differences in specification, and lack of knowledge of the design specification applicable to a particular press can very easily lead to a broken crankshaft.

2. The available force increases rapidly as the press approaches bottom dead centre. The press is designed for a 250 tonne load at 30° before BDC, but the crank/clutch mechanism is capable of exerting massive forces as α approaches zero, exceeding the design strength of the frame. These forces will only be generated if the tools are wrongly set, or if there is some obstruction to the press stroke, so great care must be taken to ensure correct setting. Squeezing type operations, e.g. coining, planishing, forging, and extrusion, require particular care.

Figure 3.36 shows a typical press crankshaft mechanism. The motor drives the flywheel through vee belts, and the power is transmitted to the crankshaft by means of the clutch plate. The clutch and the brake are operated by the same air cylinder, except that the spring loaded brake is *released* by the cylinder as it engages the clutch.

Presses are machines with a very high output rate. This means that if tool changing takes several hours (which it commonly does) large batches of parts are made to make up for the time the press is stopped for tool changing. This has lead to pressures to reduce tool changing times to avoid the pressure for large batches.

Different types of press operation make very different demands on the force profile required for the load curve. Blanking operations require a large force near the bottom of the press stroke,

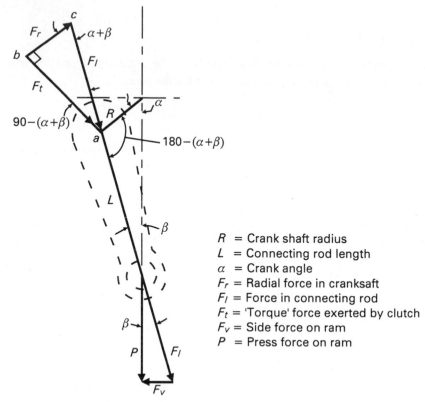

Figure 3.34 Forces acting on press crankshaft.

R = Crank shaft radius
L = Connecting rod length
α = Crank angle
F_r = Radial force in cranksaft
F_l = Force in connecting rod
F_t = 'Torque' force exerted by clutch
F_v = Side force on ram
P = Press force on ram

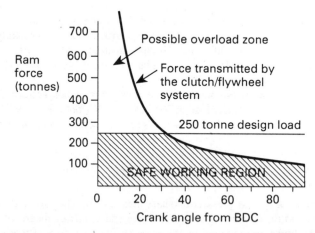

Figure 3.35 Load curve for a single action press rated at 250 tonnes 30° from BDC.

while deep drawing and extrusion operations can require a large force at a crank angle as large as 80°. It is therefore necessary to consider the use in relation to the function for which the press was designed – a 100 tonne press will not be a 100 tonne press for every application.

The press forces are also limited by the energy available in the flywheel, and operations requiring long duration application of force may demand more energy than is available.

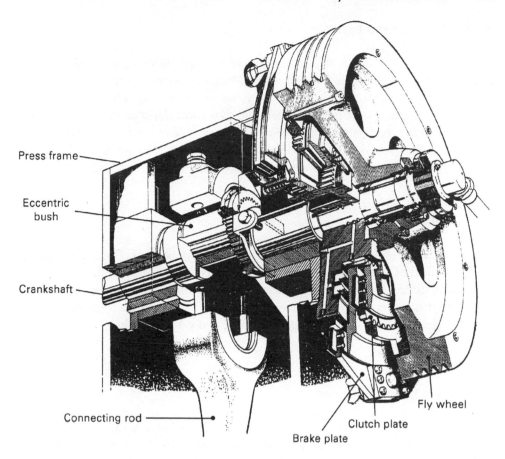

Press frame

Eccentric
bush

Crankshaft

Connecting rod

Brake plate

Clutch plate

Fly wheel

Figure 3.36 Diagram of mechanical press clutch/flywheel and crankshaft.

Presses are designed to economically produce enough energy for most operations they will be called on to perform. For economy, many presses are now designed with variable speed drive systems. It follows that a press running at half speed has only one quarter of the energy stored in the fly wheel compared to that at full speed (kinetic energy $= 1/2\ MV^2$) so slow running reduces the capacity of the press.

3.10 Hydraulic presses

Since these are powered directly by the motor pumping oil at high pressure to the cylinder powering the ram, such presses can exert maximum power at any point in the stroke. Additionally such a press cannot generate excessive forces at the bottom of the stroke. However, since there is no energy stored, the motor must be able instantaneously to provide all the power required for the operation. Hydraulic presses are intrinsically slower and more expensive than mechanical presses, and so are not widely used, but they excel when operations are of such a nature as to require a slow controlled squeezing action.

3.11 Other press types

The two basic press types used in manufacturing operations (crank and hydraulic presses) have been described above. Other types of press include toggle presses, where the action if by the application of force to the side of a toggle mechanism attached to the ram, lever presses, where the ram force depends on the mechanical advantage of a lever system, cam operated presses, where the ram is driven by the action of a follower on a cam, and many other devices. The reader is directed to specialist publications, such as *Mechanical Presses* by Makelt, for further reading.

3.12 Roll bending

Where the bends are of large radius, forming rolls may be used. The sheet is passed through a stand of three rollers, the upper of which is adjustable. The effect is to stretch one side of the sheet, producing a permanent set which forms the sheet into an even arc, the radius of which is determined by the relationship of the three rolls (Figure 3.37).

This same technique, but using grooved rolls, is used to form large radii from rod and other sections.

3.13 Pipe bending

A specialist bending application which occurs frequently in manufacturing operations is that of bending tubing. Application of normal press-bending techniques to tubing will cause the tube to collapse, due to the inability of the side wall to sustain the forces required to stretch the outside of the radius. This is overcome by restraining the side walls within a grooved former and applying the bending load via a grooved slipper progressively around the bend, so that the tubing receives the maximum support to the point where bending is taking place. With such devices radii as small as five times the tube diameter may be achieved in thinwall tube. (See Figure 3.38.)

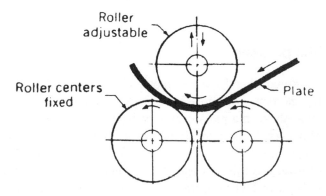

Figure 3.37 Forming rolls.

Figure 3.38 Example of complex CNC tube bending (Courtesy Addison Tube Forming Ltd).

3.14 Spinning

This is the process of shaping sheet material by forcing it against a former while it is rotating. The process is normally a cold working process, but heat may be used in some cases (Figure 3.39).

Hand spinning

Hand spinning uses a special lathe, similar to a wood lathe, except that in lieu of a tailstock there is a device to clamp the sheet to the form. The form is usually of hardwood, although metal forms may be used for large production quantities. Parts are formed using blunt hand tools resting on a tool rest similar to that shown in Figure 3.39 equipped with pins to provide fulcrums for the tools used.

Parts are formed either from sheet or from blanks previously drawn on a press.

Shear spinning

Shear spinning is a mechanised form of hand spinning, where power driven roll formers replace the hand tools to force the metal against the mandrel.

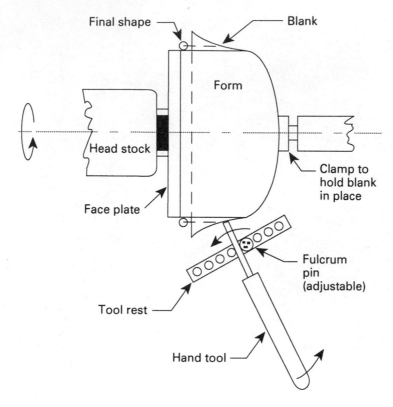

Figure 3.39 Metal spinning.

3.15 Roll forming

Roll forming machines are constructed with a series of mating rolls that progressively form strip metal as it is fed through the machine continuously at high speed. The number of rolls used is determined by the complexity of the section required.

Seam welding may be incorporated into the roll forming machine, allowing the production of seam welded tube and similar structures. The cost of the rolls is high, and thus the process is only justifiable where quantities required are large, but the speed of operation is high and the unit costs are low.

Seaming

Seaming, used in the manufacture of lightweight ducting, etc., is a not dissimilar operation. The body is roll formed, the edges being formed so as to interlock, and the final operation clinches the seams together.

A combination of roll forming, pressing, and clinching techniques not dissimilar to spinning are used in the specialist machinery used in the canning industries.

Exercises – 3

1. (a) What physical and mechanical properties control the suitability of an alloy for a deep-drawing operation?
   (b) Give a list of the chief metals and alloys which can be deep-drawn and discuss briefly the stages in a deep-drawing operation.

2. A factory producing brass sheet by the 'cold-work and anneal' cycle finds that in subsequent deep-drawing operations the sheet is subject to tears. Examination of the microstructure showed excessively large grain. Explain clearly the two ways in which this effect could be produced, and so decide the steps to be taken to prevent such losses.

3. (a) Describe the shearing action which occurs when cutting sheet metal with hardened steel tools.
   (b) Describe a method of reducing the maximum tool load during such an operation, and explain clearly why the load is reduced.

4. Figure 3.40 shows a cydindrical cup drawn from a disc. One formula for the maximum value of the drawing force (P) is $P = \pi dt\sigma (D/d - 0.7)$ where d, t and D are the dimensions shown on the diagram and σ is the yield stress of the material. The energy (W) required to draw the cup is given by

   $$W = cPh$$

   where h is the depth of the cup and c is a value depending upon the drawing conditions.
   (a) A blank of diameter 500 mm and thickness 4.5 mm is drawn in one pass to a cup of diameter 300 mm.

   Given $\sigma = 350 \text{ N/mm}^2$, $c = 0.6$, find the maximum load and the energy required to draw this cup.

   (b) A double-action press, load capacity 1500 kN and energy capacity 100 kJ when continuously stroking, is selected for the operation. Comment on the suitability of the press for this operation.

5. A cylindrical cup of inside diameter 30 mm and depth 60 mm, is drawn from brass of thickness 0.8 mm.
   (a) Calculate an approximate blank diameter.

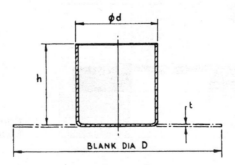

BLANK DIA D

Figure 3.40 Question 4.

(b) For the particular material the maximum ratio for successive draws (D/d) is 1.8, 1.4, 1.3 and 1.2. Calculate the minimum number of drawing operations for this cup.

(c) Explain why the drawing ratio (D/d) reduces for EACH subsequent draw.

(d) For the first drawing operation, the presses available are rated at 10 ton, 15 ton and 20 ton. The material has a tensile strength of 320 N/mm^2. Taking 1 ton force = 9.96 kN, find the lowest rated press which could be used.

Presswork problems

6. Calculate the force needed to bend up a box (see Figure 3.41) from 3 mm steel, the base is 100×150 mm and the walls are 20 mm high. The punch and die have 10 mm edge radii and the punch to die clearance is 4.5 mm. Take δy as 350 N/mm^2.

7. (a) What sort of press tool is shown in Figure 3.42?

(b) Name the main components of the tool.

(c) If the punch/die clearance = 7% of 't' give the diameters of the punches and dies.

8. (a) Describe the shearing action that occurs when cutting sheet metal with hardened steel tools.

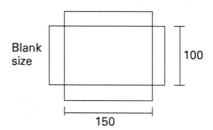

Figure 3.41 Rectangular box example (Question 6).

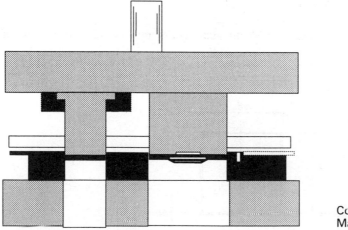

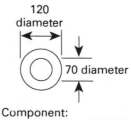

Component:
Material: 5 mm 070M20

Figure 3.42 Press tool.

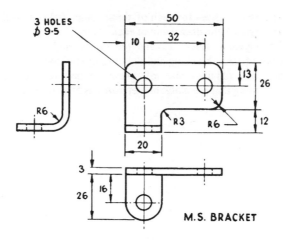

Figure 3.43

    (b) Circular blanks 150 mm diameter × 10 mm thick are being cut from steel with a shear strength of 300 N/mm². If the steel completely shears from the parent metal after 35% penetration calculate:
      (i) the punch force required (in kN);
      (ii) the energy required (in Newton Metres).
    (c) It is required to cut this part on a press with 1000 kN maximum capacity. Show how you would modify the tools to achieve this.

 9. Design a blanking and piercing tool for the component shown in Figure 3.43. Dimension the punches and dies and calculate the maximum press force required assuming a shear stress of 155 N/mm².

10. (a) What is the largest diameter that can be pierced in 1.6 mm thick steel plate of 310 N/mm² shear stress on a 250 kN press?
    (b) If the hole is the required dimension, what size should the punch be?
    (c) If the blank is to be drawn into a cup what would the cup diameter be (approximately) after the first draw?

Press mechanisms problems

11. Plot a load curve for a 250 tonne press (rated at 30° from BDC) with a stroke of 200 mm.
    (a) What is the maximum 'torque force' F_t available before the clutch will start to slip?
    (b) What force could the press exert at 10° from BDC and 1° from BDC?
    (c) What force could the press exert at 60° from BDC?

12. (a) Calculate the blank size required to produce a cylindrical component from 3 mm thick steel. The finished size is to be 150 mm outside diameter by 150 mm finished height. Calculate the blanking force if $\sigma_s = 400$ N/mm².
    (b) Plan out the drawing sequence – how many draws will be required to produce it?

(c) Calculate the drawing forces for each of these stages (take σ_y as 500 N/mm^2).

(d) Calculate the height of the drawn cup at each stage.

(e) Calculate the tangential force 'T' and the crank angle 'α' when it occurs; it is needed for each of these blanking and drawing operations.

(f) Hence specify the required size of press needed to accomplish all of these operations, assuming that the blank holder force is 50% of the punch force.

4 Machining

4.1 Introduction

The process of forming a product by the removal of material is common to all manufactured product, with only the techniques for removing the material varying. Even then, the same basic techniques are reflected in different industries.

All machining relies for its operation on a shearing action, the separating molecules of the material from the adjacent molecules by the application of force.

The basis of shearing is the application of concentrated force to a small area by means of a tool or knife while supporting the immediately adjacent material. This can be seen in the tailor's scissors, as well as in the powered guillotine. The force available may be applied over a long edge, as in the blanking of shapes from such materials as felt or paper using a shaped knife or rule and a semi-resilient support for the material, or may be concentrated at one or more points by applying shear to the blade, as in a guillotine. The application of the force may be linear, vertical as in the guillotine, or horizontal, as in a bread slicer, or rotational, as in a ham slicer. All, however, are forms of machining. Some may support the work close to the blade, while others will rely on the rigidity of the object being cut to provide the second component of the shear.

Machining operations can be split into two categories, those where the workpiece is moving while the tool is still, (referred to as 'turning') and those where the opposite applies, and it is convenient to use this categorisation when considering the processes available.

4.2 Turning

Lathe turning is generally considered to be the oldest mechanised machining process, the origins of wood-turning being lost in antiquity. The part is rotated while a tool is offered up to it (Figure 4.1). The lathe is capable of:

- Turning straight, conical, curved or grooved workpieces, such as shafts, handles, spindles, finials, etc.
- Facing to produce a flat surface on the end of a part.
- Profiling, either internally or externally, to produce concave or convex shapes.
- Producing shapes by the use of form tools.
- Boring to provide a shaped bore and internal grooves.
- Drilling to produce a hole concentric with the workpiece.
- Cutting off (or parting) to separate a piece from the stock material.
- Threading (or screwing) to produce internal or external threads.
- Knurling or embossing, to produce regular patterns on cylindrical surfaces.

Figure 4.1 Plain diameter turning on a lathe.

The basic lathe

Figure 4.2 shows a typical metal-cutting lathe, identifying the component parts. The machine illustrated is complex: for many applications, e.g. wood turning or glass working, a simpler machine is used.

The bed of the lathe may take other forms besides the dovetail form shown: it may be of round or T-section or take the form of parallel bars. Its function is to carry the headstock, tailstock, and tool carriage and maintain these in accurate relationship at all times.

The headstock is fixed to one end of the bed, and carries and drives the workpiece, either screwed or bolted to a face plate, held in a collet or chuck, or if the part is being held between centres, a pin on the headstock drive engages with a carrier clamped to the workpiece, as in the illustration. In wood turning the driven centre will incorporate spades which dig into the end grain and provide drive. The spindle of the lathe may be hollow, and may incorporate internal taper for the fitting of centres or other equipment.

The tailstock, which can slide along the bedways and be clamped in any position, supports the other end of the workpiece if required.

The carriage slides along the bedways, and carries the tool or toolrest, according to whether hand turning or a fixed tool or tools are used. Movement of fixed tools is controlled transversely by a cross slide, and in many cases by a compound slide which swivels to allow movement angular to the work axis.

Longitudinal movement of the tool comes from movement of the carriage along the bedway, controlled by a handwheel on the apron (i.e. the front of the carriage) engaging with a rack mounted along the front of the bed. Ability to power feed

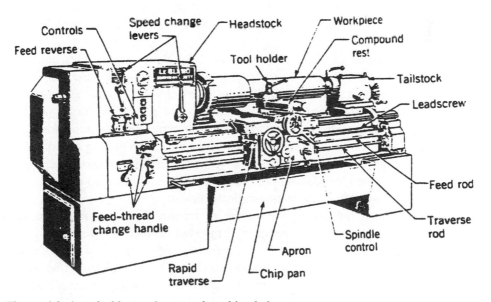

Figure 4.2 A typical heavy duty metal working lathe.

longitudinally and transversely may be provided by driving the handwheels from a feed rod running along the front of the bed, engagement and disengagement being under the control of the operator.

A separate lead screw, used for cutting threads accurately, also runs along the front of the bed, and is engaged with the carriage by closing a split nut around it.

The basic lathe requires a skilled operator, the quality of work produced depending on his care and attention. For production work it is desirable to eliminate the possibility of variation and error, and to this end mechanical stops are often fitted to the carriage motions to allow for accurate return to position, and also to disengage power feeds when the required cut is completed.

By coupling the cross slide to the spindle by means of a shaped cam it is possible to turn parts which are not round in section, but correspond to the form of the cam. Similarly by controlling the cross slide by following a shaped plate it is possible to produce repetitive longitudinal shaping. A combination of these allows the tracer lathe to produce parts with varying contours by following a rotating template. Such machines have the advantage of not requiring a high degree of operator skill, since the dimensions of the finished workpiece are dependent on the profile of the template (Figure 4.3).

The turret lathe

The next development in the provision of a production machine for use by an unskilled operator is the turret lathe (Figure 4.4, p. 108). In this the cross slide is simplified, and provided with a single inverted toolpost at the rear and a square indexing turret carrying up to four tools is fitted at the front. A numbered drum of stops is provided for longitudinal and transverse motions. All that is required of the operator is that he or she indexes turret and drum, and moves the carriage to the preset stop.

A saddle moving longitudinally is substituted for the tailstock. This carries an indexing turret designed to carry a variety of tools designed for the purpose.

A drum of adjustable stops is coupled to this indexing turret, so that as the turret is indexed the appropriate longitudinal stop comes into play. The turret saddle is advanced by a large capstan type handwheel, and the turret indexes automatically on return. The result is a very adaptable production machine not requiring any skill in its operation, provided it has been set up by a skilled setter. The capstan machine has been further automated by provision of powered feeds and powered control of the turret to provide semi-automatic or automatic operation.

Another variant of the turret lathe worthy of mention is the vertical lathe. This follows the same principles as the turret lathe, but carries the work on a large rotating table. This makes it suitable for the machining of large heavy parts. On such a machine it is common practice to provide transverse feed to the turret as well as longitudinal feed, making it a very versatile machine. In another variant it may lack

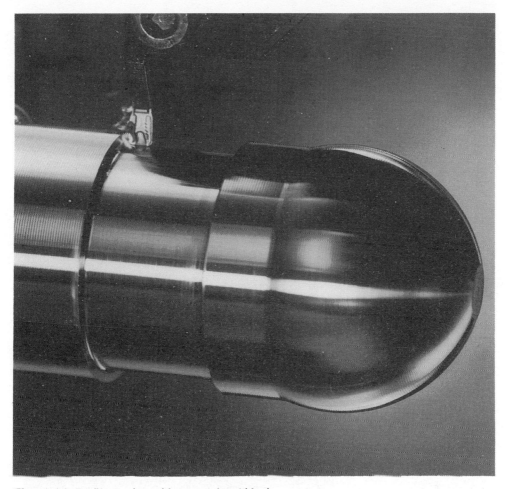

Figure 4.3 Profile turning with a coated carbide tip.

the turret, instead having a toolpost. In this state it is termed a 'vertical borer' (Figure 4.5).

The automatic lathe

The next development is the automatic lathe, where control of the machining cycle no longer rests with the operator, but is controlled by the machine itself, usually by cam actuated motions.

There are a number of types of automatic lathes, all derived from the centre lathe or the vertical lathe or boring mill. They may be broadly grouped as follows:

1. Vertical or horizontal spindle.
2. Single spindle or multi-spindle.
3. Bar, chucking, or coil fed.

Figure 4.4 Turret lathe.

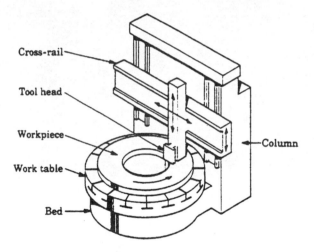

Figure 4.5 Configuration of vertical borer.

The single spindle auto

The single spindle automatic (Figure 4.6) is particularly popular as a first operation machine fed with bar stock, using a magazine type bar feeder. Notice the indexing turret, and the bar feed actuated from the main camshaft. Twin drives to the headstock give forward or reverse spindle, rotation selected by clutches operated from the camshaft. Additional vertical turning slides are cam operated from the auxiliary camshaft.

The single spindle auto differs from the capstan in that the turret revolves about a horizontal pivot instead of about a vertical axis. It will also possess several radial tool slides, used for form and parting tools.

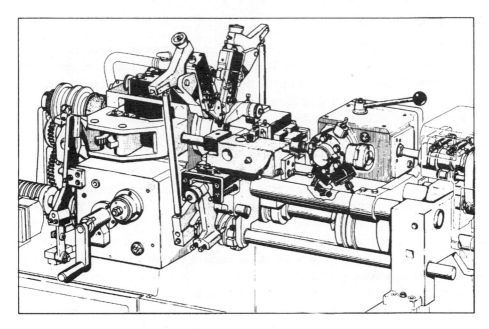

Figure 4.6 Single spindle auto.

Single spindle autos are often provided with pickoff arms also actuated from the camshaft, taking the part to auxiliary slotting or cross drilling attachments.

The sliding head auto

A variant on the single spindle auto is the sliding head auto. In this case the headstock is arranged to slide longitudinally, enabling two-dimensional profiles to be cut using simple single point tooling. The material is held in the collet of the sliding head, and passes through a steady bush close to the cutting point. It is therefore necessary on a machine of this type to ensure that the cutting takes place progressively as the material is fed out.

Second operation machines

The simplest of automatic lathes, for second operation work, closely resemble the capstan. An attendant is still required to load and unload the components, but the machine itself controls the operations. The control is either cam or sequential, and the machine will stop on completion of the work to await operator attention. One operator will normally operate several machines.

In this type of machine, as in the centre lathe, the work rotates. Each tool moves in one of two planes, in/out or longitudinal. Tooling is either single point or simple form tool, housed in roller boxes or other capstan-style toolholders. Cross slides are restricted to simple form tools or facing operations.

The multi-spindle auto

The limitations of such a machine are obvious. For mass production second operation work the multi-spindle auto (Figure 4.7) is preferred. Several spindles are incorporated in the headstock, all driven from a central drive shaft. The head is indexed, usually by a geneva-type mechanism, and machining takes place at all spindles concurrently. One spindle position is utilised for loading, usually by some form of automatic feed device. Each station is provided with a radially acting toolholder and provision for a longitudinally acting toolholder. It is normal to use complex form tools, and the machine is rigidly constructed and heavily powered to allow for this. To obtain the correct relationship between the movements these are all actuated from a single main camshaft, either directly or via auxiliary camshafts driven from it. Such machines are capable of very high output rates, while having a relatively short changeover time, since changeover comprises the replacement of a set of cams and the installation of a set of form tools. Initial tooling outlay in the manufacture of cams is relatively high, but these are long lasting and the machines are extremely reliable.

For large quantity second operation work such machines are difficult to beat.

Multi-spindle autos may also be bar-fed, with one station of the turret taken up by a feed-to-stop operation, and with one utilised for part-off. If the cutting time at the other stations is long, both part-off and feed may be carried out at the same station.

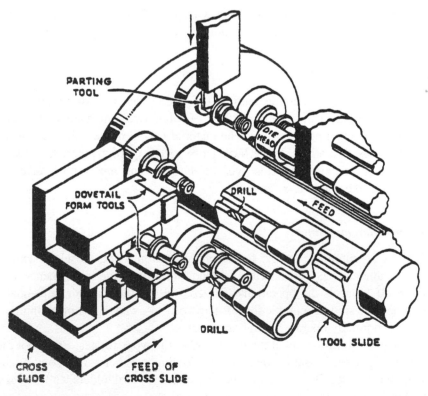

Figure 4.7 Multi-spindle auto.

The vertical auto

The multi-spindle lathe is also found in a vertical spindle form, where it resembles a rotary transfer machine, and it requires close examination to realise that the workpieces are rotating, and it is in fact a turning operation that is taking place. The vertical configuration has the advantage of easy parts loading for larger and more complex parts which cannot readily be fed automatically.

Coil fed lathes

Another form of lathe found is the lock lathe, or rotomatic. This is a clever adaptation of the turning principle to enable the turned part to be produced from coil stock. The coiled stock is fed through a rotating straightener, between grooved feed wheels and through a steady bush in the centre of a rotating toolhead. The tools are rotating about the stationary work, and are mounted on swivels which advance the tools against the workpiece under the control of cones moved along the linear axis of the spindle. Since the work is not rotating, the 'tailstock' accessories must be powered. As in the sliding head auto, profiles are generated by a combination of tool feed and bar feed.

The CNC lathe

A further development is the CNC lathe (Figure 4.8). Substitution of positional controlled drives for the handwheels of the conventional lathe allows for the

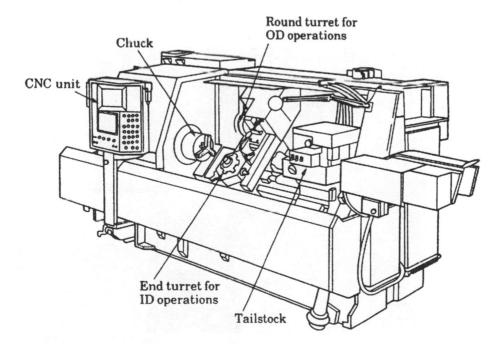

Figure 4.8 A computer numerical control lathe.

control of slide positions by computer, and opens up the possibility of generating an unlimited variety of profiles using continuous positional control of the slides.

This can then be extended by the installation of tool turrets under computer control, and a computer controlled turret on longitudinal and transverse slideways replacing the tailstock. Since provision has no longer to be made for a human operator, the machine can be made more compact and complex. The bed can be redesigned and placed at a slope to allow improved flow of coolant. The result is the versatile turning machine illustrated in Figure 4.8.

Driven spindles may be incorporated into the tool turrets, enabling milling and drilling operations to be carried out, and allowing complex parts to be produced complete in one operation.

The twin spindle CNC lathe

To enable the production of completely finished turned parts, the CNC lathe is available in a twin spindle version (Figure 4.9). The part, from chucked billet or bar fed material, is machined in the normal way, with all the facilities of the CNC turning centre. The second spindle is aligned with the main spindle, and takes the component at part-off (Figure 4.10). It then carries out a similar range of operations on the second end of the component, while the first spindle is producing the next component. The component discharged from the second spindle is thus a completely finished part.

Figure 4.9 Twin spindle operation.

Figure 4.10 Transfer between spindles.

4.3 Moving tool machining

Single point cutting

The concept of moving a single point cutting tool along a workpiece is common to a variety of industries, and although the name of the operation varies from industry to industry the machines themselves are very similar in operation. A morticer in the woodworking industry is little different in operational principles from a keyway slotter in engineering, for example, while a shaper in light engineering operates in the same way as a planer in heavy engineering. Typical machines are illustrated in Figure 4.11. The motion may not always be straight line: Figure 4.12 shows the rotary use of a single point tool, known as 'fly cutting' or 'trepanning'. Another common 'single point' use is boring, using a rotating boring bar.

Sawing

The obvious next move from the use of a single point tool is the use of a multi-point tool. Initially this was a straight line operation; early 'sawmills' were not large circular saws, as is common today, but were large straight oscillating saws, driven by water power. Such machines still exist, as they are capable of sawing very thick material, when a circular saw would require to be of a prohibitively large diameter. The principle is also still found in the fret-saw used in the woodworking industry, and the filing machine and the power hacksaw of engineering. A variant is the broaching machine (Figure 4.13), where a multi-toothed tool is pulled or pushed through material to provide a shaped hole which would be very difficult to produce by any other means (Figure 4.14).

These all incorporate oscillatory motion, and since the tool cuts in only one direction, the back stroke is ineffective and the process is relatively inefficient. For this reason, flexible blades were developed, allowing the cutting edges to follow a

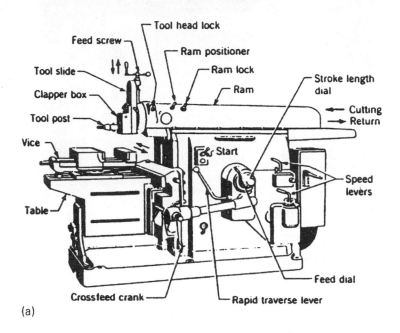

(a)

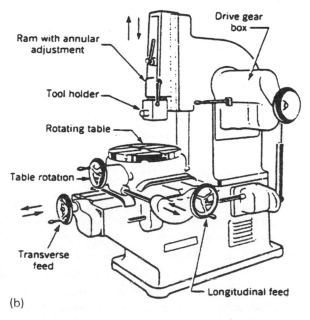

(b)

Figure 4.11 (a) Shaper; (b) slotter or vertical shaper; (c) planer.

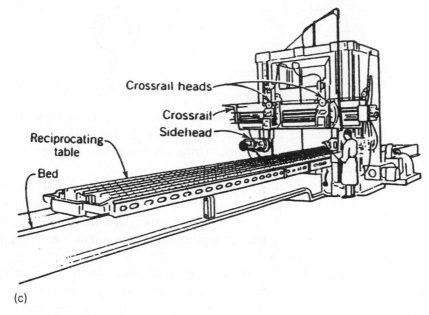

Crossrail heads
Crossrail
Sidehead
Reciprocating table
Bed

(c)

Figure 4.11 Continued.

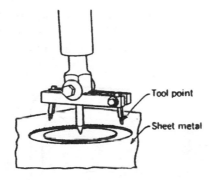

Tool point
Sheet metal

Figure 4.12 Fly cutting.

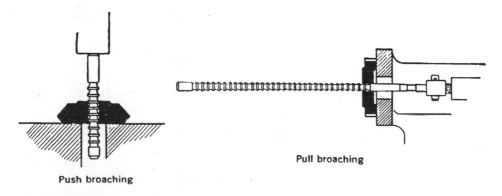

Push broaching

Pull broaching

Figure 4.13 Broaching operations.

Figure 4.14 Typical parts made by broaching.

straight path, then be curved back clear of the workpiece to return to the starting point, allowing for continuous cutting. This device, the band saw (Figure 4.15) may be found, in horizontal and vertical forms, in virtually every manufacturing industry with blades to suit all materials.

The next progression in cutting is to place the cutting teeth around the periphery of a rotating shaft or disk, and it is this form that the majority of machining is carried out. The circular saw is the most basic form and is illustrated in Figures 4.16 and 4.17 while Figure 4.18 shows another application of a saw, this time for cutting round holes in wood or sheet metal.

At this point it is appropriate to consider the cutting tooth itself. Like the lathe tool, it must have clearance for the chip produced, and front rake to provide a cutting face. There must also be some place for the chip to go, to avoid the chip locking the tool in the work, and for this reason the teeth on a saw are not in line, but are narrower than the cut which will be produced, and are staggered to provide clearance. This is referred to as 'set'. The terminology for saw teeth, and ways of achieving this set, are shown in Figure 4.19. The form of the saw tooth will vary according to the nature of the material to be cut: the cutting of cloth or the slicing of

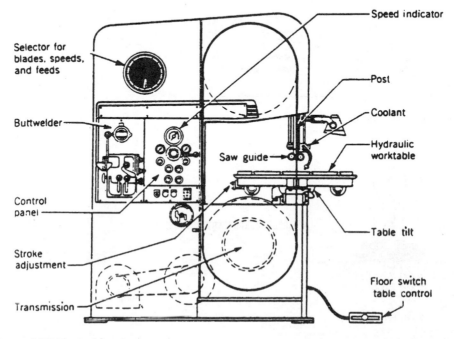

Figure 4.15 Typical bandsaw.

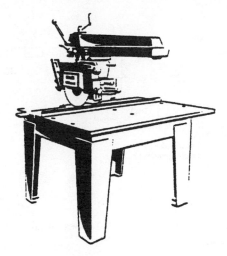

Figure 4.16 Overarm saw.

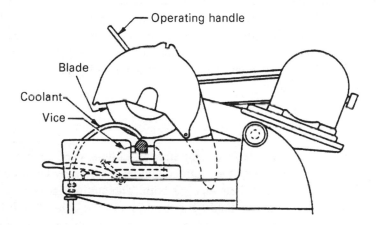

Figure 4.17 Cut-off saw.

Figure 4.18 Saw cutter.

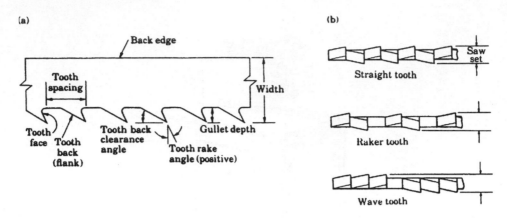

Figure 4.19 (a) Terminology for saw teeth, (b) types of saw teeth, staggered to provide clearance to prevent binding.

bread requires a cutter which is in essence a wavy-edged knife blade, sometimes without teeth as such, while steel requires a small hard tooth, and wood a larger tooth with a more pronounced set. Very hard materials may be ground; this is still a cutting action. Each particle of the grinding wheel forms a small, very hard cutting point which has the advantage, when it grows blunt, of breaking off to expose a fresh cutting point. Thus broadly similar machines, but using different cutting blades, will be found right across industry.

Milling

The term 'saw' is normally retained for narrow cutters used to cut off or slot materials, while the wider versions are termed cutters. These may be designed to cut along their periphery, or along their face, or both. The cutting faces may be linear, or may be formed. They may be held by a shank, as in the case of a drill, or they may be mounted on an arbor. Their cutting faces may be straight, or they may be inclined or spiralled to aid chip removal. A variety of cutting tools are shown in Figure 4.20.

The machines using these tools fall into two categories, according to the alignment of their cutting spindle.

Horizontal spindle machines

The horizontal spindle machine is very suitable for surfacing material. One of the commonest applications of this is in the woodworking industries, where the sawn planks are smoothed by passing them under a drum cutter. The forming of shaped surfaces is also carried out on a horizontal spindle machine, using a shaped cutter or combination of cutters. The engineering version of the horizontal spindle machine is shown in Figure 4.21.

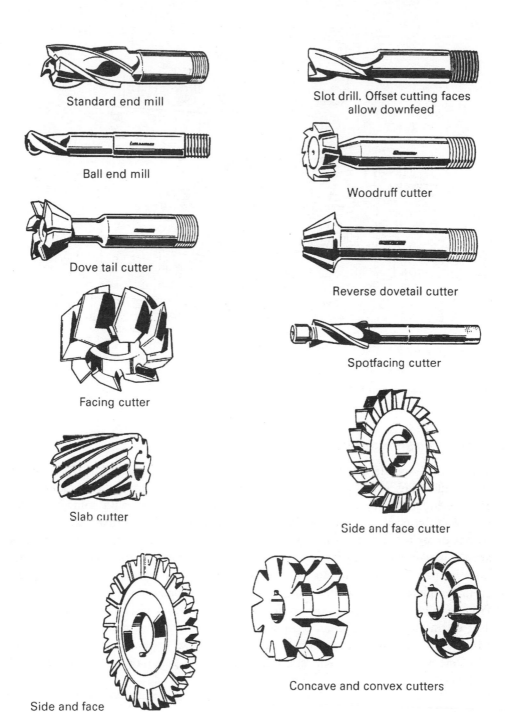

Standard end mill

Ball end mill

Dove tail cutter

Facing cutter

Slab cutter

Side and face

Slot drill. Offset cutting faces allow downfeed

Woodruff cutter

Reverse dovetail cutter

Spotfacing cutter

Side and face cutter

Concave and convex cutters

Figure 4.20 Milling cutters.

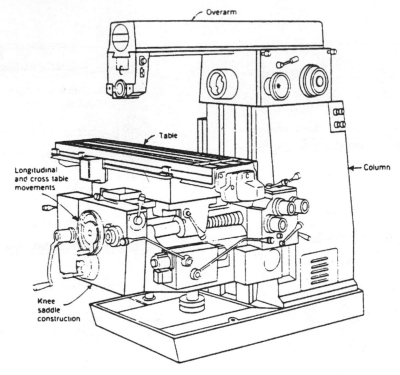

Figure 4.21 Horizontal milling machine.

Vertical spindle machines

The vertical spindle machine operates as shown diagrammatically in Figure 4.22. An engineering version is shown in Figure 4.23 and forms the backbone of engineering production, either in manual form or equipped with some form of positional control. To the engineer this machine is a vertical mill, to the woodworker it is a router, but essentially it is the same machine. Various versions are available; some with two spindles allowing two parts to be made at the same time; some with, copying attachments to allow the production of identical parts from templates; some with a table or spindle on a pantograph to allow copying from template to different scales, or to produce parts from assemblies of templates. The engraving machine is an example of the latter, allowing the production of nameplates and inscriptions from a standard alphabet in template form, with the adjustment of the pantograph ratio allowing for the generation of different sized letters.

Combination machines

Vertical milling machines may be provided with tilting heads, allowing the spindle to be adjusted at any angle from vertical to horizontal, to produce surfaces and shapes

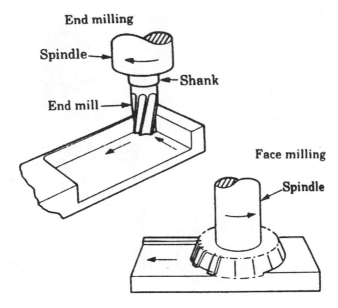

Figure 4.22 Vertical spindle milling.

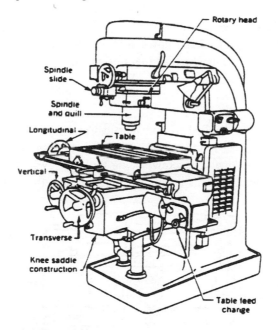

Figure 4.23 Vertical milling machine.

at odd angles on complex parts. The best known of these is the 'Bridgeport' head, where the head spindle and motor form a unit which can be angled in one, or sometimes two, planes. This is illustrated in Figure 4.24. Such machines are valuable in the tool room, allowing for complex die-sinking operations, but also have their place in production work.

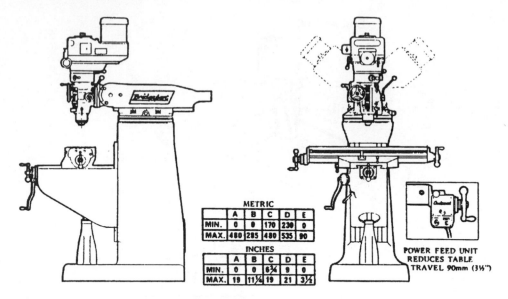

METRIC					
	A	B	C	D	E
MIN.	0	0	170	230	0
MAX.	480	285	480	535	90

INCHES					
	A	B	C	D	E
MIN.	0	0	6¾	9	0
MAX.	19	11¼	19	21	3½

POWER FEED UNIT
REDUCES TABLE
TRAVEL 90mm (3½")

Figure 4.24 Machine with Bridgeport head.

Other machines have been designed for special purposes, with spindles to suit a particular application. Figure 4.25 shows such a machine, in this case a slab mill using three facing cutters to machine an ingot of metal.

Computer control

Computer control can be applied to both vertical and horizontal machines (Figure 4.26). Many machines were retrofitted, those with vertical or Bridgeport heads being the most popular, the original leadscrews being replaced by ballscrews and resolvers fitted. Program information was manually entered into the machine's computer.

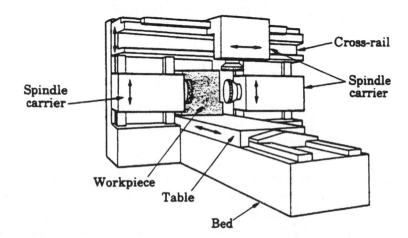

Figure 4.25 Slab mill.

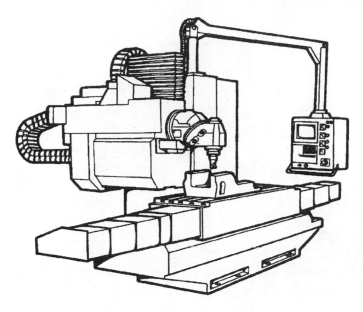

Figure 4.26 CNC milling machine.

However, to obtain the maximum gain from the use of the computer, the tendency now is to design the machine specially to take advantage of computer control. The modern machining centre (Figure 4.27) has not just computer control over X, Y, and Z axes, but also the ability to rotate the workpiece about one or more axis, and has built in arrangements for changing workpiece and changing tools within the programmed sequences. Such arrangements are dealt with more fully in Chapter 20.

Drilling

The most common manufacturing process is the provision of a hole in a component, using a rotating drill bit of some form. The commonest form is the twist drill, a device which is provided with two cutting edges, thus balancing the cutting forces and minimising flexure, helical flutes which serve to remove the debris, and margins providing a cylindrical surface which gives guidance. The material remaining between the flutes of the drill, the web, prevents the two cutting edges coming together at a point, resulting in a portion in the centre where no cutting takes place. Instead the material is forced aside to be removed by the cutting edges, this indentation accounting for the major portion of the feed force required when drilling. When drilling larger holes it is normal to use a smaller drill to remove this dead area before drilling to the final size.

When starting a drill, the presence of this chisel-edged point gives a tendency for the drill to wander, and it is necessary to restrain the tip of the drill in some way. This is done either by providing an indentation by centre-punching or drilling with a centre-drill, or by restraining the drill itself with a guide bushing close to the surface to be drilled.

Figure 4.27 Machining centre.

Figure 4.28 shows the angles used on the general purpose drill, but the geometries of the drill will vary according to the nature of material being worked. Fast spiral drills are used where the chip is small and tends to clog (e.g. ceramics) and where high cutting speeds or deep drilling are required, while slow spirals are used for softer materials where more continuous chip is formed, and for thin materials. For

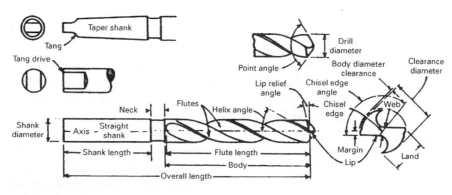

Figure 4.28 Drill nomenclature.

free-cutting brasses and thin materials straight flute drills may be used, obviating the tendency for the drill to catch the material and pull it up the helix at breakthrough. A common form of this is the spade drill, sometimes with a replaceable tip.

The alignment of the hole depends initially on the indentation of the drill point, and then once the drill has entered to its full diameter, on the guiding action of the margin. If, however, the drill point breaks through before the full diameter has been reached, the guiding action vanishes, and the workpiece will be snatched up the spiral. Thus it is necessary when drilling thin material to try to reduce the point angle to ensure that the full diameter is reached before breakthrough. However, a reduced point angle decreases the guidance of the point on starting, and increases the tendency to wander. This may be overcome by grinding the drill as a trepanning drill. Here the point is ground concave, so that cutting commences at the outside diameter instead of the core, and leaving a small indenting point in the centre to provide initial guidance. This type of drill is much used when drilling holes in brass or aluminium sheet.

A similar problem occurs when drilling soft materials, or those of a fibrous nature, e.g. wood. The soft nature of the material does not support the centre of the drill, so there is a pronounced tendency to wander, and this is exacerbated by the presence of hard and soft strata within the material. Furthermore the soft material will tend to be pushed aside by the angle of the point, resulting in a raised build-up around the hole, or tearing of the edges. This is overcome by the use of a special form of spade drill, often referred to as a wood-bit, whose cutting edge is ground with a negative point angle and has a large locating spike left in the centre.

Drill materials

Carbon steel drills are rarely found in modern industry, the majority of drills being of high speed steel. Significant increases in drill life can be achieved by the use of wear-resistant coatings. Tungsten carbide, either solid carbide for small drills, as the cutting face of spade drills or as inserts in the cutting faces of twist drills, give improved life and the ability to machine harder materials, and for extreme conditions diamond tips may be obtained.

As in any cutting action, heat is generated and for high rates of metal removal it is necessary to provide adequate cooling. It is difficult to ensure that coolant reaches the cutting faces, as the flute spiral is intended to remove swarf not carry liquid in. For normal drilling operations the cooling and swarf removal is aided by drilling in increments (pecking), withdrawing the drill at intervals to cool the drill and flush the hole. For arduous operations drills may be obtained with passages allowing the pressure feeding of coolant to the cutting face, but the cost of these limits their use to specialist applications.

Reaming

The surface finish of a drilled hole is not as good as a machined bore, and the tolerance on the hole, while adequate for most purposes, is not high. While these can both be improved by using a smaller drill then finishing with a size drill and a light cut, for high accuracy work it is necessary to introduce a reaming operation to improve the finish and provide closer tolerances.

Like drilling, reaming is a cutting action. In reaming, however, the cutting takes place as a small cut along a long face tapering gradually out to the required size. A reamer (Figure 4.29) is multi-fluted, as it relies on multiple contact with the drilled hole for its alignment and stability of cut. Reaming is carried out at a slow speed, with the reamer in a floating holder to ensure that the reamer is free to follow the line of the hole being reamed.

Special applications

Many of the holes required in industry require several concentric diameters, or concentric facing. The simplicity of the drill press operation lends itself to low-cost operations of this type.

A very common application requires that a hole should be recessed or counterbored. To achieve this with conventional drills would require three drilling operations:

1. Drill the through hole.
2. Drill the counterbore.
3. Use a flat-ended drill to provide the flat face to the counterbore.

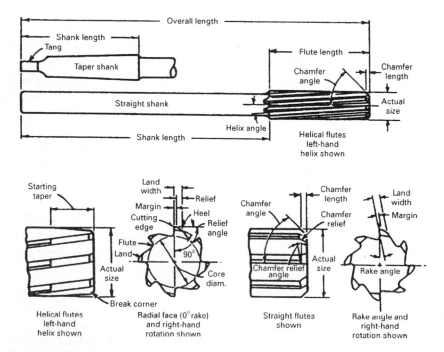

Figure 4.29 Reamer nomenclature.

This can be reduced to a single operation by grinding the drill to reduce its diameter over the front portion of its length, and to provide flat cutting faces to the step.

A similar technique is used to produce countersunk holes.

Similarly, where a face is required on a surface for a holding bolt, it may be more economic to use a drill press to machine this than to have the surface milled. In such a case a cutter would be used which carried a pilot locating in the drilled hole, or possibly the drill would be carried in the centre of the cutter, allowing drilling and spot facing in one operation.

5 Kinematics of Machine Tools

5.1 Introduction

The shape or many engineering components is derived from common geometric solids. One useful approach to the machining of such shapes is to analyse motions capable of producing them.

There are alternative ways of machining each geometric form; traditional machine tools have evolved to provide the most convenient way for machining the general run of work. Where, however, special component forms are required to be machined in large quantities, it is first desirable to consider possible generating motions and then to base manufacturing methods and equipment, including in some instances specially designed machine tools, on the generating system likely to give the lowest cost for the quality required.

5.2 Geometric form of engineering components

As a basis for analysing elements of movement of machine tools some of the common geometric figures used in engineering components will be considered.

1. *Plane*. This is the basic element of prismatic shapes, and it is used as a datum from which component dimensions are given. It is a feature which is sometimes required to a high order of accuracy, and the methods used in its production must be capable of achieving whatever degree of accuracy is required for functional purposes both as regards flatness and angular relationship with adjacent planes.
2. *Cylinder*. This appears in very many components and may have an external surface (shafts) or an internal surface (holes).
3. *Cone*. In relation to the production of components a cone is more commonly referred to as a taper, chamfer or bevel (as in bevel gears).
4. *Sphere*. The ball bearing is the most common example of the spherical shape but spherical surfaces are also used for applications such as spherical seatings.
5. *Helix*. This appears in the form of screw threads, oil grooves, drill flutes, etc.

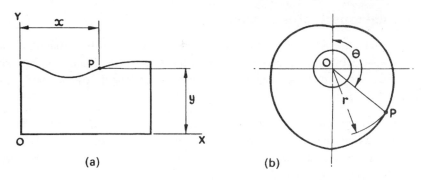

Figure 5.1 Plate cam profiles defined by: (a) rectangular co-ordinates; (b) polar co-ordinates.

Shapes more complex than standard geometric forms occur in component profiles and contours and dimensions for these may lie in two or three planes. For example, a point on the profile of a plate cam will be defined by two dimensions, each taken from a datum and expressed in either rectangular or polar co-ordinates.

These two methods of dimensioning are shown in Figure 5.1.

Similarly, relationship with a point on the flank of a tooth of a helical gear would need to be defined by dimensions in three planes, one of which would be an angular dimension. In order to produce these geometric figures on a machine tool, surplus material must be cut away as chips or swarf until the original piece is reduced to the required form and size. This requires that the machine tool be capable of reproducing controlled movements of workpiece or tool or both in such a way that geometric shapes result. Quite apart from the questions of cutting action and the power required to remove material from a component the machine must be able to perform accurately its basic functions of motion.

5.3 Kinematics in machine tools

Kinematics is the branch of science treating position and movement. Since components are machined by movement of tools and are produced to a size by the accurate relative position of tool and workpiece, kinematics is clearly important as far as machine tools are concerned.

Alignments are closely related to the kinematics of a machine both as regards straight-line motion and accuracy of rotation about a fixed axis. The machine must be capable of maintaining its alignments under conditions of:

1. Static loading – it must be strong enough to bear applied forces without undue flexure.
2. Dynamic loading – it must accept stresses set up by moving parts without the effects of vibration or deformation affecting its function.

Fundamental aspects of movement

Consider the basic problem of constraining a body, whether it be a machine slide, a cutter or an abstract mass is immaterial to our present purpose. Such a body is shown in Figure 5.2, and it may be seen that the 'pure' movements of the body can be classified as follows:

1. Linear movement along any one of the three conventional axes, i.e. XX, YY or ZZ. Thus the body has three degrees of freedom by means of axial translation.
2. Rotational movement about any one of the axes, XX, YY or ZZ, thus providing a further three degrees of freedom.

Hence it is seen that *an unrestrained body in space has six 'pure' or precise degrees of freedom.* All other movements of the body may be achieved by combining two or more of the precise movements to obtain the required motion. Thus, the path of the sphere in a given plane, Figure 5.3, is defined by relating ordinates x_1y_1, x_2y_2, etc.

> *It can be shown that a body will be constrained from movement by six locations suitably applied and which take into consideration kinematic principles. The removal of one location will then permit one degree of freedom, the removal of two locations will permit two degrees of freedom and so on.*

Figure 5.4(a) represents the location of two separate bodies – one having three ball feet, the other three radially disposed vees. Six location points occur, numbered 1 to 6, constrain movement of the body. Important aspects of kinematic design can be observed in this example:

1. The two bodies can be separated and replaced in precisely the same relationship as frequently as is necessary.
2. A high degree of accuracy in construction is not needed e.g. the diameter of the pitch circle of the feet may vary considerably and yet be accommodated by the vees.
3. Expansion of either body will not affect the precise location.

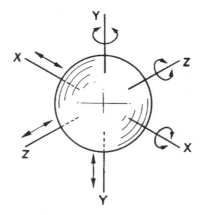

Figure 5.2 Six degrees of freedom of a body in space.

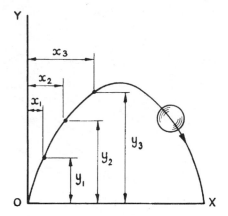

Figure 5.3 Movement of a body in one plane (without rotation).

4. The six locations are all 'point' contact, there are no redundant locations, i.e. none of the locations can be dispensed with.
5. Wear of locations will not cause slackness to occur between the mating parts.

A similar but degenerate design for locating two bodies is illustrated in Figure 5.4(b). In this case extreme accuracy is called for, and wear or expansion will cause loss of six point contact and thus uncertain location.

Notes

1. It is apparent that while the two parts of a kinematic pair are fully constrained by six restraints the parts can still be separated unless a closing force is applied. Frequently gravity is the only force necessary to close the system, in other designs some form of retention (e.g. clamping) may be necessary.

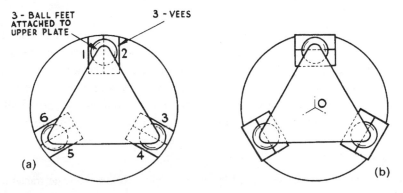

Figure 5.4 (a) Simple kinematic design for location of two bodies – the six points of constraint are numbered 1 to 6; (b) degenerate design. Constraints are tangential to a circle and do not prevent rotation about a vertical axis through O. The three ball feet must also be accurately positioned relative to the vees to ensure contact at six points.

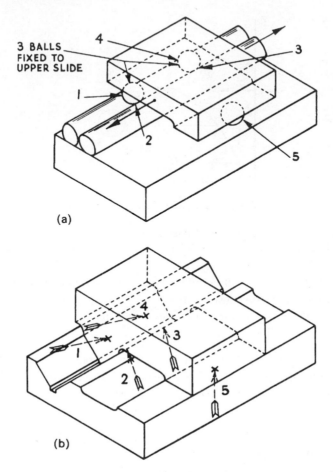

Figure 5.5 (a) Kinematic slide (point contacts with five restraints, permitting one degree of freedom (indicated by arrow); (b) slideways based on kinematic principles and having one degree of freedom.

2. 'Point' contact is not normally feasible in machine-tool applications because of the substantial forces which are invariably present. These forces must be supported over suitably large areas of contact to keep the pressure between surfaces within reasonable limits. If this were not done the pressure would be infinitely high. A comparison between a fully kinematic slide and the equivalent type of slide used on machine tools and based on kinematic principles is shown in Figure 5.5.

Example 5.1

State the degrees of freedom possessed by the parts given below, indicate transition by T, rotation by R.

1. Threaded nut on stationary screw.
2. Plug gauge in component bore.
3. Scribing block on marking-out table.

4. Arm of radial drill.
5. Single-point tool operating in lathe toolpost.

Solutions

1. $1T$, $1R$; 2. $1T$, $1R$; 3. $2T$, $1R$; 4. $1T$, $1R$; 5. $2T$.

Note

1. Two planes in contact provide four constraints, one of which is redundant. To obtain similar results using true kinematic location the scribing block, in (3) above, should be supported at three points. A simple example illustrating the existence of a redundant location is the case of the four-legged chair, which will rock on an uneven surface and a three-legged stool which will not rock.
2. A cylinder enveloped in a bore as 1, 2 and 4 has two degrees of freedom ($1T$, $1R$) and four constraints, i.e. to obtain similar results kinematically the cylinder should be supported at four points of contact as illustrated in Figure 5.6.

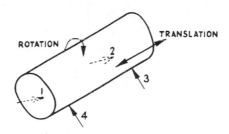

Figure 5.6 A kinematically constrained cylinder showing four points of restraint and two degrees of freedom (i.e. $1R$, $1T$).

⊓   ⊓   ⊓

5.4 Kinematics and machining geometric forms

The geometric requirements of machining are derived from two basic elements:

1. Rotation about an axis which has a fixed position.
2. Translation in a straight line.

Figure 5.7 gives the conventional symbols for linear and rotary displacement in accordance with BS 3635: *The Numerical Control of Machine Tools.*

When considered kinematically each of these motions occur in systems having five location points and one degree of freedom. This is illustrated by the rotation of a lathe spindle and by the single translatory movement of a lathe saddle shown in diagrammatic form in Figure 5.8.

For an axis of rotation to have a constant position in space, the rotating element must be exactly cylindrical (e.g. bearing surfaces of a machine-tool spindle) and there must be no axial 'float' (endwise motion of the axis).

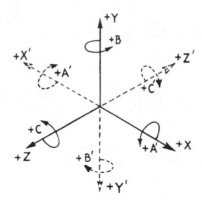

Figure 5.7 Conventional symbols in accordance with BS 3635.

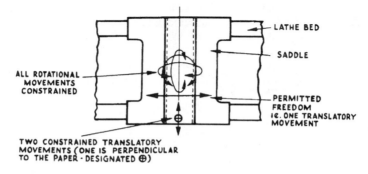

Figure 5.8 The single degree of freedom of a lathe saddle.

For translation to be in a straight line, the motion of all points of the moving element must be in two mutually perpendicular planes. Any 'cross-wind' about the line of translation produces a rotation of the sliding member. This is shown in Figure 5.9, where a cylinder rolls on two separate surfaces. If planes A and B do not lie in a common true plane, but are inclined as indicated, then the cylinder will have an additional freedom.

Symbolic notation of BS 3635

To apply the notation of Figure 5.7 in accordance with rules given in BS 3635, the following conventions must be followed:

Z-axis   The principal axis of rotation of the machine (e.g. spindle of a lathe) is designated as the Z-axis. A displacement of either workpiece or tool made parallel to this axis is a z displacement.

For machines without a rotating spindle (e.g. planing machine) the Z-axis is perpendicular to the workholding surface.

X-axis   This axis is always horizontal, parallel to the workholding surface and perpendicular to the Z-axis.

Y-axis   An axis perpendicular to the Z- and X-axes.

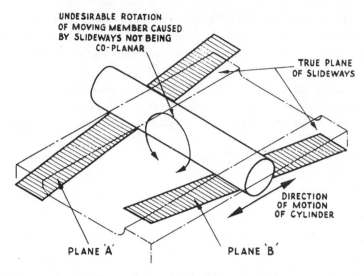

Figure 5.9 The rotational effect on a moving member caused by cross-wind in machine slideways.

Some machine tools have elements which can rotate about one or more of the above axes, such as circular tables. These rotations are symbolically designated in relation to their axes as shown in Figure 5.7. Note that the directions of displacement along the axis and the direction of the associated rotation follows the well known 'right hand screw' rule enabling +ve and −ve directions to be standardised.

Standardisation of these features has become essential since it facilitates numerical control programming. Figure 5.10 shows the application of the above rules for a particular machine tool.

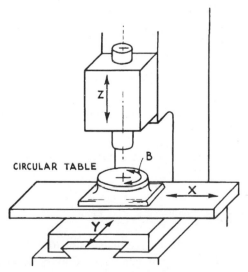

Figure 5.10 Conventional symbols for axes of single column drilling and boring machines.

Rotations A, B and C are primarily intended for feed motions under numerical control. In discussing the kinematics of generating, θ has been used to indicate rotation as an essential part of the cutting process, e.g. work rotation during turning (see Section 5.5).

In producing the geometric forms discussed in Section 5.2, machine tools use systems based on (1) or (2) on p. 133, Section 5.4, or combinations of these elementary movements. *Machine tools normally have two kinematic systems of freedom and constraints, one controlling the workpiece and the other controlling the medium through which cutting is effected, e.g. cutting tool, grinding wheel.*

These two systems are linked to obtain the relative movements necessary to produce a geometric shape. The form of this linkage may be as follows:

1. Mechanical, e.g. gear train, leadscrew and nut, etc.
2. Hydraulic, e.g. hydraulic systems of valves and cylinders as used on tracer mechanisms.
3. Electronic, e.g. punched or magnetic tape control.

Sometimes it is essential for this linkage to give a fixed and definite ratio in order to produce a required shape. In other cases the relationship is not critical, i.e. the ratio can be changed without affecting the geometry of the parts produced e.g. changing the feed rate when turning a cylinder will not affect the geometric shape of the solid body produced, but changing the traverse rate when screw cutting would most certainly affect the accuracy of the screw thread. *In this latter case the linkage ratio is said to be critical.*

5.5 Classification of generating systems

In the following examples the notation used will be as follows:

> θ_w = Rotation of workpiece.
> θ_t = Rotation of tool.
> x = Translation along XX axis.
> y = Translation along the YY axis.
> z = Translation along ZZ axis. (*Note*: displacement parallel to a spindle axis is usually designated as z.)
> $\oplus$ = Linear motion of machine element in a direction perpendicular to the plane of the diagram.

The generation of geometric shapes on a machine tool may be classified as First or Second Order Generation.

First Order Generation

In this group the separate kinematic systems of the tool and workpiece are linked with a *non-critical ratio*. This order of generation may be subdivided into Class 1 or Class 2.

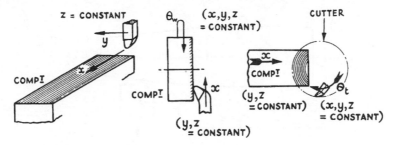

Figure 5.11 Examples of First Order, Class 1, Generation – flat surfaces. (a) Motions of translation (y and z) must be in a straight line – for convenience they are usually perpendicular to each other. The component and tool may each have one degree of freedom (e.g. planing), or the tool may have two freedoms and the component none, or the tool none and the component two freedoms. (b) and (c) The essential conditions for producing a plane surface under the conditions shown are: (i) translation (x) must be in a straight line, and (ii) axis of rotation must be perpendicular to x.

Class 1

No critical relationship exists within the kinematic system of the tool, nor within the kinematic system of the workpiece. Examples of this class of First Order Generation are the machining of plane and cylindrical workpieces, shown in Figure 5.11.

Class 2

A critical relationship exists within either:

1. The kinematic system of the tool; or
2. The kinematic system of the workpiece.

Examples of machining which fall into this group are the production of a spherical or conical surface on a lathe (Figure 5.12, p. 138), or a helix on a milling machine.

Second Order Generation

The two kinematic systems (i.e. of workpiece and of tool) *are linked with a critical ratio*, which if altered or varied, would affect the geometric form produced. Again, this order of generation may be subdivided into classes as for First Order Generation.

Class 1

Examples of machining in which the kinematic systems of tool and workpiece are linked with a critical ratio are the production on a lathe of a helix, or a spiral (as for example, the relief curve of a form-relieved cutter).

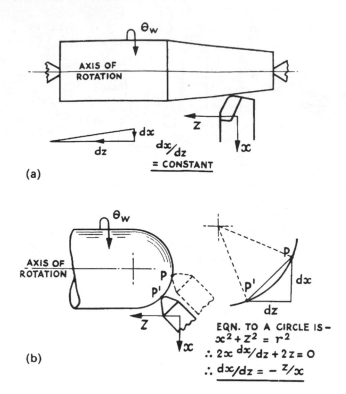

Figure 5.12 Examples of First Order, Class 2, Generation. (a) Turning a cone; (b) turning a sphere. The relationship dx/dz may be achieved by the following methods: (i) mechanical (e.g.) taper turning attachment, (ii) hydraulic, (iii) electronic.

Class 2

This is a complex form of generation in which a critical relationship exists within the kinematic system of the tool (or workpiece) in addition to the critical relationship between the kinematic system of the tool and the kinematic system of the workpiece. Examples of machining which fall into this group are the production on a lathe of a taper helix (taper thread) and the imposition of a spiral upon a helix which occurs when machining the relief curve on the teeth of a gear-cutting hob.

This classification of the generating systems employed on machine tools will now be illustrated by worked examples, some of which are problems related to conditions of alignment and serve to emphasise the connection between the kinematics of a machine tool, and its alignments.

Example 5.2

(First Order Generation, Class 1)

Illustrate methods of generating a cylindrical surface and state the essential kinematic conditions for producing an accurate form. In separate diagrams show inversion of the methods.

Solution

1. Rotation θ_w and translation z (see Figure 5.13(a)) are linked, usually by mechanical methods, in a non-critical ratio, i.e. the feed rate. Considered from a kinematic point of view, the boring of a hole by rotation of the work, θ_w, and translation of the tool, z, uses the same principle and is merely a variation (Figure 5.14(i)).

   On the other hand, machining a spigot using a boring head as shown in Figure 5.14(ii) is an inversion of the turning process, and has a variation shown in Figure 5.14(iii).

2. Translation, z, and rotation, θ_w (see Figure 5.13(b)) are linked by a hand or mechanical feed in a non-critical ratio. This method is used on slotting machines, and a variation may be seen in Figure 5.14(iv), where the grinding of the root of a spline shaft is shown.

   Producing a cylinder in this way will take longer than the methods in (1), and hence will be used only if technical difficulties prevent the use of the quicker method.

3. The essential kinematic conditions for producing a true cylinder are:
   (a) Rotation (θ_w or θ_t) must be about a constant axis.
   (b) Translation, z, must be in a straight line *parallel to the axis of rotation and lying in a plane containing the axis*.

   □   □   □

Example 5.3

(First Order Generation, Class 2)

Analyse the motions capable of generating spherical surfaces.

Solution

Spherical surfaces may be produced by tool movements using continuous co-ordinate settings based on systems of polar co-ordinates and Cartesian co-ordinates.

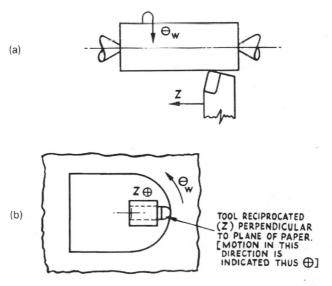

Figure 5.13 Basic methods of generating cylindrical surfaces (First Order Generation, Class 1).

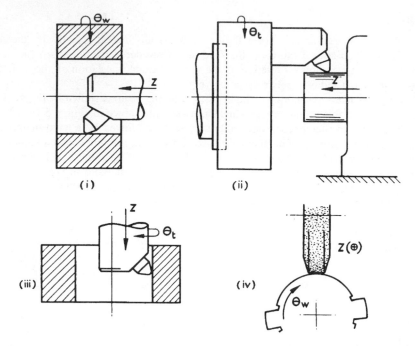

Figure 5.14 Variations and inversions of basic methods of producing cylindrical surfaces.

For both these methods, the workpiece must rotate (θ_w) about a constant axis ZZ and the locus of the tool point should follow a circular path, origin at the centre of the sphere.

1. When using polar co-ordinates the tool is maintained at a constant radius from the sphere centre by means of a mechanical linkage. The principle is illustrated in Figure 5.15(a), and any one of several alternative mechanisms might be used to achieve these conditions. For example, if on a lathe the saddle and cross slide were locked and the conventional toolpost replaced by a form of circular table capable of holding the tool the necessary movement could be achieved.
2. To produce a spherical surface using Cartesian co-ordinate settings for the tool point the relationship of the tool movement along the XX and ZZ axes must be carefully controlled, see Figure 5.15(b). When a locus of the tool point traces a circle, origin at the centre of the spherical surface, the equation to the locus will be:

$$x^2 + z^2 = R^2$$

i.e.

$$2x \cdot dx/dz + 2x = 0$$
$$\therefore \qquad dx/dz = -z/x$$

The relation dx/dz cannot be achieved satisfactorily by manual operation of the tool movement. If the work is to be performed on a lathe the relationship may be ensured by electronic control using punched or magnetic tape to feed instructions to motors controlling saddle and cross slide movement. Lathes of this type dispense with the usual handwheel controls, gear change levers, etc., and are controlled from a numerical control unit which is fed with punched tape prepared from the drawing of the component.

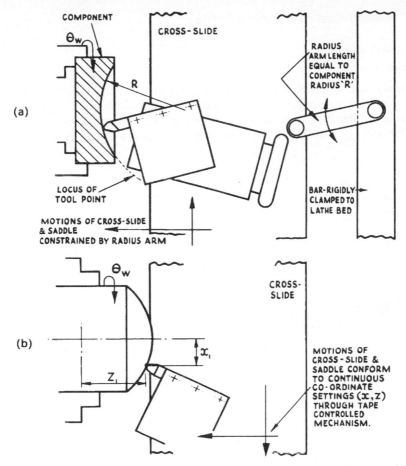

Figure 5.15 Generating spherical surfaces: (a) polar co-ordinate method; (b) Cartesian co-ordinate method.

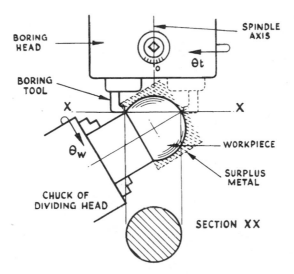

Figure 5.16 Generating spherical surfaces on boring or milling machine. Internal spherical surfaces can also be produced this way.

Alternative solution

(First Order Generation, Class 1)

In the first solution the axis (Y) about which the tool rotated was perpendicular to the axis (Z) of the workpiece. This is not an essential condition for the generation of a spherical surface, and a method is illustrated in Figure 5.16, where oblique axes intersect at the centre of the sphere. This method is suitable for use in a universal milling or a boring machine.

☐   ☐   ☐

6 Mechanics of Machine Tools

6.1 Basic features of a machine tool

A machine tool provides the means for cutting tools to shape a workpiece to required dimensions; the machine supports the tool and the workpiece in a controlled relationship through the functioning of its basic members, which arc as follows:

1. *Bed, structure or frame*. This is the main member which provides a basis for and a connection between the spindles and slides; the distortion and vibration under load must be kept to a minimum.
2. *Slides and slideways*. The translation of a machine element (e.g. the slide) is normally achieved by straight-line motion under the constraint of accurate guiding surfaces (the slideway).
3. *Spindles and bearings*. Angular displacements take place about an axis of rotation; the position of this axis must be constant within extremely fine limits in machine tools, and is ensured by the provision of precision spindles and bearings.
4. *Power unit*. The electric motor is the universally adopted power unit for machine tools. By suitably positioning individual motors belt and gear drives are reduced to a minimum.

6.2 Forces in a machine tool

Stresses which tend to deform the machine tool or workpiece are caused by the following:

1. Static loads i.e. the weight of the machine and its various parts as considered on p. 148.
2. Dynamic loads, i.e. forces induced by rotating or reciprocating masses.
3. Cutting forces as discussed in Chapters 9–11.
4. Thermal stresses caused by energy released by the machining process (these are usually the dominant distortions of the structure).

There are two broad divisions of machining operations:

1. Roughing, for which the metal removal rate and consequently the cutting force is high, but the required dimensional accuracy relatively low.
2. Finishing, for which the metal removal rate and consequently the cutting force, is low, but the required dimensional accuracy and surface finish relatively high.

It follows from the above considerations that static loads (the position of which may vary as slideways are displaced) and dynamic loads such as result from an unbalanced grinding wheel are more significant in finishing operations than in roughing operations. The degree of precision achieved in any machining process is usually influenced by the magnitude of the deflections which occur as a result of the forces acting and the thermal equilibrium of the machine structure.

Machine tools are regularly required to work to an accuracy of 0.02 mm and often to 0.002 mm. The permissible amounts of elastic flexure of the main frame and its subsidiary units must be small to achieve this degree of accuracy. The machine as a structure cannot be designed by normal stressing methods where load carrying capacity is the criterion, but must be designed to have negligible deflection and provide generous bearing surfaces so as to diminish wear. To achieve this, the section modulus must be made as large as possible without giving rise to excessive weight. This leads to the employment of very deep sections, such as shown in Figure 6.1.

Symbolically, for the given section:

Deflection, $\Delta y \propto 1/I$

where I = 2nd moment of area
    Now, for a rectangular section $I = BD^3/12$

$\therefore$ $I \propto D^3$ when thickness of casting is constant

hence

$\Delta y \propto 1/D^3$

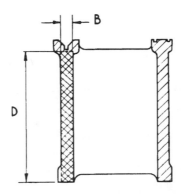

Figure 6.1 Cross-section through slideways of a grinding machine. The depth of section, D, is substantial to ensure maximum resistance to deflection.

6.3 Structural elements

Machine tool frames are frequently made in cast iron, although some may be steel castings or mild steel fabrications. Cast iron is chosen because of its cheapness, rigidity, compressive strength and capacity for damping vibrations set up in machining operations.

The cost factor deserves more than a passing mention, and the following points should be noted:

1. While the price of cast iron is about double that of mild steel plate, this includes the cost of casting to shape.
2. Cast iron is the cheapest metal which can be used to produce metal castings. Steel castings are much more expensive to produce.
3. The cost of pattern and core boxes must be spread over the number of castings produced.
4. The greater strength in tension of mild steel may permit thinner sections than those for cast iron, but this lower material cost is offset by the cost of cutting and welding.

From the above it will be apparent that welded steel fabrications offer certain advantages, particularly when the pattern cost would have to be borne by a single casting, or when time cannot be spared for pattern making. However, fabrications have a greater resonance, and cannot be so easily produced in aesthetically pleasing shapes.

A recent innovation by some companies is the use of high quality concrete structures for machine tool frames. The requirements for low distortion means that metal structures are never loaded to more than a small percentage of their possible maximum stress levels. Such metal machine frames need to be either cast or welded to shape then precision machined to produce accurate slideways. These are complex, expensive, and time consuming procedures. The development of precision 'bolt-on' ball bearing slideway units has opened up the possibility of using a composite structure, using concrete which has sufficient strength to withstand the low stress levels imposed on the frame and has excellent vibration damping characteristics. The use of precision slideways glued into the concrete structure avoids expensive machining, the resin being used both to fill any gaps between the slideways and the concrete and to lock the slideways in position once they have been precisely aligned.

To avoid massive sections in casting, carefully designed systems of ribbing are used to offer maximum resistance to bending and torsional stress. Two basic types of ribbing are:

1. Box.
2. Diagonal, as illustrated in Figure 6.2.

The box formation is convenient to produce, apertures in the walls permitting the positioning and extraction of cores. Diagonal ribbing provides greater torsional

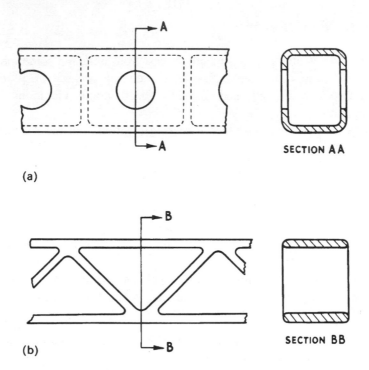

Figure 6.2 Six sections for cast structural members: (a) box ribbing; (b) diagonal ribbing.

stiffness while allowing swarf to fall between the sections, as in lathe beds. Similar stiffening techniques are used for fabricated beds.

The increasing capabilities of modern cutting tools and control systems have caused a revolution in machine tool design. Much higher feeds and speeds are used, bringing about increased power consumption and increased forces on the machine and workpiece, and producing large volumes of swarf. Motor powers are frequently in the 20–50 kW region, and 80 percent of this energy will be transformed into heat in the swarf. This, if allowed to build up on the slideways, will give rise to thermal distortion of the bed structure, and besides causing loss of accuracy could cause damage to both machine and workpiece. To assist in swarf removal, machine structures have been redesigned to ensure that swarf falls clear of the slideways, and to reinforce the structure significantly to withstand the higher forces involved. this has led to the adoption of the slant bed configuration for the majority of CNC turning machines (Figure 6.3).

Swarf removal is less easy on milling machines. Horizontal spindle machines will tend to throw the swarf clear of the workpiece, but on the more commonly used vertical spindle machines swarf removal and workpiece cooling rely on high volume coolant systems.

Support of frames

Structural members of machine tool frames are supported to conform as closely as possible with the kinematic principles outlined in Chapter 5, but for a large member

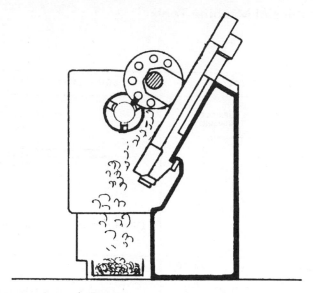

Figure 6.3 Slant bed CNC lathe structure.

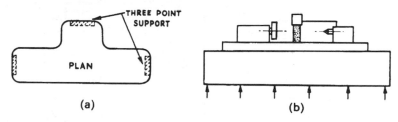

Figure 6.4 (a) Kinematic principles applied to the method used in supporting a small external grinding machine; (b) multi-point support applied to a large external grinding machine to prevent sag of machine bed.

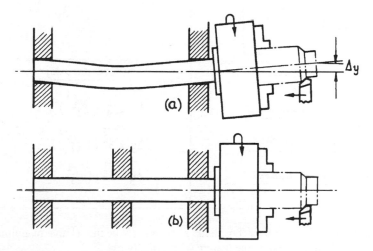

Figure 6.5 Lathe spindle considered as a beam. (a) Deflection (Δy) at tool point caused by bending of inadequately supported spindle; (b) the introduction of an intermediate bearing greatly reduces deflection of long spindles.

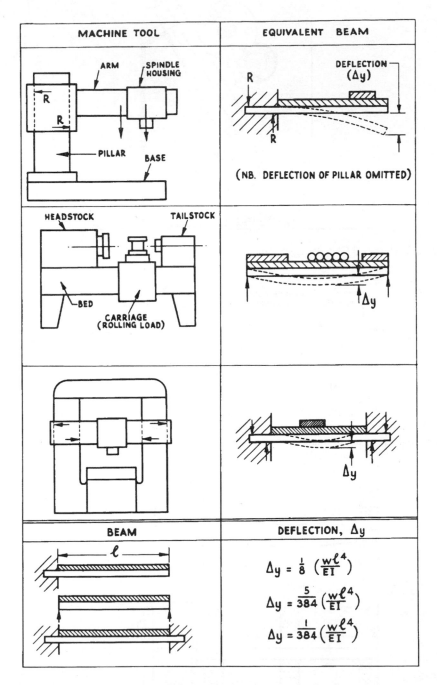

Figure 6.6 Machine-tool members considered as beams. (Note the relative magnitudes of Δy for the three basic beams shown at the bottom of the diagram).

additional points of support are needed to avoid the 'sag' caused by its own weight plus the applied loads. Such conditions can occur in the foundation supports of a machine e.g. small cylindrical grinding machines can be supported at three points, Figure 6.4(a), whereas a very large machine would need support at a greater number of points, Figure 6.4(b). In the latter case periodic checks should be made to ensure that the straightness and accuracy of slideways have been maintained.

Deflection of machine elements

Many spindles, Figure 6.5, and structural members of machine tools can be considered as cantilever, simply supported or built-in beams. Although textbook conditions rarely apply in practice, the effectiveness of these designs from the point of view of resistance to deflection can be compared by reference to expressions giving the deflection of beams. Examples are shown in Figure 6.6 together with idealised applications from common types of machine tool. Examples of modifications to the designs shown in the idealised diagrams so as to improve performance are:

1. Increase in section depth proportionate to the bending moment for the drilling machine arm.
2. Widening the legs of the lathe and so approximating the bed to a built-in beam.

6.4 Slides and slideways

The slides and slideways of a machined tool locate and guide members which move relative to each other, usually changing the position of the tool relative to the workpiece. the movement generally takes the form of translation in a straight line, but is sometimes angular rotation, e.g. tilting the wheelhead of a universal thread grinding machine to an angle corresponding with the helix angle of the workpiece thread.

Features of slideways are as follows:

1. *Accuracy of movement.* Where a slide is to be displaced in a straight line this line must lie in *two* mutually perpendicular planes and there must be no 'crosswind', i.e. slide rotation (the principles of slide movements are analysed in the previous chapter). The general tolerance for straightness of machine tool slideways is 0–0.02 mm per 1000 mm; on horizontal surfaces this tolerance may be disposed so that a convex surface results, thus countering the effect of 'sag' of the slideway.
2. *Means of adjustment.* To facilitate assembly, maintain accuracy and eliminate 'play' between sliding members after wear has taken place, a strip is sometimes inserted in the slides. This is called a gib-strip, and a simple design using this device is shown in Figure 6.8(b). In this example the gib is retained by socket-head screws passing through elongated slots and is adjusted by grub screws secured by lock nuts. Minimum slideway wear is achieved by low pressure

conditions (assuming lubrication is adequate), i.e. for a given load the greater the contact area the lower will be the wear rate.

3. *Lubrication*. Slideways may be lubricated by either of the following systems:

   (a) Intermittently through grease or oil nipples, a method suitable where movements are infrequent and speeds low.

   (b) Continuously, e.g. by pumping through a metering valve and pipework to the point of application; the film of oil introduced between surfaces by these means must be extremely thin to avoid the slide 'floating' (see p. 156, item 2). If sliding surfaces were optically flat, oil would be squeezed out, resulting in the surfaces 'wringing' together like slip gauges, and resisting movement. Hence, in practice slide surfaces are either ground using the edge of a cup wheel, Figure 6.7, or scraped. Both processes produce minute surface depressions which retain 'pockets' of oil, and complete separation of the parts may not occur at all points; positive location of the slides is thus retained.

4. *Protection*. To maintain slideways in good order, the following conditions must be met:

   (a) Ingress of foreign matter, e.g. swarf, must be prevented. Where this is not possible, it is desirable to have a form of slideway which does not retain swarf, e.g. the inverted vee.

   (b) Lubricating oil must be retained. The adhesive property of oil for use on vertical or inclined slide surface is important; oils are available which have been specially developed for this purpose. The adhesiveness of an oil also prevents it being washed away by cutting fluids.

   (c) Accidental damage must be prevented by protective guards. These can be of the rigid, sliding, 'concertina' bellows or roller-blind type.

Basic types of slideways

The basic geometric elements of slides are the flat, vee, dovetail and cylinder. These elements may be used separately or combined in various ways according to the application. The design shown in Figure 6.8(a) provides a simple, effective slide, free of unnecessary constraints, straightforward to manufacture and one which permits regrinding or rescraping of the slide surfaces. If two vees are used redundant constraints are present, and the difficulties of manufacture considerably increased, because a high degree of accuracy is required in order to obtain a perfect fit between the two sets of vees.

Dovetail slides Figure 6.8(b), are used where an upward movement of the slide must be prevented. The effect of tightening the adjusting gib-strip is to pull the sliding members into closer contact, thus eliminating 'lift' and reducing vibrational tendencies. The square-edge slide shown in Figure 6.8(c) is provided with adjustment on one side, the two retaining plates prevent lifting of the slide.

An extension of the vee-and-flat slideway is illustrated in Figure 6.8(d). Rollers retained in brass cages are interposed between the sliding members; by this means

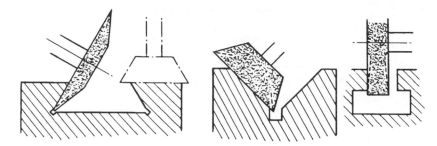

Figure 6.7 Methods employed for surface grinding machine tool slideways. Note the line contact, e.g. edge of cup wheel, between wheel and slide surface is extensively employed.

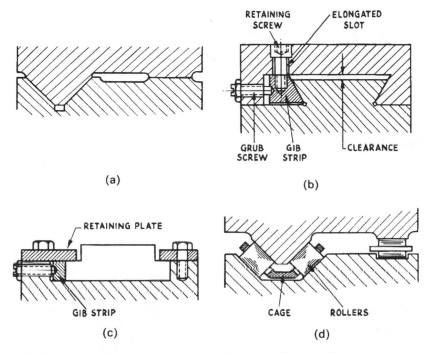

Figure 6.8 Typical machine tool slides and slideways. (a) Vee and flat; (b) dovetail (with gib strip); (c) square edge (with gib strip); (d) vee and flat using rollers.

rolling friction is substituted for sliding friction. The method has been in use for many years in the wheelhead slideways of Matrix thread grinding machines. The heavy wheelhead unit of such a machine is required to make very small displacements (often as little as a few thousandths of a millimetre) positively and with a light control. The rolling motion of this type of slideway provides the necessary sensitiveness of response and movement.

One point which should be noted in regard to roller cages, is that their linear displacement is only half the displacement of the slide. This is made clear by reference to Figure 6.9.

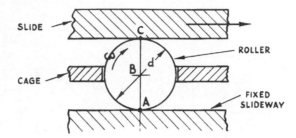

Figure 6.9 Relative movements of slide, roller and cage. Slide displacement = 2 × linear displacement of roller and cage.

Let *A* be the instantaneous centre, then

Velocity of $B = \omega d/2$ and   Velocity of $C = \omega d$

Hence in any time interval, *t*, *C* will move twice as far as *B*.

The principle of rolling slideways has been extended to the use of slide units of the roller recirculating type, Figure 6.10 (cf. recirculating ball and nut, p. 171). In addition to advantages of minimum friction and wear and the elimination of 'stick–slip' (p. 157), the length of slide travel is not limited by the use or these recirculating roller ways.

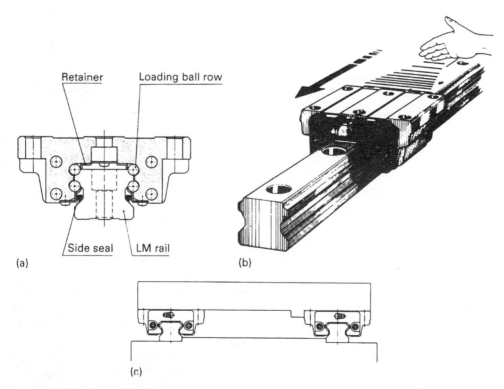

(a)

(b)

(c)

Figure 6.10 Linear ball bearing slideways (a)–(c).

Narrow guide principle

This important principle is illustrated by the vee guide and to a lesser extent by the dovetail and square-edge slides. To prevent 'jamming' of slides (cf: with the jamming of a sash window) the width between the guiding surfaces must be limited relative to the length of the guide. The application of the narrow guide principle is particularly important when the feed force of the sliding member is applied eccentrically.

Example 6.1

A slide, Figure 6.11, of weight W is caused to move by force F acting parallel to the slideway and distance e from the centre line. Obtain expressions for

1. Magnitude of force F.
2. The minimum value of e for which jamming will occur.

Solution

1. If the eccentricity of force F is assumed to slightly tilt the slide, then reaction of sides of slideway will act at A and C respectively; since $R_a = R_c$ the reactions may be represented by R.

   $\therefore$ The resistance due to friction, at each side of slideway $= \mu . R$

   Also, frictional resistance to slide motion on lower surface $= \mu . W$
   Hence, for equilibrium at the limiting conditions

   $$F = \mu W + 2\mu R \tag{6.1}$$

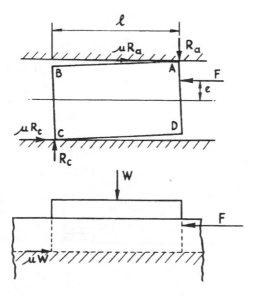

Figure 6.11 Example 6.1.

Taking moments about a point on the centre line (μR then has no moment)

$$Rl = Fe$$
$$\therefore \quad R = Fe/l \tag{6.2}$$

Substitute for R in equation 2.1

$$F = \mu W + 2\mu Fe/l$$
$$F(1 - 2\mu e/l) = \mu W$$

$$\therefore \quad F = \frac{\mu W}{1 - 2\mu e/l} \tag{6.3}$$

2. From equation 6.3 it can be seen that if $e = l/2\mu$ the denominator becomes zero and hence F is infinite,

$$\therefore \quad \text{When } e \geqslant l/2\mu \text{ the slide will jam} \tag{6.4}$$

□   □   □

A similar analysis applied to a vee and flat slide, Figure 6.12, gives the following relationships.

The distribution of weight W will be W_1, and W_2 on the flat and vee slides respectively.

Resolving W_2 to obtain forces normal to the vee surfaces, Figure 6.13

$$R_1 + R_2 = W_2/\sin \alpha$$

where α = semi-angle of vee

$$\therefore \quad \text{Frictional resistance of vee slideway } F_2 = \mu W_2/\sin \alpha$$

Also, frictional resistance of flat slideway $F_1 = \mu W_1$

$$\therefore \quad \text{Under no-load conditions, feed force } F = F_1 + F_2$$

$$= \mu(W_1 + W_2/\sin \alpha)$$

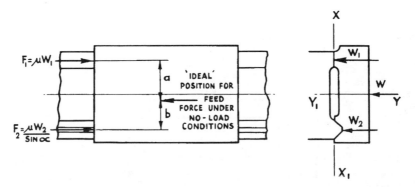

Figure 6.12 Forces acting on a simple slide.

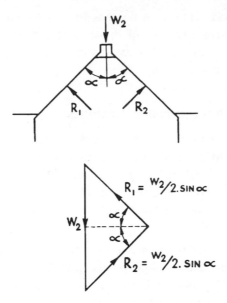

Figure 6.13 Forces on 'Vee' slide.

The ideal point of application for the feed force under these conditions is the position for which no turning moment acts on the sliding member, i.e. when

$$a \times F_1 = b \times F_2 \quad \text{(see Figure 6.12)}$$

For any other position of the feed force the vee slide must provide a restraining couple, thus increasing frictional resistance and the magnitude of the feed force.

When slideways support approximately symmetrical loads, e.g. the table slideways of a jig boring machine, the 'ideal' position for the feed force is easily determined, but where unsymmetrical loads are supported, the position is less readily found.

Figure 6.14 shows the resultant cutting force which acts on a lathe saddle. Its magnitude and point of application will vary according to the dimensions of the cut

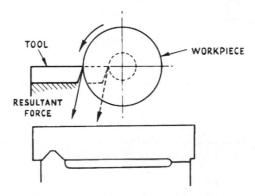

Figure 6.14 Position (and direction) of the resultant cutting force acting on the tool when turning, depends upon many factors, including size of workpiece.

and the diameter turned. Thus it is not possible to fix an ideal position for the application of the feed force because of these variations in the cutting force, and the position finally chosen, may be influenced as much by design convenience as by a consideration of the forces met in operation.

Slideways for program controlled machines

The operation of CNC machines makes functional demands which go beyond the requirements for a machine handled by an operator. One or these demands is that the response of a slide to a signal shall be positive and almost instantaneous. Because of the weight of the average slide, this requirement is not met by conventional slideway construction, and as a result slideways have been developed to keep the frictional resistance as low as possible.

1. Slides may be mounted on rollers see p. 180 and Figures 6.8(d) and 6.10.
2. In hydrostatic or aerostatic slideways, the principle of which is illustrated in Figure 6.15 the slide floats on a film of pressurised oil or air. Maximum stiffness and damping is obtained by using a very thin fluid film which must remain constant in thickness under varying conditions of loading in order to maintain accuracy in the position of the slide.
3. Low friction materials may be employed for sliding surfaces in places of the usual cast iron or steel, e.g. sintered bronze impregnated with the plastic PTFE (polytetrafluoroethylene, C_2F_4). The coefficient of friction of PTFE on ground steel surfaces is 0.04.

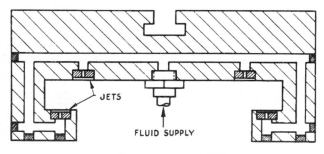

SECTION THROUGH XX

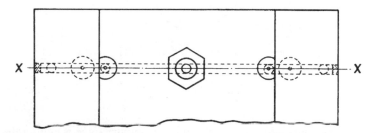

Figure 6.15 Principle of the hydrostatic slide for machine tools.

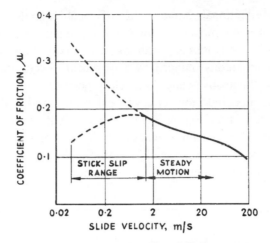

Figure 6.16 Variation in coefficient of friction during movement of a slide.

Stick-slip

An important advantage possessed by low friction slideways is uniformity of motion at low speeds and the elimination of stick–slip.

Stick–slip is the alternate sticking and slipping which occurs when partially lubricated slides operate at low speed. This produces a saw-tooth form of friction force variation in the stick–slip range illustrated by the graph, Figure 6.16. Under these conditions a machine-tool slide will remain stationary until the feed force has reached a magnitude sufficient to overcome static friction, the slide will then be accelerated by the applied force, but this movement will be immediately followed by a tendency for the slide to decelerate or come to rest. This unstable motion is repeated in rapid succession, often being so pronounced as to cause a noticeable low frequency vibration of the slide.

Stick–slip motion is one of the factors which makes fine adjustment of a slide difficult. For example, if the wheelhead of a grinding machine is to be adjusted to remove 0.01 mm from the diameter of a workpiece the slide must move 0.005 mm. If sticking occurs in the slide, no cut will result, and if the operator makes a second adjustment the sticking may be overcome but the wheelhead may move forward the total adjustment of 0.01 mm, thus removing 0.02 mm from the workpiece diameter instead of the intended 0.01 mm.

When such trouble occurs on conventional machines it can be mitigated by the use of one of the special lubricants developed for the specific purpose of reducing static friction.

6.5 Vibration and chatter

Chatter in machining operations is an objectionable manifestation of vibration in the machine, tool or workpiece; it affects surface finish, accuracy, and adversely affects the life of carbide or ceramic-tipped tools.

Research into the vibrational characteristics of machine tools is a complex matter calling for advanced experimental techniques. Investigations on a vertical milling machine carried out by Dr Tobias at Cambridge University showed the behaviour of this machine under vibrations induced by an exciting force, and also related the severity of chatter to the relative cutter/workpiece position.

Excitation of the machine table in a vertical plane produced conditions from which the resonance curve, Figure 6.17, was plotted; the relative amplitudes between table and spindle being measured by a capacitance pick-up. The vibration mode produced by an induced vibration at 73 Hz is illustrated in Figure 6.18. The thin continuous lines represent surfaces from which the amplitudes

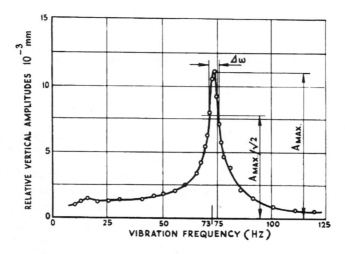

Figure 6.17 Resonance curve for vertical excitation of a milling machine (exciting force = 122 N rms) (courtesy I.Mech.E.).

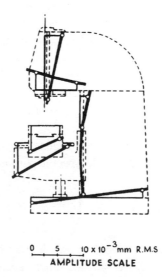

Figure 6.18 Vertical mode of vibration of milling machine (courtesy I.Mech.E.).

were measured, and the thick lines the maximum deflection. The distance of the circles from the thin base lines represents the amplitude of the dynamic deflection.

Horizontal excitation of the machine table produced vibrations represented by the curve, Figure 6.19, which shows peaks corresponding to four modes of vibration. The alignments of spindle and table at the excitation frequencies of 46, 56.5 and 78 Hz are illustrated in Figure 6.20.

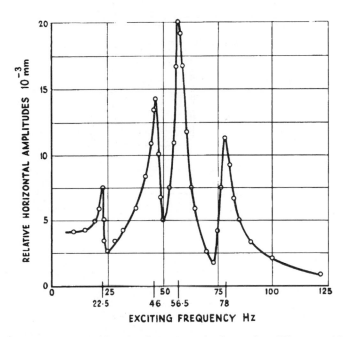

Figure 6.19 Resonance curve for horizontal excitation of milling machines (exciting force = 122 N rms) (courtesy I.Mech.E.).

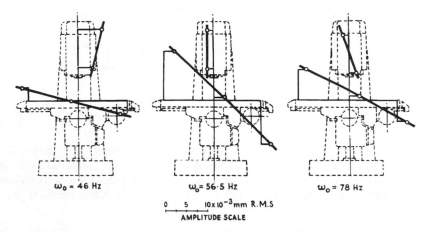

Figure 6.20 Alignment of spindle and table of milling machine in the horizontal modes (courtesy I.Mech.E.).

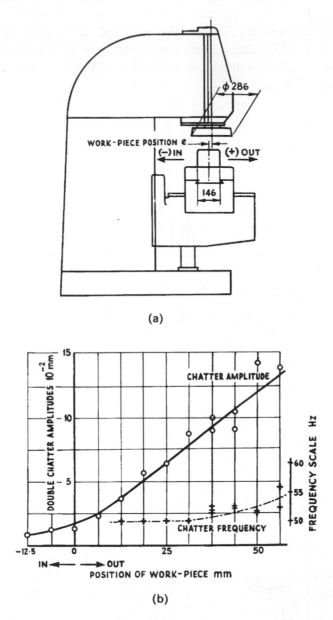

(a)

(b)

Figure 6.21 The effect on chatter of relative workpiece and cutter position. (a) Details of the milling machine set-up: no. of teeth = 32, workpiece material = CI, cutter speed = 126 rev/min, cutter feed rate = 584 mm/min, depth of cut = 1.9 mm. (b) Variation of chatter amplitude and frequency as a function of workpiece position condition stated at (a) (courtesy I.Mech.E.).

Factors determining the modes vary at different frequencies, but included among them are:

1. Stiffness of the foundation.
2. Stiffness of table and cross slides.
3. Bending and torsional deformation of knee, column and head.
4. Strength of connection between knee and column.

The behaviour of the machine tested led to the recommendations:

1. Slides, slideways and retaining strips need to be of substantial proportions to ensure stiffness, and where retaining strips are used there should be an adequate number of retaining bolts otherwise flexure may result from elasticity at this source.
2. Apertures in the column, e.g. for the motor, should be kept as small as possible as the flexure of the column of the machine under test was largely determined by such an aperture.

Effect of cutter position relative to workpiece

The conditions used to test the relationship between chatter in vertical milling and the eccentricity of the workpiece position relative to the cutter axis are shown in Figure 6.21(a). The graph, Figure 6.21(b), plotted from results obtained in the test clearly shows an increase in chatter amplitude as the table is moved outwards. The report also states that during the investigations newly sharpened cutters gave rise to increased chatter until the initial sharpness was removed. With very dull cutters chatter of a higher frequency was experienced.

For detailed findings of the investigation reference should be made to the paper 'Vibrations of Vertical Milling Machines Under Test and Working Conditions' by Dr S. A. Tobias, published by the Institution of Mechanical Engineers which has kindly given permission to reproduce the diagrams in this section.

6.6 Machine-tool alignments

Machine tool alignments are tested to ensure the machine is capable of consistently producing components to the accuracy demanded by standard tolerances. The method of testing is related to the type of machine and its functions; the tolerance magnitudes to its quality and the class of work produced on the machine.

Principles of testing

Although there are many different types of machine tool each with its own arrangement of spindles and slideways, certain elemental alignments apply to many of them; e.g. the perpendicularity of a spindle axis with a machine table. Such alignments can be tested by standard methods irrespective of the type of machine, but the inter-relationships between alignments – and their relative importance – depend on the type of machine and its function.

These factors are given the appropriate emphasis and arrangement in standard alignment test charts but in the absence of such charts *the essential alignments of a machine tool can be logically established from a consideration of the machine's construction and its method of functioning.*

Basic alignment tests

The basic alignment tests may be sectionalised under the following headings:

1. Tables and slideways.
2. Spindles.
3. Spindle-axis alignment with another feature.
4. Performance tests.

BS 3800, Methods for Testing the Accuracy of Machine Tools, gives details of suitable tests.

Some suitable tests are given in Tables 6.1–6.3. Tolerances are quoted to give the student some concept of the accuracy involved in the alignments of high-quality machine tools. It will be appreciated that tolerances will be increased for large machines and that tolerances for grinding machines will, in general, be less than those permitted on, say, horizontal boring machines. Hence the figures given for tolerances in the following section are for general guidance – when alignment tests are actually carried out, reference should be made to standard test charts such as those contained in the relevant parts of BS 4656, or to *Testing Machine Tools* by Georg Schlesinger (Machinery Publishing Co. Ltd).

Table 6.1 Table and slideways.

	Description of test	*Example of tolerance magnitude*
(a)	Straightness	±0.02/1000
(b)	Flatness	±0.02/1000
	(In the case of parallel slideways, flatness implies absence of 'twist')	
	Note: For details of the techniques employed for straightness and flatness testing see Figures 15.5 and 15.33	
(c)	Perpendicularly of two slideways	0.02/300
	(It is important to realise that this test must be carried out in *two* planes as shown in each of the diagrams)	

Table 6.2 Spindles.

Description of test	Example of tolerance magnitude*
Note: Before tests are carried out the spindle must have attained normal working temperatures	Precision
(a) Axial 'float'.	
(This may be caused by non-parallel thrust faces or incorrect bearing adjustment)	0.005 mm
	Average
	0.010 mm
	Large machines
	0.020 mm
(b) True running of location faces and diameters	Precision
	0.005 mm
(Faces must be tested to positions 180° apart)	Average
	0.010 mm

* Total indicator reading.

Table 6.3 Alignment of spindle axis.

Description of test	Example of tolerance magnitude
(a) Parallelism between axis and another feature	0.02/300*
Parallelism in one plane	
Parallelism in two planes	
(b) Perpendicularity between axis and another feature	
The tolerance for this alignment varies considerably according to the size of machine and the class of work it performs, e.g. vertical milling machines, axis perpendicular to the table	0.02/300
Perpendicularity of axis to slideway is tested by setting a straight-edge perpendicular to the spindle using the 'turn-round' method, as shown by positions *P* and *Q*; errors of perpendicularity will be observed as the cross slide is moved to traverse the straight edge past the dial gauge	
(c) Alignment between spindle axis and some other axis	0.03 mm

* Some tolerances are directional in character, e.g. headstock spindle of a centre lathe is permitted to slope upwards towards the free end of the test mandrel. Where these conditions apply, the direction of the slope is such that it tends to counter the normal deflection or wear of the machine (the direction of the slope is always stated in the general description of the alignment being tested).

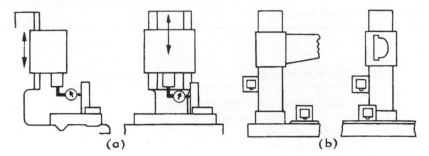

Figure 6.22 Testing perpendicularity of slideways in two planes. (a) Dial indicator method, (b) spirit level method.

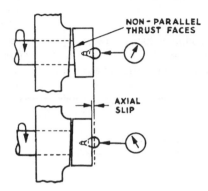

Figure 6.23 Testing for axial float.

Note

1. Machines must be satisfactorily supported and level before alignment tests are commenced, otherwise twist or deformation may be induced in the frame or its members. A spirit level is used for this purpose, and the test should be made in both transverse and longitudinal directions.
2. The sensitivity of a spirit level suitable for testing machine tool alignments is 0.04 mm/1000 mm per division.

If a location face has been machined with the spindle in position and the dial indicator touches the spindle on the same point no error will be apparent, hence two readings are taken as stated in Table 6.2(b).

The errors encountered in location faces and diameters are illustrated in Figure 6.25(a) and (b), and may be classified as follows:

1. Inclination (α) of spindle axis in relation to the axis of rotation.
2. Eccentricity (e) of the spindle axis.

Laser alignment testing

Laser interferometry uses the wavelength of light as the unit of measurement. Basically it involves measuring the relative displacement between two optical

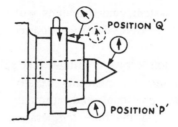

Figure 6.24 Testing concentricity of location diameters and true running at location faces (faces must be tested at two positions P and Q, separated by 180°.

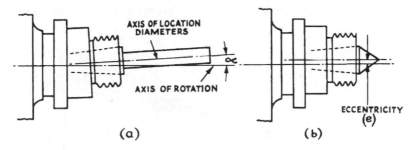

Figure 6.25 Types of error in concentricity. Inclination (a) and eccentricity (b) of an axis relative to the axis of rotation.

elements by splitting the light beam into two separate paths and counting the fringes – the wavelengths – which indicate the displacement between two bodies. The light beam is split using polarisation techniques; half is reflected back into the laser head with a retroreflector, the second half continues on its path to be returned by another reflector positioned on the object being measured. The difference between the moving and stationary beams induces a display which gives the corrected displacement values of one optical element in relation to another.

Interferometer systems can make measurements in three distinct modes; linear, angular, and straightness.

Linear measurements show positional accuracy, repeatability, and velocity. Acceleration and deceleration can be measured without additional software, as can vibration with the use of FFT (Fast Fourier Transform). In terms of linear, positional, and repeatability calibration, software is available covering all the major national and international metrology standards.

The second group of measurements – angularity – uses a different range of optics and a laser configuration which emits two parallel beams of light. Of a body's 21(7) degrees of freedom laser interferometry systems can measure 18(6), roll being, as yet, unmeasurable.

Machine tool manufacturers can now carry out all development, build, quality, and final test runs with one laser measurement system. Drive control velocity functions and finally automatic scale or lead screw compensation can be measured with a laser interferometer system, and machines will usually have compensation software systems to enable the CNC control to correct errors in the accuracy of leadscrews, etc.

Environmental compensation allows continual and automatic compensation for variations in air temperature, air pressure, relative humidity, and material temperature. It also allows continual and automatic normalisation of linear displacement measurement to a machine temperature of 20 °C, based on the average reading of up to three material temperature sensors and on a manually entered thermal expansion coefficient. This is especially recommended when carrying out linear measurements in workshop conditions, where environmental conditions are continually changing, and it considerably improves the resulting accuracy.

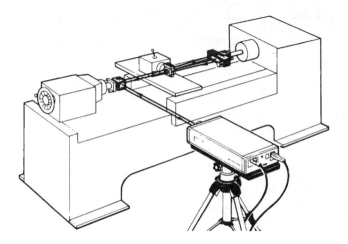

Figure 6.26 Parallelism measurement using an alignment laser.

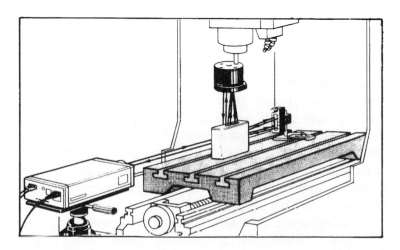

Figure 6.27 X–Z axes squareness measurement set-up.

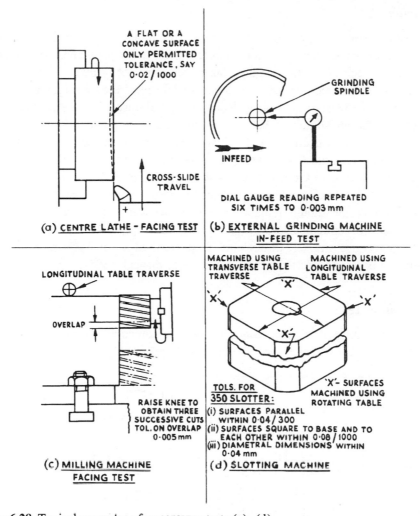

Figure 6.28 Typical examples of acceptance tests (a)–(d).

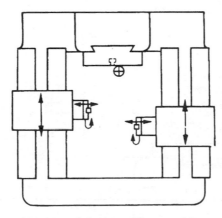

Figure 6.29 Alignment requirements of duplex milling machine.

Application

Parallelism measurements comprise two sets of straightness measurements performed with a common straightness reflector reference. Both linear parallelism and rotational measurements can be undertaken using a laser interferometer system.

Linear parallelism measurements are made to determine the misalignment between two nominally co-axial axes or plane. Rotational parallelism measurements are made to determine the misalignment between a rotational axis and a nominally co-axial linear axis. Such measurements are particularly pertinent to lathes. A set-up measuring the straightness of travel and out of parallelism of a lathe carriage axis relative to the spindle's axis of rotation is shown in Figure 6.26.

Deviations of perpendicularity can be assessed with suitable reflectors. The straightness reflector should remain fixed and provides the optical straight edge. Measurements of straightness are made through the pentaprism which bends the optical straight edge precisely through 90°. In the case of vertical squareness the large retroreflector is used to fold the optical straight edge through 180° (Figure 6.27).

Performance tests

At the present time performance tests are subject to agreement between the purchaser and the maker of a machine tool, the agreement preferably being made during the

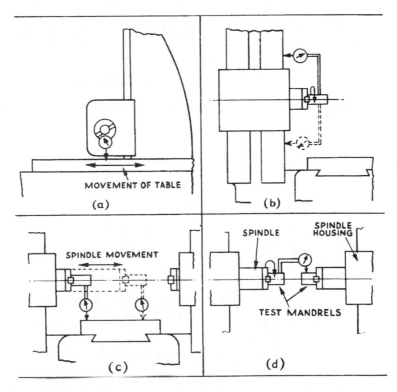

Figure 6.30 Some of the alignment tests for the duplex spindle milling machine.

negotiations which precede the placing of an order. An agreement made at such a time not only avoids possible disagreement at a later date but also gives the manufacturer an opportunity to ensure the machine will reach the required standards.

Certain well known performance tests are applied to standard machine tools, some of these are illustrated in Figure 6.28.

Example 6.2

Explain with the aid of diagrams the principal alignment checks for a duplex spindle surface milling machine of the type shown in Figure 6.29.

Table 6.4

1. Slideways and worktable

Description of test	Tolerance
(a) Flatness of worktable in longitudinal and transverse directions	±0.04/1000
(b) Rise and fall of worktable longitudinally	0.03/1000
(c) Slideways of column perpendicular to table in plane of spindle axes	0.02/300
(d) Slideways of column in plane perpendicular to spindle axes	0.02/300

2. Spindles

Description of test	Tolerance
(a) Internal tape of spindles runs true:	
(i) near spindle nose	(i) 0.01 mm
(ii) 300 mm from spindle nose	(ii) 0.02 mm
(b) Location diameter of spindle nose runs true	0.01 mm
(c) Location face of spindle nose runs true	0.01 mm
(d) Axial float of spindles	0.02 mm

3. Spindle alignments

Description of test	Tolerance
(a) Spindles square with slideways of upright	0.02/300
(b) Spindles square with slideways of table	0.04/300
(c) Spindles parallel with worktable	0.02/300
(d) Axial movement of spindle slide parallel with worktable	0.04/1000
(e) Spindles co-axial in vertical plane	0.01 mm
(f) Tee slots square with spindle (turn-round method)	0.05/1000

Solution

Note
Problems dealing with alignments are best approached from a consideration of the general to a consideration of the particular. In other words, plan the overall sequence of the tests before considering the detail testing.

In this type of machine it is clear that a basis for the tests is provided by the table and its slideways. The next most basic feature is the slideway of the upright columns, and these should be tested next, followed by tests of the spindles. Once the accuracy of the fundamentals has been determined, the relationships between the spindles and the other machine features should be checked. A detailed analysis of the testing procedure is shown in Table 6.4 and some alignment tests are illustrated in Figure 6.30.

□   □   □

6.7 Straight-line motion

Some common methods for obtaining straight-line motion are illustrated in Figure 6.31.

The thread form used for leadscrews, Figure 6.31(b), is usually acme or square thread: the efficiency of this type of screw is low, as can be seen from the following example.

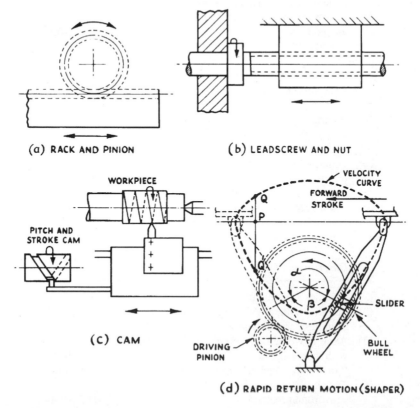

Figure 6.31 Conversion of rotary motion to linear displacement using mechanical methods.

Example 6.3

Determine the efficiency or a 5 mm pitch acme screw, thread angle 29°, 44.5 mm effective diameter when the coefficient of friction is 0.05.

Solution

From mechanics, efficiency for an acme screw thread is:

$$\eta = \frac{\tan \alpha (1 - \mu \sec \theta \tan \alpha)}{\tan \alpha + \mu \sec \theta}$$

where

$$\theta(\text{semi-angle}) = 29°/2$$

α (helix angle) is given by

$$\tan \alpha = \frac{5}{44.5\pi} = 0.03577$$

$$\sec \theta = 1.033$$

$$\therefore \quad \eta = \frac{0.03577\{1 - (0.05 \times 1.033 \times 0.03577)\}}{0.03577 + (0.05 \times 1.033)} = 0.4084$$

$$\therefore \quad \text{Efficiency, } \eta \approx 41\%$$

Note

The recirculating-ball screw and nut, Figure 6.32, greatly improves upon the performance of an acme thread, the minimum efficiency being above 90 per cent. The assemblies comprise a shaft with a semicircular helical groove and a nut with a corresponding helical groove. A

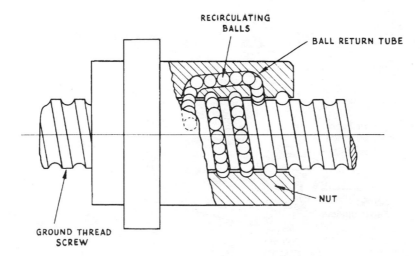

Figure 6.32 Recirculating-ball screw and nut. The minimum efficiency of this mechanism is greater than 90 per cent.

stream of balls fits into the groove, so forming a connection between the nut and screw. When in operation, the balls at the end of the stream are deflected into a return channel, thus ensuring a continuous and recirculating flow of balls when either screw or nut is turned. With a single nut assembly, backlash is as little as 0.02 mm; with a double nut and axially preloaded balls, backlash can be completely eliminated.

□   □   □

Figure 6.31(d) shows the basic features of a rapid return motion for shaping machines with a velocity curve superimposed upon it. The radial position of the slider on the 'bull' wheel may be varied and determines the length of stroke. The forward stroke occurs while the bull wheel rotates at constant speed through angle α, and the return stroke during rotation through angle β. The ram velocity at any position P is determined from the velocity curve by length PQ for the forward stroke and PQ^1 for the return stroke.

Spindle systems

Spindle systems have to be made so that vibration caused by the metal cutting process is not transmitted through the bearings, thus causing resonance.

Most systems today have a two stage gearbox and the gear change method is relatively simple (Figure 6.33). The two gear systems are engaged permanently, but are not attached through to the drive shaft. This is done by means of clutches which pull in on command from the control. Special precautions have to be taken to ensure that when the spindle is running both clutches do not engage at the same time. When stopped this method is often used to lock the spindle.

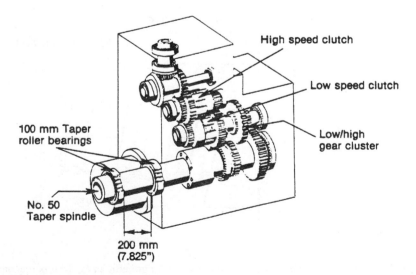

Figure 6.33 Spindle system showing a two speed gearbox.

6.8 Machine tool spindles

The material used for machine tool spindles is normally a low-carbon alloy steel, heat treated to give a case hardened surface. Such a spindle possesses resistance to wear combined with a tough core for strength in torsion. Where high precision spindles with maximum stability are required, e.g. spindles for external grinding machines, Nitralloy steel is used. The hardness of the case achieved by the nitriding process is greater than that obtained by carburising and hardening, and since nitriding is carried out at a temperature (500–550 °C) which is below the critical range, the structural equilibrium of the steel is not affected and secular changes due to retained austenite cannot occur. (See Figure 6.34.)

Allowance must be made for the thermal expansion of spindles. This may be done by preventing axial float by means of bearings situated close together at one end of the spindle and allowing the other end to expand freely in bearings which provide radial location, but permit axial movement, e.g. cylindrical roller bearings. The spindle of a lathe illustrating these principles is shown in Figure 6.35.

Example 6.4
A machine tool spindle is supported between bearings 560 mm apart. The lowest temperature at which the machine is used is 10 °C and the working temperature of the spindle is found to be 45 °C, while the average temperature of the spindle housing casting is 35 °C. Determine

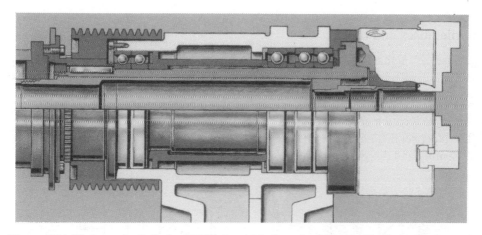

Figure 6.34 High speed spindle for CNC lathe with hollow chuck operated by hydraulic draw tube.

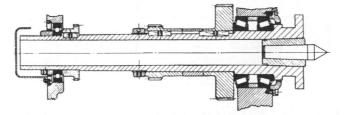

Figure 6.35 Bearings and drive gear arrangement on spindle of lathe.

the necessary allowance in the bearings for expansion. (α_s for spindle $= 0.000\,015$, α_n for housing $= 0.000\,01$, per degree C).

Solution

$$\text{Expansion of spindle} = \text{Length} \times \alpha_s \times \text{Temp change}$$
$$= 560 \times 0.000\,015 \times (45 - 10) = 0.294 \text{ mm}$$
$$\text{Expansion of housing} = 560 \times 0.000\,01 \times (35 - 10) = 0.140 \text{ mm}$$
$$\text{Difference in expansion} = 0.294 - 0.140$$
$$= 0.154 \text{ mm}$$

It is clear that 0.154 'play' would be unacceptable in a machine tool spindle and the provision for expansion mentioned above must be made without allowing axial 'float' in the bearings.

□   □   □

In heavy duty spindles and where a high torque is transmitted the final drive gear should be positioned as close to the cutting force as the design will permit, in order to keep torsion of the spindle to a minimum. By reference to Figure 6.36, it can be seen that if the final drive gear is positioned at A the spindle length between the gear and the cutting force is as small as possible. If the gear was situated at B, where $l_2 = 4l_1$, the total twist of the shaft would be quadrupled, since it is proportional to the length l.

Spindle noses and tapers

The nose of a machine spindle is bored out to receive and locate tools or tool holders; the wheel spindle of an external grinding machine has an external taper to locate the wheel adapter. The advantages of the taper for location purposes are:

1. Accurate location ensuring concentricity with spindle journals.
2. The taper seating eliminates all 'play'.
3. Assembly and disconnection of tapered parts is easily performed.

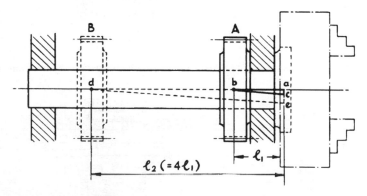

Figure 6.36 Effect of final drive gear position on the torsional rigidity of a lathe spindle. Angular deflection $ae = 4 \times$ angular deflection ac.

The tapers used may be classified into two groups:

1. *Self-holding tapers*, e.g. Morse taper and 5 per cent metric taper (BS 1660). These are used where parts are to be retained in position and withstand the application of a moderate torque by means of the friction between the taper surfaces e.g. taper shank drill.
2. *Self-release tapers*, e.g. taper ratio 7/24 (BS 1660). Where the torque is too great to be transmitted by the friction of a taper, positive drive by means of two keys is employed, and the taper angle increased to avoid difficulty in releasing the mating parts.

Because of their widespread international use the original Morse taper and the self-releasing taper of $3\frac{1}{2}$ in/ft($=7/24$) provide the bases of the current metric standard (BS 1660: 1972).

Holding and releasing forces for taper fits

Figure 6.37 shows the forces acting on a taper spindle when inserted in a taper bore.

Let P = force to drive in;
  Q = force to drive out;
  S = total normal thrust of shank on socket;
  R = total normal reaction of socket on shank;
  F_1 = frictional resistance when driving in;
  F_2 = frictional resistance when driving out;
  F_3 = frictional resistance of taper acting tangential to mean diameter;
  θ = semi-angle of taper;
  t = taper ratio change of diameter/length;
  μ = coefficient of friction.

$$F_1 = F_2 = \mu R \tag{6.5}$$

By resolving forces horizontally

$$P = F_1 \cos \theta + R \sin \theta$$

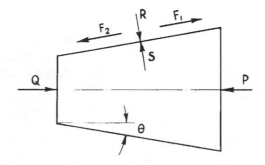

Figure 6.37 Forces acting on a taper spindle when inserted in a taper bore.

Substituting from equation 6.5 and rearranging

$$P = R \sin \theta (\mu \cot \theta + 1)$$
$$Q = F_2 \cos \theta - R \sin \theta$$

(6.6)

Substituting from equation 6.5 and rearranging

$$Q = R \sin \theta (\mu \cot \theta - 1)$$

(6.7)

Now

$$Q = P\left(\frac{Q}{P}\right)$$

∴   From equation 6.6 and equation 6.7,

$$Q = P\left[\frac{R \sin \theta (\mu \cot \theta - 1)}{R \sin \theta (\mu \cot \theta + 1)}\right]$$

$$= P \frac{(\mu \cot \theta - 1)}{(\mu \cot \theta + 1)}$$

hence, by division,

$$Q = P\left(\frac{2P}{(\mu \cot \theta + 1)}\right)$$

(6.8)

From equation 6.8 it follows that Q increases as the taper semi-angle (θ) decreases and that $P = -Q$ when $\theta = 0$, i.e. shank is parallel.

Similarly, for a shank and socket with a known coefficient of friction there will be a value of θ at which the taper will become self-releasing.

The value of θ at which a taper becomes self-releasing, i.e. the value at which the releasing force, Q, just equals zero, is determined as follows.

Equation 6.7 may be rewritten as $Q = R \ (\mu \cos \theta - \sin \theta)$ and since for self-releasing tapers, $Q = 0$, but $R \neq 0$,

$$\mu \cos \theta = \sin \theta$$

$$\therefore \mu = \tan \theta$$

(6.9)

The taper ratio (t) is $2 \times \tan \theta$ and substituting from equation 6.9 a taper is self-releasing when $t = 2\mu$. If we take $\mu = 0.146$ for steel tapers in steel sockets (0.15 is considered an average value for μ, steel on steel), then the taper ratio for self-releasing action,

$$t = 0.292 \text{ or } 7/24$$

(equivalent to $3\frac{1}{2}$ inches/ft on diameter.)

This is the value specified for self-release machine tapers in BS 1660.

Torque transmitted by self-holding tapers

The maximum torque which can be transmitted through the friction of a self-holding taper will be limited by the accuracy in matching of the sleeve and shank. For accurately fitting tapers the actual transmitted torque is reasonably close to the value shown in Figure 6.38.

Maximum value of transmitted torque,

$$T_{max} = F_3 \times \frac{\text{Mean diameter}}{2}$$

$$= \mu R \left(\frac{D+d}{4} \right)$$

Rearranging equation 6.6 and substituting for R,

$$T_{max} = \frac{\mu P}{\mu \cos \theta + \sin \theta} \left(\frac{D+d}{4} \right) \tag{6.10}$$

From equation 6.10 it will be noted that the maximum torque transmitted can be increased by:

1. Increasing mean radius, R_m.
2. Increasing value of P.
3. Decreasing value of θ.

All three of these expedients are used in the design of taper shanks for cutting tools; θ is given a low value for self-holding tapers, P is increased by the provision of draw bars, R_m is increased for the shanks of large diameter drills and other heavy duty cutters.

Example 6.5

The following data is taken from a series of practical tests on drilling.

> Material, cast iron. Cutting speed 20 m/min. Diameter of drill, 24 mm.
> Feed = 0.33 mm/rev. Axial thrust = 4500 N.
> Torque = 37.6 Nm.

Determine whether a No. 2 or a No. 3 Morse taper shank is suitable for this drill. (Take taper ratio as 5 per cent for both sizes and $\mu = 0.15$.)

Solution

$$\tan \theta = \frac{0.05}{2}$$

$$\therefore \quad \theta = 1°26'$$

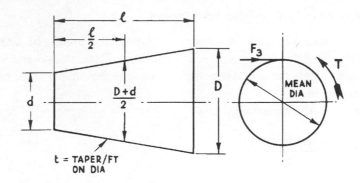

Figure 6.38 Frictional torque transmitted by a taper.

From equation 6.10

$$T_{max} = \frac{\mu.P}{\mu.\cos\theta + \sin\theta} \times \frac{(D+d)}{4}$$

$\therefore$ Mean diameter, $\dfrac{D+d}{2} = \dfrac{2 \times T_{max}(\mu.\cos\theta + \sin\theta)}{\mu.P}$

$$= \frac{2 \times 37.6 \times 10^3 (0.15 \cos 1°26' + \sin 1°26')}{0.15 \times 4500}$$

$$= 19.5 \text{ mm}$$

Now

No. 2 Morse taper $d = 14$ mm, $D = 17.78$ mm

No. 3 Morse taper $d = 19$ mm, $D = 23.825$ mm

$\therefore$ A No. 3 Morse taper is required.

☐   ☐   ☐

6.9 Rolling bearings

The great advantage of rolling bearings lies in the substitution of rolling friction (the theory of which is discussed on p. 180) for sliding friction. The coefficient of friction for rolling bearings is $\mu = 0.001 - 0.0015$, although needle roller bearings are a little less efficient and $\mu \approx 0.004$. Other advantages of rolling bearings include reduced torque for starting from rest, reliability, ease of lubricating, easy replacement and low maintenance costs – although to obtain these advantages correct design and assembly is essential.

Selection of bearing

Basic considerations in the selection of bearings are as follows:

1. Direction of load relative to bearing axis: the bearing may be required to meet axial thrust, radial load or a combination of both.

2. Magnitude and type of load: ball bearings will sustain considerable loads, but for severe conditions and shock loads roller bearings with their line contact are to be preferred.
3. Shaft stiffness: rigid bearings may be used for stiff well-aligned shafts, but shafts subject to flexure or misalignment require self-aligning bearings.
4. Speed of rotation: speed and load are inter-connected; for a given life, speed must be reduced as load is increased, and vice versa, i.e.

$$\text{life} \propto \frac{1}{(\text{speed})^a(\text{load})^b}$$

Basic types of bearing

Basic types of bearing are illustrated in Figure 6.39.

1. *Single-row radial ball bearings.* This type of bearing will accept pure thrust in either direction, pure radial load or a combination of both; the maximum thrust load is less than the maximum radial load. In common with other rolling bearings, the permissible load decreases with increasing speeds, and its life is reduced by elevated operating temperatures.
2. *Single-row angular contact bearings.* These bearings will accept a greater thrust load than (1) above, but in one direction only; in consequence, they are used in pairs. The angle of contact (α) may be selected according to the amount of thrust the bearing is to sustain.
3. *Single-row thrust bearings.* The applied load should be pure thrust only, as a journal load affects the concentricity of the rings. The balls are subject to centrifugal force, and very high speeds must be avoided.
4. *Cylindrical roller bearings.* These bearings are used for applications where a high radial or shock load is encountered. The inner race may move axially relative to the outer race, and this property is utilised in machine tool spindles to permit expansion due to temperature effects without loss of radial location.
5. *Needle roller bearings.* For a given bore size, less room is taken up by needle roller bearings than by the standard cylindrical type, although needle rollers are less efficient. The bearings are used for the lower speeds and as bearings for oscillating motions.
6. *Taper roller bearings.* This type of bearing has a wide field of usefulness, especially in connection with machine tool spindles, gearshafts, etc., where high axial and radial forces are combined. The angle α, Figure 6.39(f), can be selected according to the relative magnitudes of the axial and radial forces, the usual range being from $7\frac{1}{2}°$ to $25°$, the steep angle bearings being suited to heavy thrust conditions.

In all cases basic bearing types have variants to suit particular conditions. Makers' catalogues not only give details of the wide range of types and sizes which are available but also provide data relating speed, load and life factors.

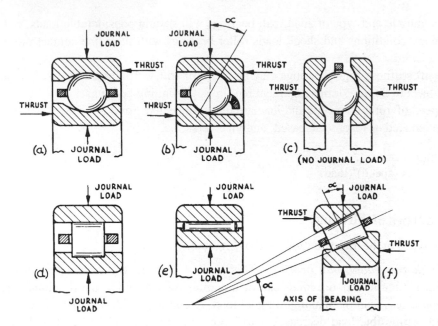

Figure 6.39 Basic types of roller bearings. (a) Single-row ball bearing; (b) angular-contact bearing; (c) single-row thrust bearing; (d) cylindrical-roller bearing; (e) needle-roller bearing; (f) taper-roller bearing.

Rolling friction

If an inelastic ball or roller rolls over an inelastic plane surface, Figure 6.40(a), the ball is in point contact and the roller in line contact: under such theoretical conditions the stress is infinite; stress = load/area, but when area = 0, stress = ∞. However, the materials used for rolling bearings are elastic and deform under load, Figure 6.40(b); this enables them to withstand loads without fracturing due to excessive stress.

Since the deformation of rolling bearings in practice is of significant size, it is clear that relative movement between the inner and outer races can take place due to this factor, and, additionally, that energy is absorbed by the process of deformation. This energy is an important factor in total energy absorbed by 'rolling friction'.

A further aspect of the deformation process relates to endurance: in use, the metallic crystals of the bearing are subjected to repeated stresses which, even if they do not reach the ultimate strength of the material, can eventually lead to failure by fatigue of the metal.

In dealing with rolling bearings it is desirable to know their load carrying capacity and, in precision applications, to have a concept of the magnitude of the deformation which can occur in bearing elements. These matters have been treated in research carried out by Hertz, Stribeck, Palmgren and others.

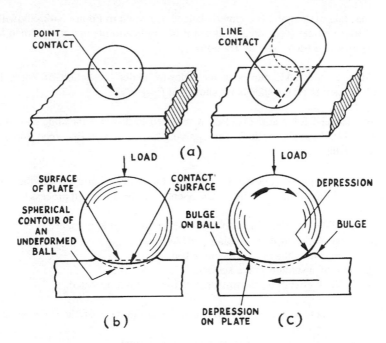

Figure 6.40 Contact conditions between a plane and a rolling element. (a) Inelastic elements; (b) elastic elements (stationary); (c) elastic elements (rolling).

Exercises – 6

1. Analyse the effect of dynamic forces induced by rotating and reciprocating masses in the operation of a machine tool. Discuss the relevant factors and their effects in regard to one of the following: (a) centre lathe; (b) planing machine; (c) external cylindrical grinding machine.

2. The wheelhead slide shown, Figure 6.41, is supported on rollers retained in a cage. If the centre of gravity of the wheelhead is 145 mm from the left-hand end of the slide, and the position of the rollers when the slide is at the extreme right-hand end of its movement is as shown, find the maximum slide movement (L). If the rollers are spaced at 43 mm intervals, what is the maximum number of rollers the slide will accommodate?

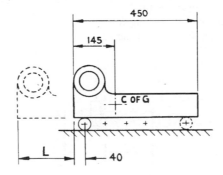

Figure 6.41

3. A machine slide of height h is counterbalanced as shown in Figure 6.42. Show that it can be raised by a vertical feed force if its point of application is not greater than $h/2 \cot \phi$ from the centreline (ϕ = angle of friction).

4. A vee slide, Figure 6.43, moves in its slideway under the action of force P. If the vertical force acting on the slide is W, show that $P = \mu W \sqrt{2}$.

5. Determine the feed force at the leadscrew required to displace the slide shown in Figure 6.44 ($\mu = 0.12$). Find the position for applying the least possible feed force, and the magnitude of this force.

6. Illustrate the principle of a device to eliminate the effect of backlash in the table screw and nut on a milling machine. Why is this essential in the down-cut process?

7. How are the following relations established in machine tool alignment?
   (a) Squareness of two or more machine surfaces.
   (b) Flatness and straightness of a machine bed.
   (c) Alignment of a spindle with a surface.
   The general characteristics of the instruments used should be stated.

8. (a) What equipment would you use to check the straightness of the slideways of a large slideways grinding machine?
   (b) How would you check the vertical slide for squareness?
   (c) How would you check the bed for cross-wind, and
   (d) What accuracy could be achieved in each of the above measurements?

9. (a) What principal alignment checks are necessary for a first-grade centre lathe and what tolerances should be met?
   (b) What equipment is necessary to check the alignments?

10. The vertical and horizontal cutting forces acting on the boring bar shown in Figure 6.45 are 900 N and 270 N respectively. If the tool is set at a radius of 19 mm determine:
    (a) Torsional deflection due to the 900 N component. (*Note*: torsional deflection,

$$\theta° = \frac{584.T.L}{G.D^4}, \text{ where } T = \text{torque, Nm, and modulus of rigidity.}$$

$G = 82 \times 10^9 \text{ N/m}^2$, D and L in mm.

    (b) Deflection as a cantilever due to:
       (i) vertical component of cutting force;
       (ii) horizontal component of cutting force.

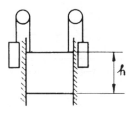

Figure 6.42

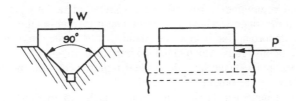

Figure 6.43

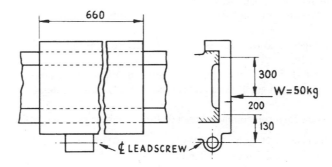

Figure 6.44

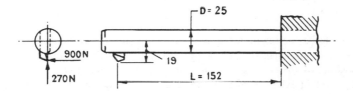

Figure 6.45

Hence determine the final position of the tool point and the size of hole which will be produced by the boring tool ($E = 206 \times 10^9$ N/m^2).

11. The spindle speed-change gear for a drilling machine has to comply with the following specification:

    Max diameter of drill   = 12 mm
    Min diameter of drill   = 3 mm
    Max culling speed of smallest drill = 30 m/min
    Min cutting speed of largest drill   = 20 m/min
    Six spindle speeds.
    Motor speed   = 2800 rev/min

Make a line diagram of a suitable layout for the spindle drive and determine the gear ratios.

12. Make a diagrammatic sketch suitably annotated of a typical transmission from the headstock of a lathe, through the feed box to the rack for sliding, the saddlescrew for surfacing and the leadscrew for screwcutting.

13. (a) The design requirements for the main spindle of a lathe are that it must be constrained in a manner which will ensure that it is rigid in all directions, except that it is very important for it to rotate about a fixed axis. Enumerate the minimum number of constraints which must be provided in an ideal pneumatic bearing arrangement and indicate how these ideal requirements are modified to provide a design which will be satisfactory in service.

(b) What arrangements can be made to ensure that the main spindle of a lathe with a very wide speed range will:

(i) not be subject to serious angular oscillations during heavy machining operations, which may include intermittent loading;

(ii) not be subject to high-frequency vibrations during high-speed finish machining?

14. Show that a taper ratio of 7/24 is the most suitable taper to be used for a 'non-stick' taper fit. ($\mu = 0.145$.)

Make a neat sketch of the type of arbor end and spindle nose used on a milling machine having a 'non-stick' taper and add notes explaining the design.

15. A cutting test on cast iron conducted on a drilling machine gave the following results:

Drill diameter = 19 mm   Spindle speed = 302 rev/min
Feed/rev   = 0.3 mm   Torque   = 21 Mn
                Axial thrust = 3400 N

Using an average value of 5 per cent for the taper, and allowing a safety factor of 1.25 on the torque and thrust values, determine a suitable size Morse taper shank for this drill. ($\mu = 0.15$.)

16. Analyse the forces acting on the spindle of a conventional vertical milling machine, and by reference to this analysis draw a cross-section through the spindle to show the type and disposition of the bearings you suggest to meet the conditions.

Clearly explain what provisions are made in the design to allow for temperature variations of the spindle due to speed and load conditions.

7 Control of Machine Tools

7.1 The need for automatic control

In those processes of manufacturing which include machining operations, several types of machine tool may be involved sequentially, e.g. turning, milling, grinding. Manual control of such operations becomes uneconomic as the required output rises and alternatives of an automatic or semi-automatic (manual loading and unloading) nature have long been available. These trends are now further advanced by numerical control (NC), applicable to more complex machining, and by in-process measurement with feedback to compensate for tool wear or other minor variables affecting a critical dimension.

Manually controlled machining operations involve three distinct steps: planning, machine setting, machining and, for one-off jobs, these may be done on the shopfloor by a skilled craftsman. For automatic or semi-automatic operation a company generally has separate groupings; a production planning department performs the operation planning, skilled tool setters prepare the machines and machine operators/loaders tend the machines while they produce the work.

From early times machine tools were fitted with devices to reduce manual labour, e.g. automatic traverse of slides. A considerable range of mechanical hydraulic and electrical devices have contributed to the development of automatic operation and control, some of which are treated below.

7.2 Mechanical control

Figure 6.31 (a), (b) and (c) illustrate common mechanisms used to propel slides. By linking these to some device which engages and disengages them at the required points within a cycle, a simple automatic form of operation is achieved.

Figure 7.1 shows a trip and clutch mechanism used on some automatic lathes. The angular setting of the trip dog on the disc gives control of operation timing. The arrangement shows how a trip-displaced lever can operate a spring loaded clutch which incorporates a device for disengaging the same clutch after a short period of action.

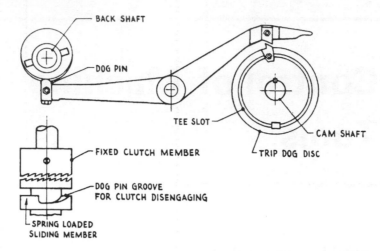

Figure 7.1 Trip and clutch mechanism.

Such mechanically controlled clutches can be used to drive other items of an automatic lathe such as the geneva indexing turret mechanism shown schematically in Figure 7.2. When the index plate is driven its rotation has two consequences. By the partial rotation of the connecting rod about D, the sub-slide is rapidly retracted and then returned to its former position. Simultaneously, pin A, after most of the slide retraction has occurred, engages slot B of the geneva plate and rotates it to position C, thus indexing the tools mounted within the turret. In small high-speed single spindle automatic lathes the entire action can be completed in as short a time as 0.5 seconds.

The turret feed is driven by a geared quadrant driving a rack, the quadrant being part of a cam-operated lever mechanism as shown in Figure 7.3.

The mechanism in Figure 7.3 gives the typical method for cam-operated slides where a specially designed and manufactured plate cam is used. The cam, its timing

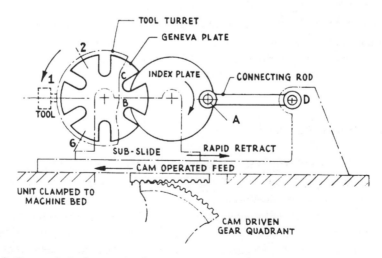

Figure 7.2 Geneva indexing mechanism.

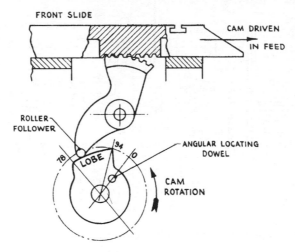

Figure 7.3 Cam mechanism for advance of slide.

location fixed by means of a dowel as shown, brings the slide into action at 78/100 of the cycle, and advances the slide at a feed rate fixed by the rate of cam rise until the operation element is completed at 94/100 of the cycle.

The equivalent cam for operating the turret could of course be six-lobed to make use of all six tooling stations on the turret.

7.3 Single spindle bar automatic lathe (SS Auto)

This machine often called a *screw machine* because of its widespread use for rapid manufacture of turned and die-threaded small screws from bar-stock, is representative of the majority of the ideas incorporated in mechanically controlled automatic machines.

Figure 7.4 gives a schematic arrangement of an SS Auto. Mechanical control of this machine depends upon the drive to the back shaft on which clutch mechanisms of the type shown in Figure 7.1 and other devices are mounted. The back shaft is linked by change gears and a worm reduction gear to the front shaft and the turret cam shaft. These shafts are arranged to make one complete revolution while the headstock spindle makes the number of revolutions needed for machining the part. The front shaft rotates the front and rear slide cams and, via trip mechanisms, operates collet opening, spindle speed changes, turret indexing etc. Data are given in the machine handbook for preparing an operation layout with timing based on headstock spindle revolutions, and for the design of the special cams where the unit used is 1/100 of the cam circle. Such a machine is automatic in operation except for the manual loading of bar stock as it is consumed.

Single spindle *chucking* autos operate on similar principles but require manual loading and unloading unless specially designed equipment is fitted. Some of these machines can be set-up from standard parts, thus avoiding the expense of special

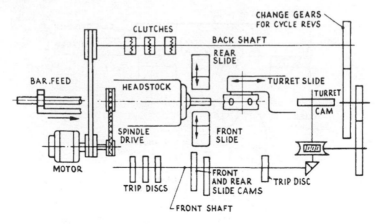

Figure 7.4 Layout of drives on single spindle automatic.

cams, but this method normally causes a larger proportion of idle time and consequently a lower rate of production.

7.4 Multi-tooling

By skillful planning, cycle times for the above machines can be reduced by overlapping operations. Two different developments of the automatic lathe exploit this principle.

Simultaneous cutting by a number of tools is a feature of *multi-spindle* automatics, the principle of which is shown in Figure 7.5. The headstock incorporates a number of equi-spaced work spindles each with its own cross slide. The central toolblock has a tooling position for each spindle serving the function of the turret of the SS Auto. Traditionally, cams are used to control the slide and central toolblock motions and are either specially designed or selected from a standard range at some sacrifice of optimum output.

As the headstock indexes, spindles move round to consecutive tooling positions, the cycle being completed by parting-off the job at spindle 5 and feeding bar to a stop. Ideally, given the same speed and feed rates, the machine should turn a part in 1/5 of the time required on the SS Auto. These have evolved into multi-spindle CNC lathes which allow much faster change over times between jobs, as well as much shorter delivery times for new jobs. Designing and manufacturing a set of cams for a new component may take several months for a cam controlled automatic lathe and a six spindle lathe can easily take eight hours to change over to the next job.

Programs for a new job on a CNC lathe should only take a few days to write (even allowing for the queue of other jobs) and the standard tooling should be immediately available. Thus lead times should be minimal and the change over time between jobs should be an hour or so (less if well organised). This means that batch sizes will be considerably reduced and cam controlled machines are now limited to long running jobs which rarely need resetting (Figure 7.6).

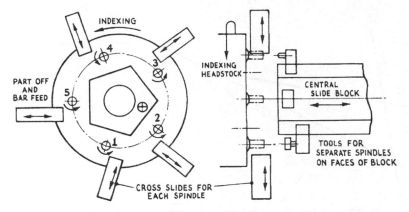

Figure 7.5 Arrangement of multi-spindle automatic lathe.

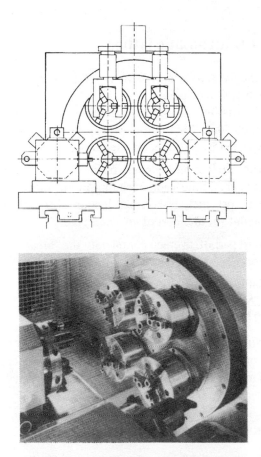

Figure 7.6 CNC lathe with four working spindles. When the part is to be changed, the entire spindle disc turns 90° or 180°. The part can be turned automatically after the first operation.

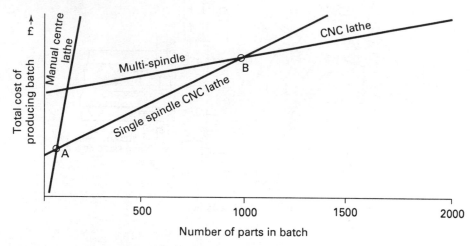

Figure 7.7 Comparison break-even points for various lathes.

7.5 Economics of automatic lathes

Figure 7.7 shows, tentatively, the approximate relationships between various types of lathes in terms of break-even points.

Note that the initial costs increase with the complexities of special equipment and setting of the machine, while high output with low direct labour cost reduces the price per piece as presented by the slope of the graph.

1. Cost of getting into production includes: planning and tool design, manufacture of special tools and cams, time spent in setting-up the machine etc.
2. Machine operating cost includes: direct labour cost, supervision, overheads, amortisation of capital (or hire cost of machine) etc.

For a straightforward part of average complexity the 'break-even' points would be in the following region.

Manual centre lathe	1–5 parts
Single spindle CNC lathe	5–1000 parts
Multi-spindle CNC lathe	>1000 parts

Realism when investing in new equipment requires the whole range of work to be considered, because only very high utilisation will justify the capital costs involved in advanced types of machine.

7.6 Advantages of numerical control

As an alternative to the control of machine tools by mechanical, hydraulic, or other conventional means, numerical engineering provides considerable attractions in

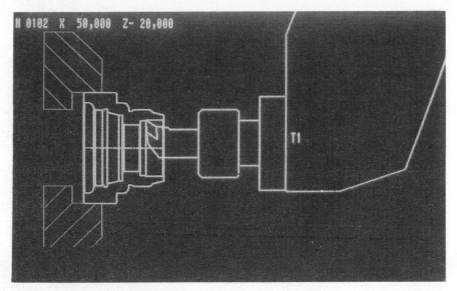

N 0102 X 50,000 Z- 20,000

TI

Figure 7.8 Graphic simulation.

suitable manufacturing applications. Numerical control, as applied to machine tools and other equipment, is versatile in that it can be employed for a wide variety of processes and used on components of widely differing characteristics. It is flexible, for whilst it may be use for high production quantities, it is of particular value in the production of small and medium-sized batches of components, since the need for jigs and fixtures – and the very high cost these involve – is almost eliminated. The computer is a very powerful tool in the preparation of programs for NC, and the development of mini-computers has enabled machine control units (MCU) of compact size to be dedicated to individual machines. The MCU will usually incorporate extended facilities, e.g. keyboard, visual display unit (VDU), memory etc. which can display graphics of the program and is thus invaluable in realising the full potential of a machine tool. These are termed CNC controls (Figure 7.8).

7.7 Analysis of the functions of a CNC machine tool

Application of CNC to a machine tool enables the functions normally performed by an operator in conventional situations to be taken care of by the CNC system. For present purposes these functions may be separated into two groups.

1. *The primary function* – namely the displacement of the machine slides to maintain a relationship between the cutting tool and the workpiece which will result in the desired geometric shape of the component to the required degree of accuracy.

    The conventions adopted to identify slide displacements are illustrated in Figure 7.9. In addition to the primary linear displacements X, Y and Z, certain

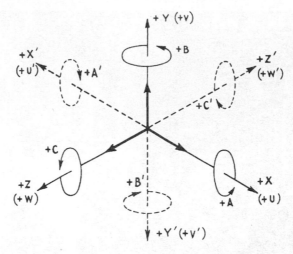

Figure 7.9 Conventional notation for linear translation along machine axes (notation for secondary displacements shown in brackets) and for rotational displacement about each axis.

types of machine may possess secondary slides operating in directions parallel to those of the primary slideways e.g. CNC lathe with twin turrets (Figure 7.10). The Z axis of a CNC machine is always in line with the axis of the principal spindle of the machine whether that axis is vertical or horizontal. The X axis is always horizontal and parallel to the work-holding surface, and the Y axis is perpendicular to the X and Z axes.

2. *The secondary functions* – that is, those supporting functions which are necessary for the normal operation of the machine. The functions will vary according to the type of machine or equipment, e.g. the requirements for the operation of, say, a CNC lathe will differ from those of a CNC electro-discharge machine. Common examples of secondary functions include some or all of the following:

(a) Spindle, start/stop/reverse.
(b) Cutting fluid, on/off.
(c) Select desired spindle speed.
(d) Select desired feed rate for slides.
(e) Index/rotate circular table.
(f) Exchange cutting tools.
(g) Lock slides when desired position is reached.

The extent of the range of secondary functions is determined not only by the type of machine, but also by the degree of sophistication considered necessary in the machine for a particular application, e.g. the desirability or otherwise of providing automatic tool changing. Many extra functions are offered by machine tool manufacturers as additional options which the machine user can select – at extra cost – according to the nature of the work the machine is to perform. Clearly, before adding to the purchase cost of a machine by demanding a wide range of extra

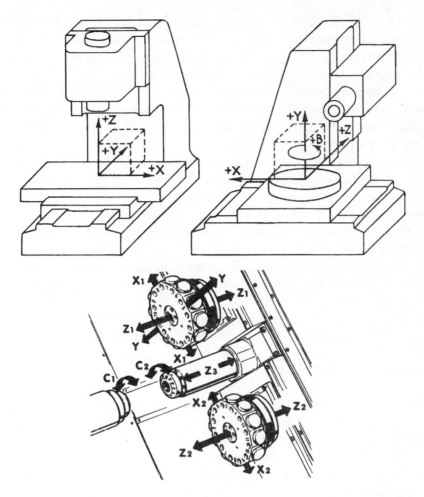

Figure 7.10 Conventional notation for slide displacements in some common types of CNC machines; lower illustration shows multiple axes.

options, an investigation into the particular needs of the manufacturing organisation must be carried out.

7.8 Inputs to the machine control unit

Operating instructions to a CNC machine are conveyed through the machine control unit. The operating instructions are devised by the process planning engineer (or programmer) in the form of a part-program and are then converted into a physical medium acceptable as an input by the MCU – e.g. punched tape or are transmitted via an RS232 link from a computer.

Originally, MCUs were of the 'hard-wire' type, i.e. they were constructed to initiate a series of set functions when requested to do so by the input of a particular code specified on the tape; for example, G81 is an alpha-numeric code which signifies a complete sequence of machine movements needed to perform a drilling or reaming operation. The MCU would recognise this signal and the control system would initiate the necessary responses from the machine tool.

CNC machine control units possess storage, i.e. memory, facilities. This facility can enable the contents of the tape to be put into store on its first run through the tape reading head. Thereafter the operation of the machine tool can be controlled from the memory each time a component is machined without re-running the tape. The tape may then be retained as a permanent record, and for use on future occasions if repeat components are required. Taking the information direct from the memory of a CNC unit is faster than reading and re-reading tape each time a component is machined, and errors or hold-ups due to damaged tape are eliminated. It also gives control back to the machinist, who can now edit the program on site to correct minor errors.

CNC machine tools do not necessarily need the program information input to be in the form of punched tape. Input data may be dialled straight into the memory by means of a keyboard incorporated in the MCU or the input data may be received by the MCU from a separate computer located at a distance from the machine tool. This method of control is termed DNC (distributed numerical control) and the computer may concurrently monitor several machine tools.

Other benefits of CNC include the following:

1. *Editing*. The data on the tape can be modified by overriding during which new information is keyed into the MCU by the operator or programmer. Feed rates can be optimised, over-runs eliminated, cutter paths shortened etc.
2. *Optimised tapes*. Once the program has been optimised by editing, the modified program may be recorded on a floppy disk, or sent back to the shop computer via the communications link.
3. *Display*. The visual display unit (VDU) provides a medium of communication between machine and operator, with most machines providing viewing of the program to facilitate debugging of the program and operation of the machine.
4. *Diagnostics*. A series of standard test procedures are held in store, and may be called up to check the operational status of various parts of the system. A few manufacturers even offer remote diagnostics checks via telephone connection directly between machine controller and their service department.

7.9 Program preparation

The use of CNC machine tools increasingly transfers responsibility for the machining process from the shopfloor to the process planning office, and requires more detailed work at this stage than when conventional machines are employed for batch production. In addition to specifying the normal sequence of operations CNC programming also necessitates the definition of workpiece geometry, calculation of co-ordinate positions and points of intersection, specification of tool types and sizes, spindle speeds, feed rates, and miscellaneous functions such as cutting fluid supply on/off, etc. The component geometry will be defined by means of a CAD system, and can produce a part geometry file for the machine programmer to use as a basis for the CNC program by direct transfer to the program generating the machine code, although CNC programs may also be compiled directly by the programmer from 'paper' information, typing the information directly into a computer and editing the program on screen. Pencil and paper programming and punched tape inputs are virtually obsolete.

The complete information for producing the component is entered on a part-program sheet. There are several different classes of programming, i.e. manual and differing types of computer-based methods. According to which method is used, it will be necessary for the programmer to ensure the appropriate procedure is followed in preparing the program (see Figure 7.11).

A simple example of a single line taken from a manually prepared program (i.e. a program not requiring the use of a computer to prepare the tape) is shown in Figure 7.12(a). The complete program comprises many such lines to take the operation of the machine step by step through the process of part manufacture.

This program is reproduced in coded format on punched tape, each line of the program forming a corresponding block of information as shown in Figure 7.12(b). The format for the punched holes, in this example, uses the ISO system. This system is computer compatible and employs track No. 8 for obtaining even parity i.e. each

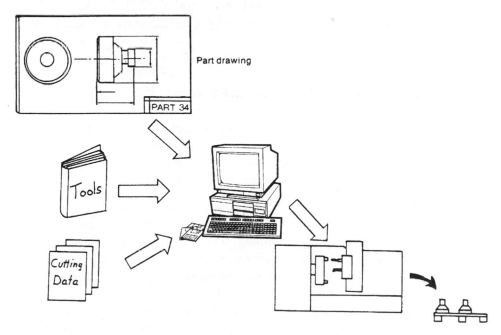

Part drawing

PART 34

Tools

Cutting Data

Figure 7.11 Programming a part for a CNC lathe.

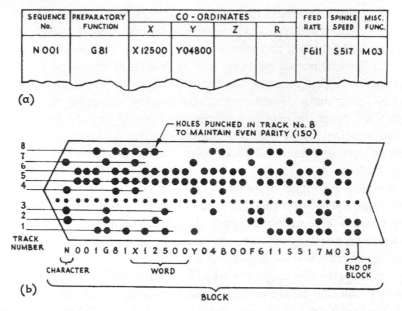

SEQUENCE No.	PREPARATORY FUNCTION	CO - ORDINATES				FEED RATE	SPINDLE SPEED	MISC. FUNC.
		X	Y	Z	R			
N 001	G 81	X 12500	Y04800			F611	S 517	M 03

(a)

(b)

Figure 7.12 (a) A line of information on a part program; (b) the corresponding block of information on a punched tape using ISO format (even parity).

line across the tape must contain an even number of holes otherwise an error is present. Writing programs in such detail is time consuming and costly; the method is suitable for simple components on relatively basic CNC machines but more sophisticated programming methods are essential for advanced work. These methods will be described later.

7.10 Classification of CNC machine types

Consideration of the geometry of components produced by CNC machine tools (see Figure 7.13) shows that the basic requirements in slide control are for positional, linear and contouring capabilities. Examples of machine types and equipment falling into these categories are shown in Figure 7.14. Note that some processes, e.g. welding, can fall into more than one group.

1. *Positional control.* Two orthogonal slides move rapidly to a fixed position at which machining takes place, e.g. drilling, boring, hole punching. The path taken by the slides to reach the desired position should be the quickest possible route, but is otherwise of no importance unless obstructions, e.g a clamp, need avoiding. This class of slide control is referred to as 2D.

2. *Linear motion control.* The slides can be displaced at rapid rates for positioning purposes, as for (1), but they are also capable of being controlled to move at feed rate, individually or concurrently, to enable straight cuts to be made. The

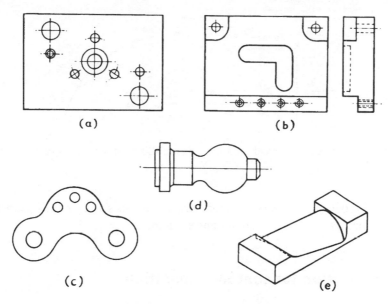

(a)   (b)

(d)

(c)   (e)

Figure 7.13 Examples of workpieces produced on CNC machine tools. (a) Positional; (b) line-motion control; (c) contouring (2D); (d) contouring (2D); (e) contouring (3D).

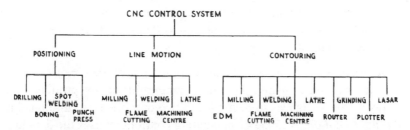

Figure 7.14 Classification of types of control for slide displacements on a range of CNC machine tools and equipment.

designation is 2L or 3L for linear control in two or three axes respectively, and may be extended to 4L, etc.

3. *Contouring Control.* The machining of some components requires that slides move concurrently in a non-linear relationship, and need control systems able to accept and process detailed and complex data. These data are necessary to define the cutter path required to produce the desired geometric shape. This process involves controlling the velocities, accelerations and decelerations of the slides along their respective axes. A simple example of constantly changing slide velocities when machining an arc is illustrated in Figure 7.15. The commonest example of this type of control is designated $2\frac{1}{2}$ D; i.e. the spindle in the Z axis has a linear control system (L), but the other two axes, X and Y, have a continuous type control system (2D) capable of dealing with non-linear data.

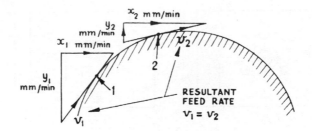

Figure 7.15 To maintain a constant feed rate when milling a curve it is necessary that slide velocities in X and Y are varied, i.e. X to X_2, Y to Y_2, etc.

Continuous control of motion for contouring processes is extended to three axes (3D) or more for complex machines and equipment.

7.11 Interpolation for contour generation

To produce smooth curves when contour machining using CNC, points on the surface are followed in a sequence of slide movements made possible by a system of interpolation. The control system of a CNC machine ensures related slide motions at the velocities necessary to achieve the desired geometrical shape of the component. Slopes and circular arcs are among the most common features needing this facility, and the two common interpolation systems used are linear and circular interpolation, defined in BS3635 as follows:

1. *Linear interpolation.* A mode of contouring control that causes a slope or straight-line operation, using data in a single block to produce velocities proportional to the distances to be moved in two or more axes simultaneously.
2. *Circular interpolation.* A mode of contouring control that uses the data in a single block to produce an arc of a circle, the velocities of the motions in two axes used to generate this arc being varied by the control system.

Example 7.1
Examine the differences in the linear cut vectors when using (i) chords, (ii) tangents, (iii) secants, to approximate a 25 mm external radius through a 90° arc, if the maximum variation from a true radius is not to exceed 0.025 mm. What will be the length of the cut vector in each case?

Solution
The three methods of approximating the arc are shown in Figure 7.16.

    R = true radius of the arc
    ϕ = angle subtending a cut vector
    t = tolerance, i.e. maximum permitted deviation from true arc.

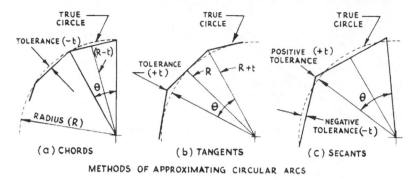

METHODS OF APPROXIMATING CIRCULAR ARCS

Figure 7.16 Three methods of approximating a circular arc by cut vectors: (a) chords; (b) tangents; (c) secants.

The tolerance t is negative for the chordal method, positive for the tangents, and both positive and negative for the secants.

1. $\theta = 2$ arc $\cos[(R - t)/R] = 2$ arc $\cos[(25 - 0.025)/25] = 5.125°$

    $\therefore$ Number of cut vectors in $90°$ of arc $= 90°/5.125° = 17.56$

    Length of cut vector $= 2\sqrt{[t(2R - t)]} = 2\sqrt{[0.025(50 - 0.025)]} = 2.235$ mm

2. $\theta = 2$ arc $\cos[R/(R + t)] = 2$ arc $\cos[25/(25 + 0.025)] = 5.122°$

    $\therefore$ Number of cut vectors in $90°$ of arc $= 90°/5.122° = 17.57$

    Length of cut vector $= 2\sqrt{[t(2R + t)]} = 2\sqrt{[0.025(50 + 0.025)]} = 2.237$ mm

3. $\theta = 2$ arc $\cos[(R - t)/(R + t)] = 2$ arc $\cos[(25 - 0.025)/(25 + 0.025)] = 7.245°$

    $\therefore$ Number of cut vectors in $90°$ of arc $= 90°/7.245° = 12.422$

    Length of cut vector $= 4\sqrt{(Rt)} = 4\sqrt{(25 \times 0.025)} = 3.162$ mm

$\square$ $\square$ $\square$

7.12 Displacement of machine tool slides

To displace a machine tool slide by means of a control system, a rapid response by slides and rotating drive members to the input signals is essential, see Figure 7.17, and yet, in order to withstand heavy loads and forces in the cutting of metal, rugged machine construction is needed.

To meet these conflicting requirements, traditional machine tool design is modified and developed, e.g. frictional forces are reduced by the substitution of rolling for sliding friction in slideways and leadscrew nuts, thus reducing loads and enabling components in the drive system, e.g. clutches, gears, shafts, to be reduced in size. By such means, inertia in the system is diminished, drive systems of smaller power are used and the cost of components of sophisticated design minimised.

In the control of slide elements, a distinction must be made between *drive systems* for physically displacing the slide, and *feedback systems* for monitoring the slide position (closed loop systems). This distinction is illustrated in Figure 7.18, and examples of open and closed loop control are shown in Figure 7.19. A transducer is

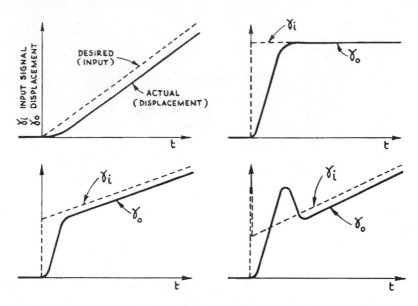

Figure 7.17 Examples of input signals and actual response conditions.

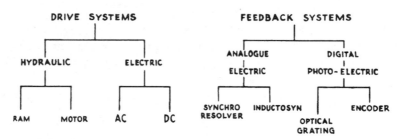

Figure 7.18 (a) Types of drive systems used to displace machine tool slides. (b) Examples of analogue and digital methods for providing feedback on slide position.

required to monitor displacement of slides. It may receive its input from a rotating member of the drive system, e.g. leadscrew, in which case accuracy may be affected by leadscrew errors, wind-up, axial slip etc.; or from a linear transducer, e.g. optical gratings or laser interferometers, and may monitor displacements directly from the slide itself. In all cases, the system must be capable of monitoring slide displacement (and velocity in the case of contouring machines) over long distances of travel and to a high degree of accuracy.

Stepping-motor drive system

A stepping-motor drives a leadscrew either directly, or indirectly through gearing, in a series of incremental movements. The principle of its design is illustrated in Figure 7.20, which shows three sets of co-axial stator laminations (s_1, s_2, s_3) arranged with their slots in exact angular correspondence. Three sets of rotor

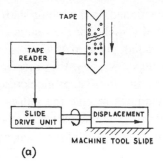

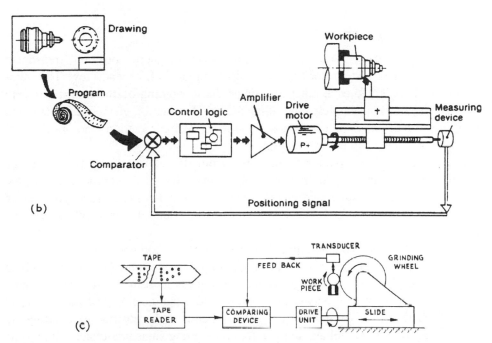

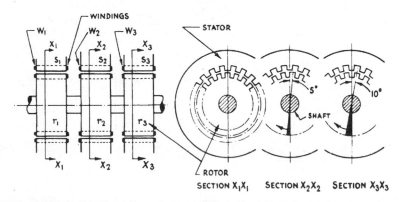

Figure 7.19 Open and closed loop sections for machine tool slide controls. (a) Open loop; (b) closed loop control system; (c) closed loop based on in-process workpiece measurement.

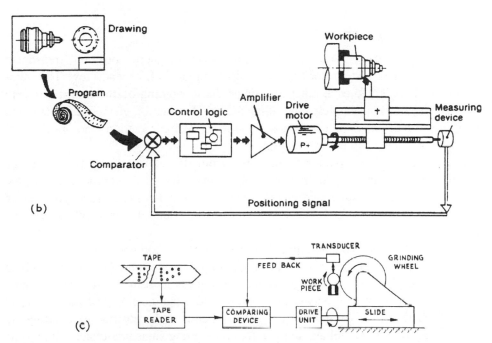

Figure 7.20 Principle of operation of stepping-motor; note the relative radial displacement of the rotor slots.

laminations (r_1, r_2, r_3) are positioned on the motor shaft to coincide axially with the stators, but are disposed radially such that their slots are displaced relative to each other by one-third of the radial pitch of the slots, i.e. in the diagram, $360°/24 \div 3 = 5°$. The stator windings (W_1, W_2, W_3) receive pulses from the supply in sequence, causing the rotor to turn in 5° steps clockwise or anticlockwise according to the order in which the windings receive the pulses. The speed of rotation is a function of the pulse frequency. The magnitude of the angular displacement (and hence the displacement of the machine slide) is dependent on the total number of pulses received.

The synchro-resolver

A synchro-resolver (Figure 7.21) is an electro-magnetic position transducer comprising a stator and a rotor connected to an output shaft or leadscrew. The stator has two coils at 90° to each other, the single phase winding of the rotor turns with the output shaft. By accurately controlling the angular rotation of the rotor relative to the stator, a machine slide can be displaced to its desired position via a mechanical linkage, e.g. gears, clutches and leadscrews.

The positional input data are converted from digital to analog form and the resulting a.c. voltage signal is resolved into two sinusoidal components of 90° phase difference, i.e. sin θ and cos θ, where θ represents the angular position of the rotor necessary to give the required slide position of the machine via the leadscrew drive.

In the resolver, the voltage induced in the coil of the rotor varies sinusoidally as the rotor turns through 360°; when the angle turned by the rotor (ϕ) corresponds to the desired angle (θ), the voltage magnitude reduces to zero (i.e. the null). Thus, $\theta - \phi$ is the angular error between the desired and actual positions of the rotor, and when $\theta = \phi$ the error is zero and the machine slide is in the position required by the input signal. Due to the voltage in the rotor coil varying sinusoidally according to its

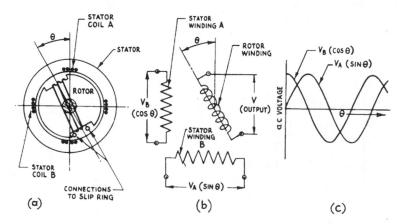

Figure 7.21 Principle of operation of synchro-resolver. (a) Cross-section; (b) relationship between rotor position and voltages of stator windings; (c) phase difference in stator voltages.

angular position, a null will occur every 180° and a phase sensor is employed to identify the correct phase change.

Since a null also will occur for each 180° turn of a leadscrew directly connected to a rotor, it is necessary to connect several resolvers together through gears and clutches to extend the range of the system.

The Inductosyn

Effectively, the Inductosyn (Figure 7.22) is a linear resolver. It employs the principle of inductive coupling between two conductors to provide a signal analogous to the linear position of a machine tool slide. It has two windings, the long (scale) winding extending the length of the slideway, the short (slider) winding is usually fitted to the slide. A gap of 0.2 mm separates the two elements.

As illustrated in Figure 7.22(a) the windings are in 'hairpin' formation; the scale winding is single phase and analogous to a resolver rotor, the slider has two sets of windings, one set displaced $P/2$ relative to the other. When an a.c. signal is applied to the scale winding the voltages induced into the two slider windings will be displaced by 90°, i.e. they will be proportional to $\sin \phi$ and $\cos \phi$ respectively.

If S = linear displacement of the slider,

P = spacing of the poles,

then $\phi = \dfrac{2\pi S}{P}$

If a line aa_1 is drawn, cutting the two curves shown in Figure 7.22(c), two voltages will be identified, which, taken together, will be unique for the curves illustrated and hence may be used to identity the position of the slider relative to a pole in the scale winding. Alternatively, if these same two voltages are fed to the windings of the slider, a voltage will be induced into the scale winding which will be at a maximum for the same unique position.

These principles provide the basis of a system to identify the position of a machine tool slide, but, as in the case of the synchro-resolver, because the identified position is unique only so far as each cycle winding is concerned, it is also necessary to:

1. Provide means of identifying the cycle.
2. Establish a range sufficient to accommodate the full linear displacement of the slide.

The rotary encoder

The basis of this type of transducer (Figures 7.23 and 7.24) is an encoder disc mounted at the end of the slide leadscrew. The disc is divided into segments which are transparent/opaque for use in photocell scanning techniques, or alternatively, conductive/non-conductive for scanning by commutator-type brushes.

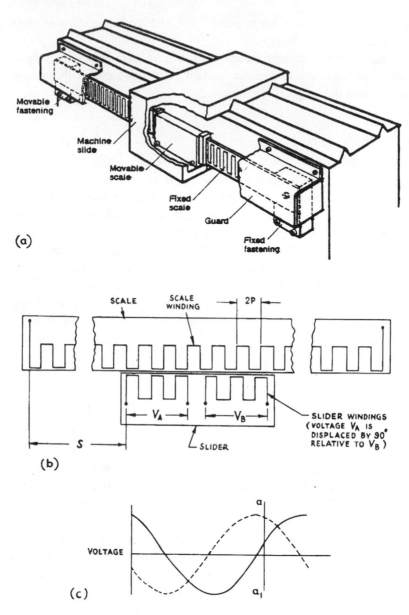

Figure 7.22 (a) A linear Inductosyn; (b) arrangement and relationship of scale and slider printed circuits; (c) phase differences in voltages V_A and V_B of slider windings.

The coding of the disc is analogous to the coding used on punched tape. If the disc segments are 'straightened out' as shown in Figure 7.23(b, ii) it can be seen that the unshaded areas correspond to a hole in punched tape. By scanning the disc as it is rotated via the leadscrew, pulses are fed into the system. After suitable decoding, these output signals are compared with the input for correction to take place if a difference, i.e. error, exists between the desired and the actual slide positions.

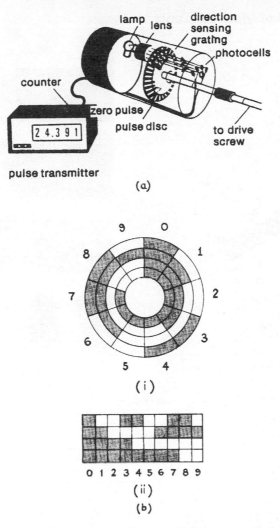

pulse transmitter

(a)

(i)

(ii)

(b)

Figure 7.23 (a) Rotating pulse generator; (b) rotary shaft encoder, (i) disc, (ii) development of coded segments demonstrating the analogy between coding for discs and coding for punched tape – unshaded areas corresponding to holes in punched tape.

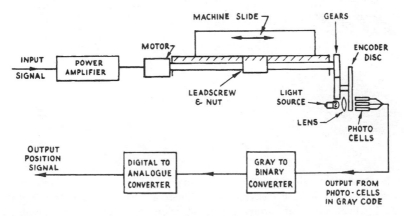

Figure 7.24 Principle of the rotary shaft encoder system as applied to a CNC machine tool.

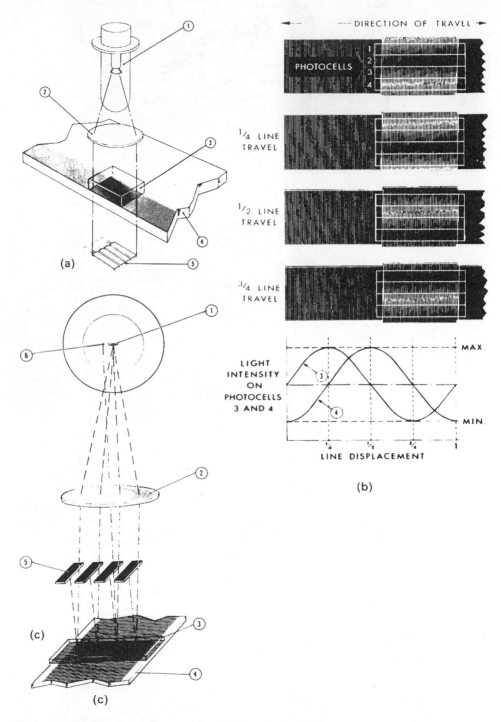

Figure 7.25 Principle of linear and optical gratings, line and space transmission type. (a) Optical arrangement: 1 – light source; 2 – collimating lens; 3 – index grating; 4 – scale grating; 5 – silicon photocell array. (b) Light intensity wave forms caused by fringe displacement due to movement of machine slide.

Optical gratings

The principle of one method employed in the application of linear gratings for monitoring machine tool slide displacement is shown in Figure 7.25. The lines of the short index grating are inclined at a slight angle to the lines of the scale grating which extends for the length of the slideway.

The intersection of these two sets of lines creates an interference effect, giving rise to the Moiré-fringe pattern. When one scale is displaced relative to the other, the dark interference fringes pass across the width of the grating at right angles to the direction of travel; see Figure 7.25(b). The dark fringes give rise to a variation in the light intensity received by the photocells, as shown by the waveforms in the diagram. The de-energising of each photocell in turn generates pulses in direct proportion to the magnitude of slide displacement. The order in which the photocells are de-energised provides the information necessary to discriminate in the sense of direction of slide movement. Once again, the pulses provide the output signal which can be de-coded and compared with the position input signal for the slide of the machine tool.

Tool length measurement

Obviously there are a wide range of tools used on CNC machines. All these tools must be sent to the work surface without crashing into solid metal, so the machine needs to know the exact length of each tool, and usually also the width or diameter of the tool (Figure 7.26). There are a variety of ways of determining this length, and of entering it into the machine memory (Figures 7.27 and 7.28) such as:

1. Manual on machine – measure tool length using machine axis system.
2. Manual off machine – measure tool length using height gauges system, then type into machine.

Table 7.1 gives a comparison of tool length measurement techniques.

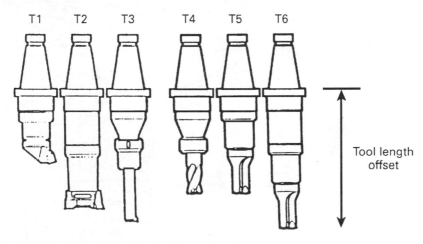

Figure 7.26 A selection of tools of different lengths.

208   Control of Machine Tools

Table 7.1 Comparison of tool length measurement techniques.

Method	Details	Comments
Manual on-machine	'Wind' tool down onto step gauge, take readings from axis calculated reading and type into memory	Machine stopped Time consuming – 6–10 minutes per cutter Error prone – measuring and typing errors
Manual off-machine	Set tool in fixture, measure length with height gauge, calculate length, type into memory	Faster – can be done while machine is running, 2–4 minutes per cutter Typing errors
Tool pre-setter – measure off machine	Set tool in special measuring machine, measure lengths (a) type into machine memory (b) send to machine electronically via wire	Faster – machine running Typing errors – 2 minutes per cutter No errors – 1 minute per cutter
Tool pre-setter with memory chips	Set tool in measuring machine, records length onto memory chips in tool	Machine running. Very fast, no errors. Sizes read when machine loads tool – 2 seconds per tool
Tool length measuring probe	Tool is loaded into spindle and driven onto length probe	Tool length measured at time of use – length can be checked to see if tool broken – 6 seconds per tool

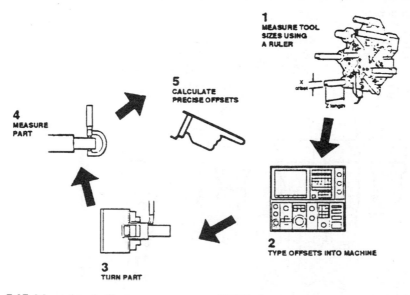

Figure 7.27 Manual tool offset measurement on CNC lathe.

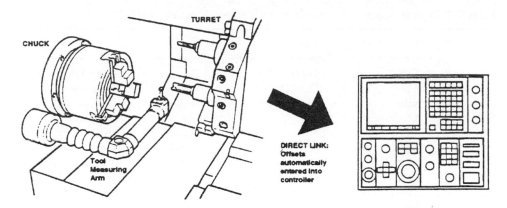

Figure 7.28 Total length measurement using tool setting probe.

7.13 Manual programming

The procedure adopted in basic (manual) programming is to set down by hand, in ISO machine codes, each detail of the required process, specifying the sequence of operations, co-ordinates of each point, machine functions, feed rates, spindle speeds, etc. The program is then typed into the machine computer, which normally has provision for viewing the programmed cutter path to check for error before cutting commences.

The programming procedure uses a low level language, and is slow and cumbersome to write. Machine manufacturers adapt the ISO code to suit their own machines, and as the format for a machine must be followed exactly, it is often not possible to transfer a program from one machine to another apparently similar machine.

Programming instructions use these main types of commands to drive machines:

1. 'G' codes are motion control codes for example: G0 means 'move at full speed', G01 means 'move at controlled speed', G02 means 'move in clockwise arc'.
2. 'M' codes are supplementary control codes, for example: M00 means 'stop program here', M02 means 'end of program', M03 means 'turn spindle clockwise'.
3. Particular instructions are specified by identifying letters, for example: X200.0 means 'X coordinate 200 mm', F150 means 'feed at 150 mm per minute', S2000 means 'spindle 2000 revolutions per minute'.

These codes are tabulated in Tables 7.3, 7.4, and 7.5.

The example below gives a brief introduction to the programming principle, but for further detail the student should refer to the machine manufacturer's handbook.

Example 7.2   Manual programming

A simple component requiring machining by drilling, milling, boring, and counterboring is shown in Figure 7.29. Given the following information regarding the code, the program should be self-explanatory, and it is shown in Table 7.2.

Table 7.2 Typical manual CNC program for the plate shown in Figure 7.29.

Line	Code	X dim	Y dim	Z dim			Comment
N0	G0	X100.0	Y100.0	Z20.0	T1	M06	Load tool 1, 5 diameter drill
N5	G0	X10.0	Y40.0	Z3.0			Move to 3 mm above hole 1
N10	G81	X10.0	Y40.0	Z−15.0	F200	M08	Drill hole 1, 15 deep, coolant on
N15		X10.0	Y10.0				Drill hole 4, 15 deep, coolant on
N20		X45.0	Y10.0				Drill hole 3, 15 deep
N25		X45.0	Y40.0				Drill hole 2, 15 deep
N30	G0	X100.0	Y100.0		T2	M06	Load tool 2, 8 diameter c/bore drill
N35	G0	X10.0	Y40.0	Z3.0			Move to 3 mm above hole 1
N40	G81	X10.0	Y40.0	Z−3.0	F200	M08	Counterbore hole 13 mm deep
N45		X10.0	Y10.0				C/bore hole 4
N50		X45.0	Y10.0				C/bore hole 3
N55		X45.0	Y40.0				C/bore hole 2
N60	G0	X100.0	Y100.0		T3	M06	Load tool 3, 15 diameter drill
N65	G0	X40.0	Y25.0	Z3.0			Move to 3 mm above hole 5
N70	G81	X40.0	Y25.0	Z−18.0	F200	M08	Drill hole 5, 15 diameter, 18 deep
N75	G0	X100.0	Y100.0		T4	M06	Load tool 4, 16 diameter boring bar
N80	G0	X40.0	Y25.0	Z3.0		M08	Move to 3 mm above hole 5
N85	G85	X40.0	Y25.0	Z−12.0	F150		Bore hole 5 to 16 mm diameter
N90	G0	X100.0	Y100.0		T5	M06	Load tool 5, 12 diameter end mill
N95	G0	X−8.0	Y25.0	Z3.0			Move to 3 mm above point 6
N100	G0			Z−4.0		M08	Move to Z−4.0
N105	G1	X15.0	Y25.0		F200		Mill along to point 7
N110	G0	X25.0	Y58.0	Z3.0			Move rapid to point 8
N115	G0			Z−4.0			Move to Z−4.0
N120	G1	X25.0	Y40.0		F200		Mill along to point 9
N125	G0	X100.0	Y100.0	Z3.0			Move to tool change point ready to load tool 1 for the next part
N130						M02	End program

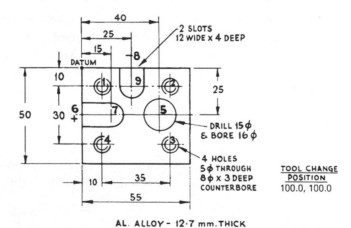

AL. ALLOY - 12·7 mm. THICK

Figure 7.29 Aluminium plate, 10 mm thick, datum at X0, Y0, top face at Z = 0, tool change position assumed to be X = 100, Y = 100; tool 1 = 5 mm drill, tool 2 = 8 mm c/bore drill, tool 3 = 15 diameter drill, tool 4 = 16 diameter boring bar, tool 5 = 12 diameter end mill.

Table 7.3 The significance of the code letters in a lathe system. Agrees with ISO Standard ISO 2539 in all essentials.

N1–N999	Block number (order number)
G0, G1, G2, etc.	Preparation function. Specifies how the control system is to work
X	Motion (position) in lateral direction
Z	Motion (position) in longitudinal direction

(X and Z can be specified 0.001 to ±9999.999 mm)

I	X-co-ordinate for centre of circle
K	Z-co-ordinate for centre of circle

(I and K can also mean thread pitch mm/varv)

F (4 digits)	Feed rate 0.001–9.999 mm/varv or mm/min
S (4 digits)	Setting speed r/min or cutting rate m/min
T (4 digits)	Tool indexing to specified number and connection of specified compensation
M (4 digits)	Miscellaneous machine functions

Table 7.4 M functions according to draft ISO standard.

M00	Program stop
M01	Optional (planned) stop
M02	End of program
M03	Spindle CW
M04	Spindle CCW
M05	Spindle OFF
M06	Tool change
M07	Coolant no. 2 ON
M08	Coolant no. 1 ON
M09	Coolant OFF
M10	Clamp
M11	Unclamp
M12	
to	Unassigned
M18	
M19	Oriented spindle stop
M20	
to	Permanently unassigned
M29	
M30	End of data
M31	Interlock bypass

Table 7.5 Preparatory functions (G functions) according to draft ISO standard.

G00	Point-to-point, positioning	G54	Linear shift
G01	Linear interpolation	G55	Linear shift
G02	Circular interpolation, arc CW	G56	Linear shift
G03	Circular interpolation, arc CCW	G57	Linear shift
G04	Dwell	G58	Linear shift
G05	Unassigned	G59	Linear shift
G06	Parabolic interpolation	G60	
G07	Unassigned	to	Unassigned
G08	Acceleration	G62	
G09	Deceleration	G63	Tapping
G10		G64	
to	Unassigned	to	Unassigned
G16		G67	
G17	XY Plane selection	G68	Tool offset inside corner
G18	ZX Plane selection	G69	Tool offset outside corner
G19	YZ Plane selection	G70	Inch data
G20		G71	Metric data
to	Unassigned	G72	
G32		to	Unassigned
G33	Thread cutting, constant lead	G73	
G34	Thread cutting, increasing lead	G74	Move to home position
G35	Thread cutting, decreasing lead	G75	
G36		to	Unassigned
to	Permanently unassigned	G79	
G39		G80	Fixed cycle cancel
G40	Cutter compensation/tool offset cancel	G81	
G41	Cutter compensation – left	to	Fixed cycle
G42	Cutter compensation – right	G89	
G43	Tool offset positive	G90	Absolute dimension
G44	Tool offset negative	G91	Incremental dimension
G45		G92	Preload registers
to	Unassigned	G93	Inverse time feed rate
G53	Linear shift cancel	G94	Feed per minute
		G95	Feed per spindle revolution
		G96	Constant surface speed
		G97	Revolutions per minute (spindle)
		G98	
		to	Unassigned
		G99	

Canned cycles

Machines have a range of 'canned cycles' built into their memory, to save programming time. Two cycles are shown in Figure 7.30.

The drilling cycle (a) moves the cutter rapidly to the position required, and then feeds to depth and moves back up out of the work at high speed ready to drill the next hole. This can be seen in use in Table 7.2.

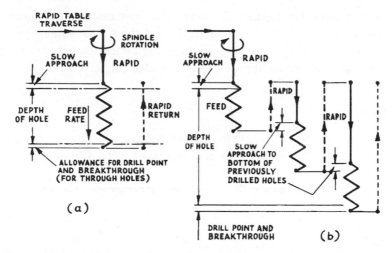

Figure 7.30 Drill cycle diagrams. (a) Nominal drilling cycle (cycles identical for drilling and reaming); (b) peck-feed (woodpecker) drilling cycle for clearing swarf when drilling a deep hole.

The peck drilling cycle (b) is used when drilling deep holes, where the swarf is likely to seize onto the drill and to break it, especially with very small drills. The diagram shows that the drill 'pecks' into the work, having a series of up and down motions to lift it out of the hole and clear the swarf as the hole gets deeper.

Both of these examples reduce many lines of G0 and G1 instructions into far fewer program instructions.

Cutter compensation

All modern CNC machines can use 'cutter compensation' or work surface programming. This is essential to produce accurate workpieces, since none of the feedback systems currently in use actually measure the final size of the workpiece. They measure the position of the machine table, but the large forces involved in metal cutting inevitably cause distortion of the machine structure and the workpiece. The structure is obviously designed to be extremely rigid but the tools used are usually the weakest part of the structure. They invariably deflect and they have some manufacturing errors, so a milling cutter which is nominally 20 mm diameter may actually be 19.95 mm diameter. Obviously tools also wear and the 19.95 mm cutter may wear down to 19.91 mm, this wear will increase the forces acting upon it and hence cause some additional bending. These factors combine to cause inevitable differences between the sizes in a program and those produced on the component. Cutter radius compensation allows the machinist to take measurements of the workpiece, and to adjust the cutter size by typing a corrected dimension into the machine memory.

A major advantage of this facility is the opportunity to machine a part using a wide range of tools. The program now defines the shape of the workpiece and the cutter size is given by the offset. Thus if the 20 mm cutter mentioned above wears out an 18 mm cutter could be used merely by changing the offset value. Cutter compensation is very convenient in milling, but Figure 7.31 shows that it is essential for turning. Every turning tool has a nose radius, however small, and the compensation system has to correct the errors caused by the constantly changing dimension between the program mid-point P, and the actual position of the cutting edge (Figure 7.32).

Subroutines

Very often it is necessary to program a particular shape several times in a program, e.g. when taking several cuts to depth in a pocket, or when drilling, countersinking,

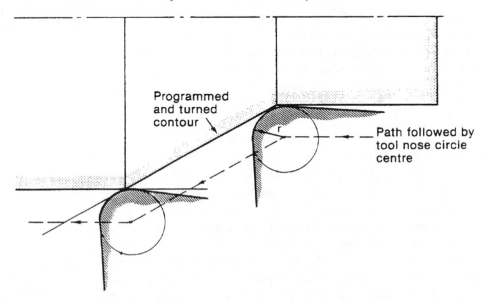

Example of choice of programmed point (P)

Example of contour error depending on position of P

Programmed path

Contour error

Turned contour

Figure 7.31 The nose radius gives a contour error (most commonly in connection with motions not parallel to an axis) since the point P is the controlled point in the machine. This contour error can be avoided by means of time consuming manual calculations or by means of an automatic nose radius compensation function in the system.

Programmed and turned contour

Path followed by tool nose circle centre

Figure 7.32 Comparison of required contour and tool path.

and tapping a series of holes. For example (Figure 7.33):

Subroutine 1 – hole centres
Subroutine 2 – pocket shape

Program structures

A sample program structure is shown in Figure 7.34.

Macro programming

Very often common features are produced on components e.g. rectangular pockets, and it is relatively easy to write a library of standard programs which are expressed in terms of 'variables' as shown in Figure 7.35.

Graphical programming

Many machine tools now offer a menu of common shapes, and the operator can 'reply' to a set of questions to create a part program. This has now provided a way of avoiding the use of G and M codes (see Figures 7.35 and 7.36).

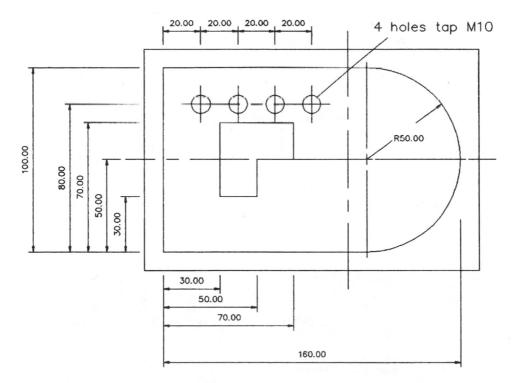

Figure 7.33 Holes and pockets example.

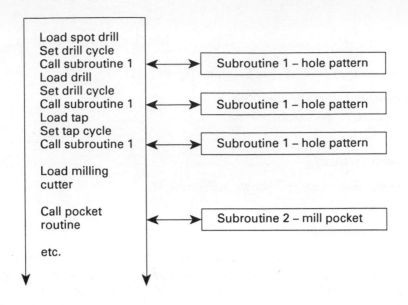

Figure 7.34 Program structure.

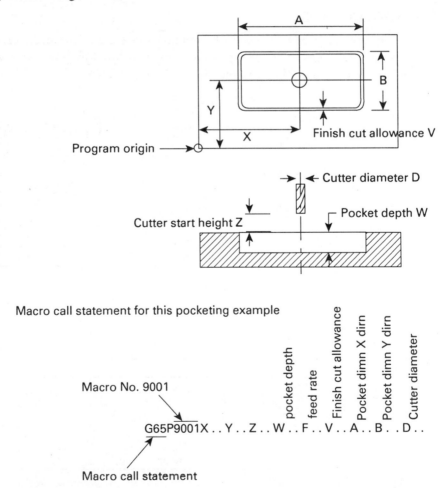

Figure 7.35 Rectangular pocket example.

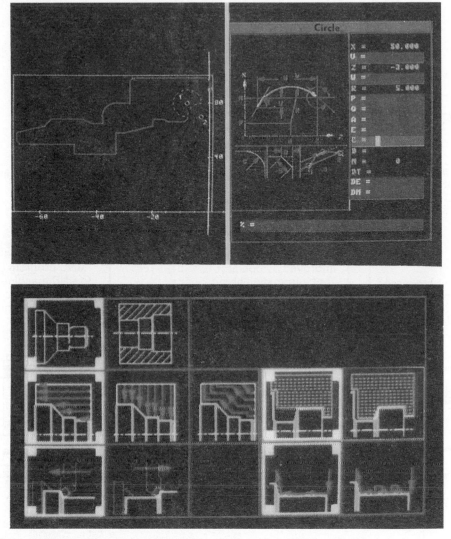

Figure 7.36 (a) Geometry definition; (b) selection of roughing method.

7.14 CAD/CAM links

The widespread use of CAD systems has led to major developments in the generation of CNC programs. The computer 'knows' the geometry that the designer has defined, i.e. dimensions on a drawing are there for human benefit; they are redundant as far as the computer is concerned. It stores the values of the equations of the lines and circles etc. in its memory. This means that the geometry can be transferred to the CAM program which adjusts the sizes produced by the designer to produce a new set of geometry to define the cutter path needed to make the part. Someone thus has to define which cutter, what feed rate, and what spindle speed will

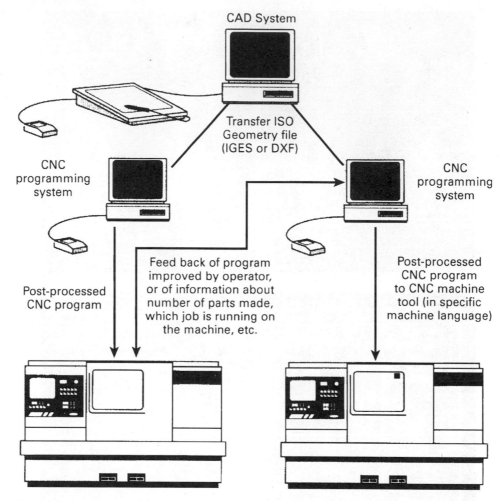

Figure 7.37 File transfers.

be used. These are technology decisions, usually made by the manufacturing engineer.

When choosing which tool to use, the engineer has to choose from many possibilities and to allow for such factors as the power available, the strength of the workpiece material, how the workpiece will be clamped, and the amount of material to be removed.

It is now common practice to download the geometry specified by the designer and then to process this into a machining program. This can save considerable amounts of time, and is less likely to lead to errors than if the programmer has to do many calculations and retype the part dimensions into the computer. It does however place the prime responsibility for the product shape where it should be – upon the designer. The drawing must now be free of errors, because it will be followed faithfully by the CAM system.

Unfortunately the machine tool builders have not agreed a universal standard for the machine language (the G and M coded etc.) and this leads to many problems on the shopfloor of the factory – one cannot, for instance, transfer a Fanuc program to a machine with a Siemens controller. This has forced companies to use 'post processors' to translate from the ISO CL file CAM output, into the particular machine tool language. Similarly CAD systems have their own individual formats and the INTERNATIONAL GRAPHIC EXCHANGE STANDARD (IGES) has been established to enable companies to exchange CAD information across different systems. A widespread shortened version of this is used by AutoCAD, the DXF (drawing exchange file) format, and most CAM systems will accept a DXF or IGES file as an input. File transfers are illustrated in Figure 7.37.

The editing facilities offered by the CAM software can enhance the utilisation of the machine considerably – programmes may be copied to enable several parts to be made from one piece of material and kits of parts may be made at one setting on the machine.

7.15 Machine tool probing systems

The original touch trigger probe was designed by David McMurtree, who then founded Renishaw Metrology to make and sell probes for co-ordinate measuring machines. These were soon adapted to suit CNC machines, by being made more robust and being protected from surface damage.

The probe is a spring loaded plunger which has a kinematically designed six point location. A circuit is connected through these contact points. The probe can be lifted off the contacts by any machine overtravel and reseats itself to an accuracy of better than 1 micron when it is released. The instant at which the circuit is broken can be used to trigger a reading from the CNC machine's axis measurement system, but for additional accuracy an enhancement is made to the circuit. As the probe begins to move the resistance of the circuit starts to increase and a suitable threshold resistance value is taken as the 'measure now' signal just before the probe lifts off its seating. Figures 7.38 and 7.39 illustrate these points.

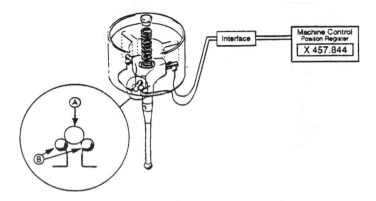

Figure 7.38 Probe operating principle.

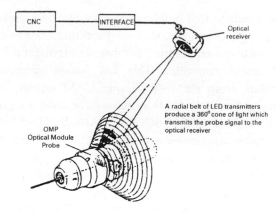

Figure 7.39 Transmission system.

The probe, in conjunction with the machine's CNC system, thus provides a highly sensitive system of sensing the workpiece size – (or the tool size) providing the machine with a sense of touch. Many ways of using this information have been developed, which can be classified as (see Figures 7.40 to 7.45):

Tool measurement	Measure tool lengths
	Check tool breakage
	Check tool wear
	Measure tool diameters
Work measurement	Measure workpiece size
	Identify the workpiece
	Check the workpiece loading
	Adjust tool offsets
	Find position of workpiece

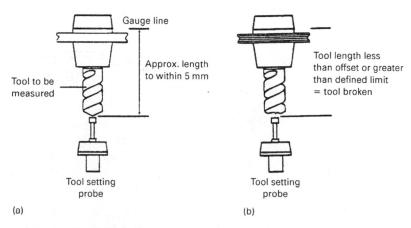

Figure 7.40 (a) Tool setting (machining centres), (b) tool breakage detection.

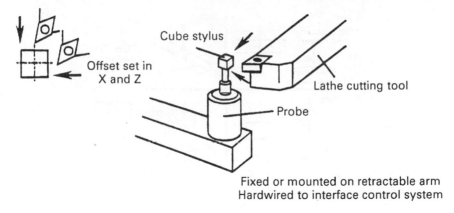

Cube stylus

Offset set in
X and Z

Lathe cutting tool

Probe

Fixed or mounted on retractable arm
Hardwired to interface control system

Figure 7.41 Tool setting on lathes.

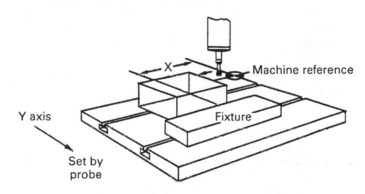

X

Machine reference

Y axis

Fixture

Set by
probe

Figure 7.42 Jobs set-up using a simplified fixture.

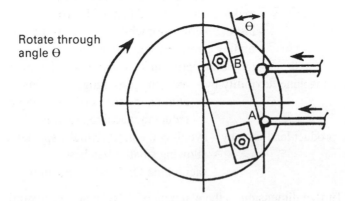

Rotate through
angle Θ

Θ

B

A

Figure 7.43 Automatic job set-up on rotary table.

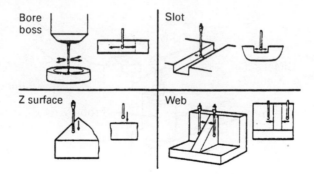

Figure 7.44 Typical probing cycles.

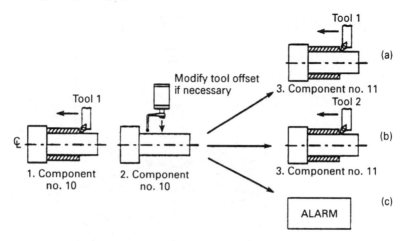

Figure 7.45 Automatic tool wear compensation – lathe.

These can be applied in various ways:

1. To reduce machine set-up times – simpler/no fixtures
                             – quicker tool setting
                             – more accurate tool setting.
2. To improve quality by – making parts RIGHT FIRST TIME
                            – inspection of work while it is on machine
                            – finding errors before they occur
                            – providing optimum location of part datum on *this* casting.
3. To improve machine flexibility by – reducing job change over times
                                    – eliminating the need for complex fixtures.
                                    – reducing machine downtime.
4. To reduce product lead time by – avoiding the need to make special fixtures
                                – allowing smaller batches
                                – facilitating ONE-HIT machining.

These add a further dimension to the planning of CNC machine operations. Planning engineers can now revise their jig and fixture requirements, and the programmer has to add probing cycles to the CNC program. Frequently a standard probing subroutine will be suitable, but it can often be worthwhile writing special macro programmes for a particular purpose. These allow the machine to make decisions from the probing results and to take corrective actions to improve quality – while the operator is at the end of the machine shop, or even when the operator has gone home and the machine is unattended!

Tool monitoring

There are a large number of factors influencing efficient machine operations, and it is frequently necessary to monitor a process to ensure that high quality parts are being produced, or to optimise the performance of the process.

Several systems are available, and Figures 7.46 to 7.51 summarise the various sensors used.

Power monitoring is frequently used, it is relatively cheap and it is easy to retrofit to any machine. Basically, it measures the motor current and is reliable for high power operations, the steady state power requirements for a rough turning operation, or for drilling, say, a 20 mm diameter hole, can easily be established.

When, however, low power is taken from the motor, e.g. for a light finishing cut, or a 2 mm drill, the power needed to overcome machine friction and to accelerate the chuck and machine spindle can easily swamp the small signal arising from the actual cutting operation. In these situations force sensing systems become more

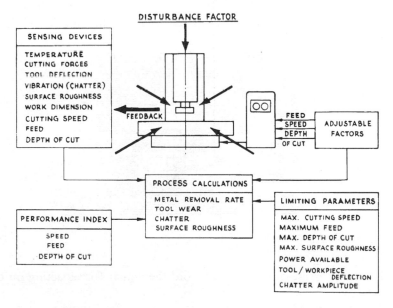

Figure 7.46 Application of adaptive control to a machining process. The effect of factors disturbing the process are monitored and corresponding kinematic or geometric corrections introduced into the system.

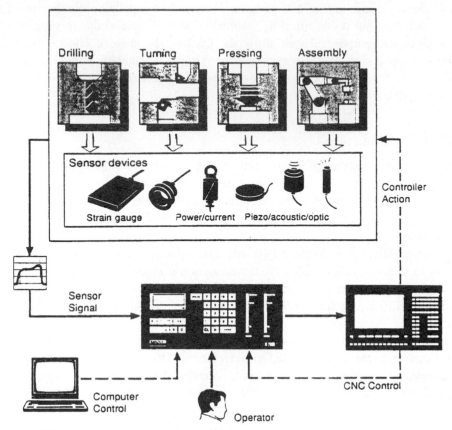

Figure 7.47 Monitoring of various processes via a range of sensing devices.

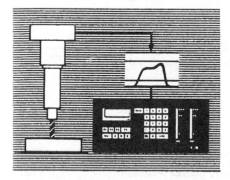

Figure 7.48 Force monitor used in drilling.

reliable, the sensor being positioned to record the actual forces acting on the tool itself.

These signals can be used to stop the machine if the force is higher or lower than the pre-set value, usually giving a warning signal of tool damage or breakage. Force sensing can be valuable in drilling operations to sense the end of the drill cycle as

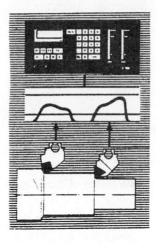

Figure 7.49 Force monitor used in turning.

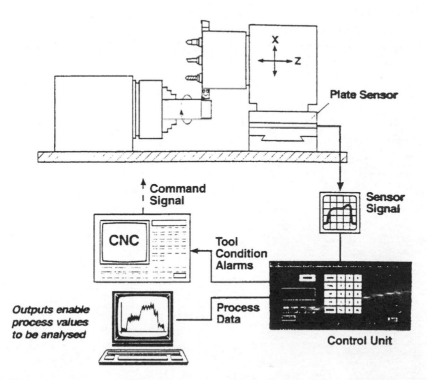

Figure 7.50 Force monitoring on CNC machines. The sensor collects cutting process data and sends it to the control unit. Data is processed by the control unit to determine tool condition and machining process values. Alarm signals are sent to the CNC for appropriate action.

☐ Guarantees high workpiece quality

☐ Minimises scrap and damage

☐ Reduces downtimes

☐ Cuts indirect costs.

■ High sensitivity and fast reaction

■ Measures relevant forces only. Uninfluenced
by external force components

■ Choice of measurement of X or Z axes independently
or combined

■ Provides a record of each cutting process

■ Only a single sensor is required for monitoring
collision, breakage and wear

Figure 7.51 Benefits of tool monitoring.

the drill breaks through the workpiece, and to reduce the forces applied by the machine to avoid damage to the work or the cutter.

Power monitoring is often used in adaptive control conditions, however, when you can pre-set the machine to apply say 10 kW to the workpiece, and the feed rate or spindle speed can be adjusted automatically if a hard or soft workpiece is encountered.

8 Introduction to Cutting

8.1 Introduction

Metal cutting by the use of a hard cutter to cut a softer workpiece has been used for thousands of years. The simplest form is a chisel cutting wood or metal. Metal cutting on a lathe is illustrated in Figure 8.1.

This process is brought under control on a machine tool by using a tool post to guide the tool along a rigidly held workpiece. The cutting edge has a precise geometric form determined by the needs of the process. The cutting edge is ground to shape on most high speed steel tools, e.g. drills and milling cutters, but nowadays an increasing number, around 50 per cent, of tools use disposable carbide or ceramic cutting edge inserts having an edge formed to control chip flow during cutting. The growth of CNC machine tools means that unmanned operation is now commonplace, and the chip formation (Figure 8.2) must not damage the workpiece or tangle around the tools on the machine.

In the actual cutting of the metal the tool deforms some of the workpiece material, and then separates it through plastic deformation. Large stresses build up as the layer of metal which is just about to become the chip approaches the cutting edge. This elastic and plastic deformation of the metal takes place as it approaches and exceeds the yield strength of the material as it moves past the tool face. Chips vary considerably with the type of workpiece material, but if the material is sufficiently tough the process resembles a continuous flow of plate-like elements sheared successively from the workpiece. Figure 8.3 shows the extent of the severe deformation occurring as the chips are formed.

There are obviously extremely high contact pressures generated between the chip and the tool top face. These are well in excess of the yield strength of the workpiece material.

The character of the motion of the chip along the tool face requires further consideration. Friction comes partly into the process as the metal is forced onto the tool at great pressure and high temperature. According to some research the chip and tool join together, and the bottom face of the chip flows over a layer of work material which is effectively welded to the tool face, and is stationary. The flow zone

Figure 8.1 Metal cutting on a lathe.

is thus created when the surfaces seize, but the relative chip/tool motion continues.

The pattern of movement is very dependent on the actual workpiece material being cut, and upon the actual speeds and feeds in use.

The contact zone between the chip and tool (Figure 8.4) can be divided into three zones in which different reactions take place:

1. *Sticking*. The flow zone is a region of molten material at high temperature.
2. *Adhesion and diffusion.*
3. *Abrasion*, where with higher temperatures diffusion and adhesion increase.

Under certain conditions (mainly slow speeds with soft materials) successive layers of flow zone materials build up on the cutting edge and form an unstable layer of soft materials on the cutting edge. Material is pressure welded on to this layer, and the shape of the tool is altered. Eventually this breaks off and a new cycle of welding the built-up edge starts all over again (Figure 8.5). Unfortunately the fragments of the built-up edge will probably weld to the workpiece, and a very rough surface will result. Sometimes parts of the tool face will be broken away when the build up fractures, altering the characteristics of the tool.

Often there is a particular temperature/speed range which promotes build-up and growth, and certain tool/workpiece material combinations are more prone to the

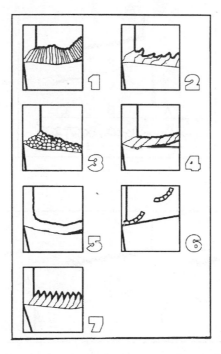

Figure 8.2 Throughout this chapter, seven basic types of material-related chip formation will be referred to: (1) continuous, long chipping, such as most steels; (2) lamellar chipping, such as most stainless steels; (3) short chipping, such as most cast irons; (4) varying, high force chipping, such as most super alloys; (5) soft, low force chipping, such as aluminium; (6) high pressure/temperature chipping, such as hard materials; (7) segmental chipping, such as titanium.

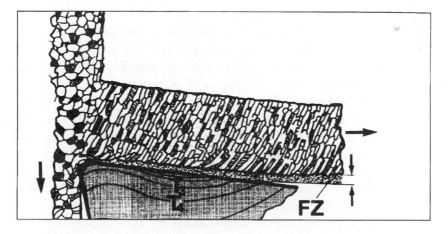

Figure 8.3 Flow zone.

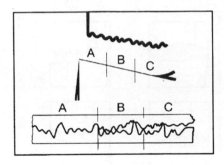

Figure 8.4 Chip contact areas.

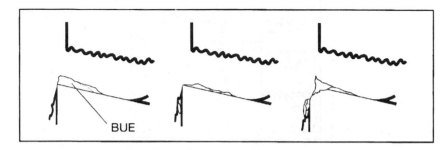

Figure 8.5 Built-up edge formation.

problem than others. Higher speeds soften the layer and replace it with a flow zone. Larger positive rake angles also tend to remove build-up.

8.2 Chip formation

The chip form is greatly influenced by the materials being cut, varying from continuous forms to crumbling material. The deformed chip is in different segmental forms, but usually held together in ductile materials.

When comparing a medium carbon content steel and an alloyed steel with different mechanical properties, the first is more deformed and has a larger initial curve. Because of the lower material strength to start with, which is reduced further with deformation, the unalloyed steel is broken more easily than the stronger alloy steel. Chip breaking is softer for the unalloyed steel also when the feed rate is high. The deformation of the alloyed steel will be less also at higher feed rates, which leads to the recommended chip breaker area being positioned towards higher feed rates. The segments will be compressed, depending on how the chip is subjected to pressure through the size of the rake angle – greater compression leading to a thick chip means less strength and thus, in some cases, the chip self-breaks at certain lengths. Chip breaking, however, cannot be resolved generally by just selecting various combinations of rake angles and feed rate values.

There are other machining process factors involved which can be adversely affected by combinations which take into account only breaking the chip. Cutting forces, tool strength, temperatures, vibrations are examples of factors that become important when designing the means with which to form chips at various cutting data. Chip formation thus becomes a technology area of its own, an area that has seen considerable development, especially during the past two decades.

In Figure 8.6, orthogonal cutting (A), the cutting speed direction or axis of rotation of the workpiece material being cut is at right angles to the main cutting edge. This is an over-simplified view of the cutting process, employed only in a few operations such as facing and plunging. Most metal cutting is oblique (B) – the cutting direction is not orthogonal but at a certain angle relative to the main edge. This changes the geometrical conditions considerably and the chip flow direction is altered. Also there is more than one edge and nose radius involved, as well as variation in speed across the width of the chip. Instead of a watch-spring type chip, as in a typical parting operation, there are various forms of comma or helical shapes to the chip.

The entering angle of the tool affects the chip formation in that the chip thickness is reduced and the width increased with a smaller angle. Chip formation is softer and smoother with a smaller entering angle (60–45°). The direction of chip flow is also changed, usually advantageously, with the spiral pitch being increased. The shape and direction of chips change with the nose radius on the cutting edge. This is an important part of the cutting edge geometry, seen from above the tool. Just as the entering angle affects the chip flow direction so will the nose radius in relation to cutting depth. Small cutting depths mean a comma shaped chip cross-section with a small angle in relation to the cutting edge. A larger depth leads to less influence from the radius and more from the actual entering angle of the edge with an outward directed spiral as the result (Figure 8.7). The feed rate also affects the width of the chip cross-section and the chip flow.

Chip formation starts with the initial curving and is affected by the combination of cutting data – especially the feed rate and depth of cut, rake, the type and condition of the material and also the size of the nose radius. A square chip cross-section usually means excessively hard chip compression while a wide, thin band-like chip is formed in unsuitably long strands. When the chip curve becomes smaller for a thicker chip, the chip/tool contact length becomes longer with more deformation and pressure as a result. Excessive thickness has a negative influence on the machining process.

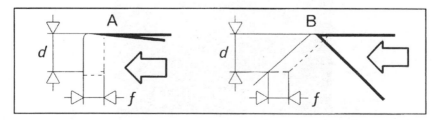

Figure 8.6 Orthogonal and oblique metal cutting.

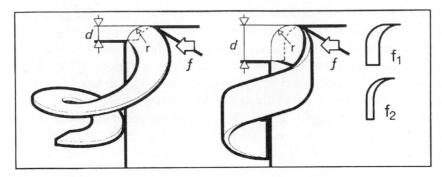

Figure 8.7 Effect of nose radius/cutting depth on entering angle and chip formation.

If the feed rate is increased to above what the insert geometry has been designed for, the chip will pass over the chip forming geometry, with the effect that machining is performed with a negative instead of positive geometry, with balanced chip breaking.

Curve, direction, helix and shape of the chip is designed into the ability of the cutting edge. Each insert type has a geometry developed to provide satisfactory chip formation within a certain area of feed rate and cutting depth and material types. Controlled chip formation is imperative throughout the various areas of modern metal cutting. A finishing insert (Figure 8.8), working mostly with its nose radius, will have the geometry concentrated to the corner of the insert while a heavy roughing insert will have geometry right across the rake face (Figure 8.9). Some inserts are capable of providing satisfactory chip formation across a broad intermediate range, having incorporated combinations of chip breakers, ranging from the corner radius and across the insert face.

There are different ways for a chip to break:

1. Self-breaking, Figure 8.10(A).
2. Breaking when the chip is stopped by the tool, Figure 8.10(B).
3. When the chip is stopped by the workpiece, Figure 8.10(C).

There are advantages and disadvantages with all three alternatives. Self-breaking has been mentioned already, and achieving a suitable direction of chip is one of the

Figure 8.8 Light chip formation.

Figure 8.9 Heavy chip formation.

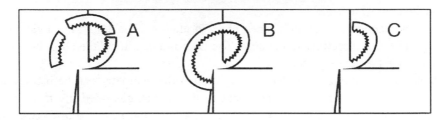

Figure 8.10 Three ways to break a chip.

more important factors here. Breaking against the tool may be negative if chip hammering takes place against the insert edge. Breaking against the workpiece may be negative if the chip affects workpiece quality or lands up in the cutting zone again. Uncontrolled swarf should always be avoided as it can very quickly lead to tool breakage, inferior results, machine downtime and operator injury.

Short chipping materials need little or no chip formation while some long chipping materials need chip breakers designed into the insert geometry to deform the chip into breaking. The initial curving of the chip is in most cases not sufficient to break the chip into required lengths.

Various forms of ground and later pressed indexable insert chip breakers were developed before today's modern inserts. The modern indexable insert is a complex combination of angles, flats and radii to optimise chip formation through cutting action, contact length, chip breaking, etc.

Most inserts have positive rake angles, combined with being inclined negatively in the toolholder, to promote good chip formation and positive cutting action. Negative primary lands, of varying lengths depending upon the working area of the geometry, are applied to strengthen the cutting edge. Small cutting depths, in many cases, involve only, or to a great extent, the nose radius. As shown this gives a comma shaped chip cross-section and a small angled chip flow direction in relation to the main edge.

The stagnation zone, that appears in front of the flow zone, cushions the material flow against the primary land, giving rise to overall positive cutting action. In this way, the primary lands need not deteriorate the cutting action if applied in the right amount for the edge design in question. A sharp edge, however, needs only minimised or no land, with only some edge rounding.

Chip control is, thus, one of the key factors especially in turning and drilling. Milling creates a natural chip length thanks to the limited length of cutting edge engagement. In drilling and boring, chip control is vital because of the limited space inside holes being machined. Also in modern high-performance drilling, chips have to be of exact form so as to be evacuated efficiently from the cutting zone – any congestion, quickly leads to tool breakdown.

Chip formation, then, is affected by several factors. The workpiece material plays a large role. Type, strength, hardness structure, shape and size all affect chip formation.

The cutting data directly influence the size and shape of the chips, especially feed and cutting depth and to some extent cutting speed. Cutting fluid application also affects the chip formation.

The tool geometry also influences chip formation. The entering angle affects length, width and direction of the chip. The nose radius affects the chip, how much depends upon the depth of cut.

The shape of the edge also determines whether the geometry can be incorporated on a double-sided insert or whether it has to be on a single-sided insert. A certain amount of support surface has to be incorporated for the insert to have two sides. Inserts A and B, in Figure 8.11, are double-sided while C is single-sided. For finishing with insert (A) the usual priority factors for the area for satisfactory machining economy are:

1. Chip control.
2. Surface texture/accuracy.
3. Cutting edge strength/forces.
4. Tool life/predictability.

For semi-finishing to light roughing (B):

1. Chip control.
2. Productivity/security.
3. Cutting edge strength/forces.
4. Tool life/predictability.

For rough machining (C):

1. Chip control.
2. Metal removal rate.
3. Cutting edge strength/forces.
4. Tool life/toughness.

The general principle is that when cutting data is selected for an operation, cutting depth should be maximised and then feed maximised within the recommended

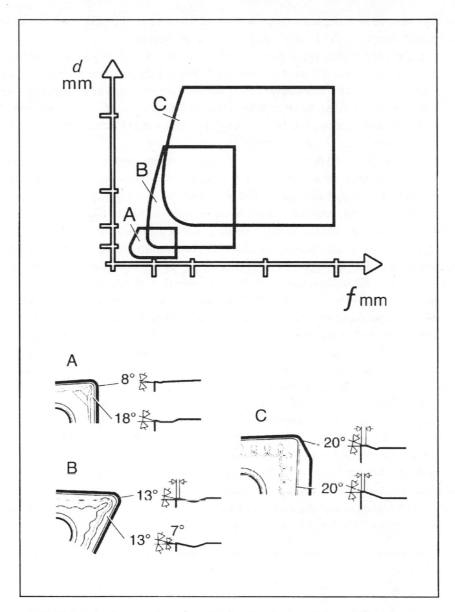

Figure 8.11 Chip breaking envelops for: (A) finishing inserts; (B) semi-finishing inserts; (C) roughing inserts.

application area. Finally, the cutting speed should be established in accordance with recommendations for the tool material relative to workpiece material, conditions, and power.

Chip control is put to the test when copy turning or profiling in CNC machines. The cutting edge is applied to cuts that vary along the profile of the component so as to change entering angle, direction of cut, cutting depth, feed and speed. The chip breakers have to be designed to cope with great flexibility. The effect is demonstrated with a shaft having typical variations in diameter and tapers.

The insert used in this application is a double-sided one with good chip control at smaller depth of cuts, but over a wide feed range. Representing a common semi-finishing operation the forged, alloy steel working allowance varies between one to three millimetres. As the insert machines round the profile of the component, the cutting depth and feed varies. In so doing the values move across the area of the possible d/f diagram for the geometry.

Chip forms, sizes and directions change but the cutting geometry controls the chip flow during the rapid transition along the component, as can be seen in Figure 8.12. The chip breaker design determines the ability of the cutting edge to form chips at varying feed rates and cutting depths. Some designs limit the edge to a small area, with light feeds and small depths, others cover a combination of chip breaking ability, at the nose and across the rake face. A corner chip breaker on the nose radius will form the chip at the lightest cuts while, as the feed is increased, the main cutting edge takes over. At the heaviest cut, chips are formed by the design on the rake face.

Figure 8.13 shows the range of chip shapes produced by the various combinations of feed rate (f) and depth of cut (d). Fine feed rates show thin continuous chips, while increasing feed rates promotes chip breaking. This leads to what is termed the 'chip breaking envelope', the acceptable cutting zone for that tool and material. The manufacturing engineer's job is to select the correct combination of tool, feed, and speed to cut within this envelope. In this way, an area of cutting data is established for each insert design where the chip formation is acceptable. Chips outside this area are usually either in the form of long thin strands or thick, over-compressed chips. Consequently, the geometry at the nose radius of the cutting edge influences the cutting action in various workpiece materials and different feeds. The design of the main cutting edge, at greater cutting depths, must take into consideration also the strength of edge.

Geometry design is often a compromise to accommodate the best of several factors, for instance: cutting edge strength versus power requirement optimization versus versatility, toughness behaviour versus tool life, etc.

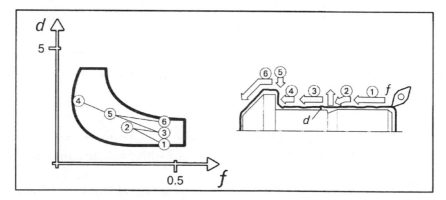

Figure 8.12 Variations of d/f combination throughout the application area during different cuts.

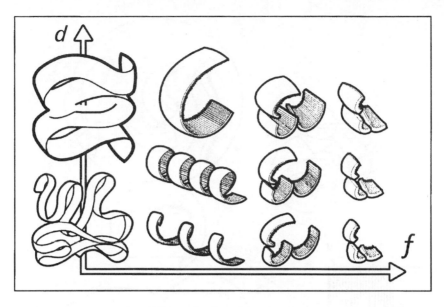

Figure 8.13 Chips produced at different feeds and depths of cut.

8.3 Machinability

Chips may be cut from some materials with relative ease and from others only with the greatest difficulty. This difference may be attributed to the 'machinability' of the respective materials.

Machinability is made up of a combination of five criteria (Figure 8.14):

1. Wear resistance.
2. Specific cutting pressure.
3. Chip breaking.
4. Built-up edge formation.
5. Tool coating character.

Attempts which have been made to represent the relative machinability of various materials numerically are only partially successful.

The most significant variables indicating machinability are tool life and the quality of surface finish produced. Conditions of the material which determine machinability are composition, heat treatment and microstructure. The measurable mechanical properties of hardness, tensile strength and ductility give some indication of the machining properties to be expected, but cannot distinguish between, say, the machinability of a free-cutting mild steel and of an austenitic stainless steel having somewhat similar mechanical test properties.

Some significant facts relating to machinability are given below:

1. *Hardness*. Steels up to 300 HB do not present great machining difficulty unless large amounts of alloying material are present. Steels up to 350 HB are

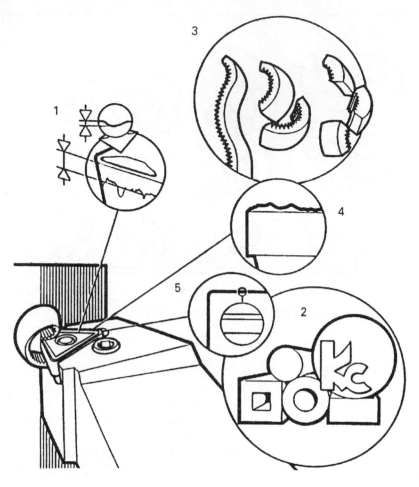

Figure 8.14 Factors contributing to machinability.

machinable with the superior grades of HSS, and with carbide tools. It is possible to hob splines on shafts of medium alloy steels following their heat treatment to somewhere near the above degree of hardness.

Some materials, notably manganese steel (manganese content 12–14 per cent) and alloys of the Nimonic series, have pronounced work-hardening properties which influence machinability. Indentation hardness tests may give values below 200 HB, which does not suggest any difficulty will be encountered during machining. Plastic deformation of the chip, however, involves 'working' of the material, and hardness may increase considerably from this cause. The increase of cutting power, due to work-hardening, gives rise to the generation of more heat; the resulting fall in tool life reveals an increased resistance to machining as compared with other materials of equivalent hardness. An extensively worked manganese steel may rise in hardness value to 550 HB, in which state machining becomes impracticable. A dull cutting tool operating at too small a feed may fail to cut a chip and will then work-harden the surface of the material so that subsequent penetration with a reconditioned tool is impossible.

The machinability of plain carbon steels falls steadily as the carbon content rises. Surface finish is better on the low-carbon steels if these have been cold drawn.

2. *Micro-structure*. Banded structures, as shown in Figure 8.15, do not machine well. Surface finish is poor due to tearing of the ferrite-rich areas.

Fine-grained materials take a good surface finish but have an increased resistance to machining. High-carbon steels may be machined at greater metal-removal rates if annealed or spheroidised; the resulting surface finish is then not very good, so that the treatment is generally confined to roughing operations.

3. *Composition*. High-nickel–chrome alloy steels are hard to machine due to their toughness, a combination of high strength with relatively high ductility. Under such conditions the amount of work to be done on the material necessary to cause plastic shear resulting in chips must be large.

The 18.8 type of austenitic stainless steel is difficult to machine because the ductile austenite transforms to the harder martensite under cutting stresses. For such conditions the cutting edge of the tool must be kept beneath the surface of the workpiece; rubbing, without producing a chip, may so work harden the surface that it is almost impossible to get a tool to penetrate the hard skin formed.

4. *Free machining properties*. Inclusions of a weaker insoluble material, such as manganese sulphide in a mild steel, considerably increase machinability both in regard to metal removal rates and the resulting surface finish. The inclusions give rise to local weaknesses which, under the cutting forces, increase stress concentration and thus cause failure at lower force values. Such inclusions cause a small reduction of mechanical properties only, and the free machining steels are of considerable economic importance in bar turning on automatic lathes.

Numerous manganese sulphide inclusions occur in steels having higher than normal amounts of sulphur (0.2 per cent in place of 0.04 per cent) to which manganese of about five times this amount has been added to neutralise the sulphur. The addition of the manganese is necessary to prevent the formation of the very weakening ferrous sulphide at the grain boundaries. The microstructure is illustrated in Figure 8.16. The introduction of small amounts of lead, 0.15–0.35

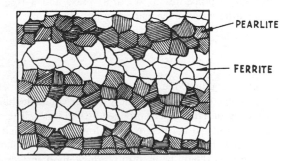

PEARLITE

FERRITE

Figure 8.15 Banded structure in a medium-carbon steel.

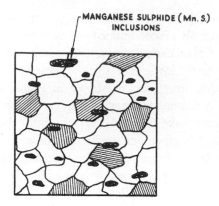

Figure 8.16 Inclusions in a free machining low-carbon steel.

per cent in steels, 1.0–3.0 per cent in brasses and bronzes, is effective in improving machinability for similar reasons.

Machinability ratings

In their *Manual on Cutting Metals*, the American Society of Mechanical Engineers (ASME) give numerical ratings indicating relative machinability. AISI Steel No. B1112, carbon 0.13 per cent max, manganese 0 70–1.00 per cent, sulphur 0.16–0.23 per cent in cold-drawn condition, UTS 82 000 Ibf/in^2 (566 N/mm^2), 179 HB, is rated at 100. By comparison a $3\frac{1}{2}$ per cent nickel, $1\frac{1}{2}$ per cent chrome steel, heat treated to 375 HB is rated at 24; a leaded brass 82 HB is rated at 200; and a phosphor bronze 140 HB is rated at 40. The ratings are based on actual cutting tests and, as instanced by the values for the alloy steel and bronze, do reflect common experience in cutting such materials.

The Manual contains excellent information on microstructures suitable for machining and on many other special aspects of machinability.

Example 8.1

Compare the relative machining properties of a 70/30 brass and a free machining brass. Mention chip types, and any differences in the cutting conditions required.

Solution

70/30 brass: in the annealed condition, the elongation is approximately 70 per cent; in the cold-worked condition the elongation is approximately 10 per cent. Machining is only likely to be required to dress the edge of pressings or produce drilled holes. The material would be machined in the worked condition and before any annealing process which may be required. The chip will be continuous and chip disposal difficult. A rake angle of about 25° is suitable; a cutting lubricant and a fine feed should be used to produce good surface finish. Tool faces should have a very good surface finish to minimise built-up edge.

Free machining brass: 58 per cent Cu, 39 per cent Zn, 3 per cent Pb, is a probable composition and the bar is likely to be extruded. Ductility is represented by an elongation of approximately 30 per cent, but the material machines easily due to the lead inclusions. It should be cut dry, using a small or zero rake angle, at moderate speeds and feeds. Discontinuous chips, very easily cleared from the tools, make this material suitable for turned work on bar automatics.

□   □   □

8.4 Tool wear

All cutting tools wear during machining and continue to do so until they come to the end of their tool life. The life of a cutting edge is counted in minutes and today tool lives are often less than the old, established mark of fifteen minutes, but often quite a bit more as well. It is the productive time available during which the edge will machine components which are acceptable within the limiting parameters. In the early days of man and tools, the tool life parameter was simply when the tool could not cut any more. Today, the usual parameters are surface texture, accuracy, tool wear pattern, chip formation and predicted reliable tool life. The one applied depends upon the type of operation, finishing or roughing, and often the amount of manual control and supervision involved.

The cutting edge of an insert in a finishing operation is worn out when it can no longer generate a certain surface texture. Not a lot of wear is needed along a very small part of the insert nose for the edge of an insert to need changing. In a roughing operation wear develops along a lot longer part of the edge and considerably more wear can be tolerated as there are no surface texture limitations and accuracy is not close. The tool life may be limited when the edge loses its chip control ability or when the wear pattern has developed to a stage where the risk for rapid edge breakdown is imminent.

The heat (H) generated during metal cutting directly influences the wear development in various ways. The process involves the conversion of the power from the motor (K) to other forms during the work of parting the chip form from the workpiece. This plastic strain condition involves deformation (H_d) and friction (H_f), the absolute majority of which is turned into heat energy. The rest of the energy, normally only around one per cent, is retained as elastic energy (Ee) (Figure 8.17).

Most of the heat energy is taken away from the cutting zone by the chip flow (C). As regards temperature, this is usually highest close to the chip face, in the tool (T), where up to 1600 °C can occur. Some heat remains in the tool and some in the workpiece (W). The temperature (t) is lower in the chip and workpiece, although the flank temperatures can get very high. It is the cutting speed that has the greatest influence on the heat generated and thus the wear development according to the mentioned wear mechanism. The cutting speed and heat will mostly influence the type of wear and tool life depending upon the interplay between tool material, workpiece material and machining conditions.

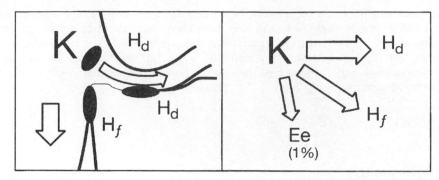

Figure 8.17 Metal cutting involves conversion of energy.

Tool wear is the product of a combination of load factors on the cutting edge. The life of the cutting edge is decided by several loads, which strive to change the geometry of the edge. Wear is the result of interaction between tool, workpiece material and machining conditions. The main load factors (Figure 8.18) are:

A. Mechanical.
B. Thermal.
C. Chemical.
D. Abrasive.

Metal cutting generates a lot of heat on the chip face and flank of the insert. Thermal load is considerable on the tool material and in some operations, such as milling, there is also a dynamic factor when edges leave and re-enter the workpiece.

The chip forming process means that a fresh metal interface is continually produced and forced at very high pressure and temperature along the tool material (Figure 8.19). The zones produced make it an attractive environment for diffusion and chemical reactions of metals.

Various types of very hard particles occur in most workpiece materials, often comparable in hardness to the tool material itself. These are bound to achieve a grinding, abrasive effect on the tool.

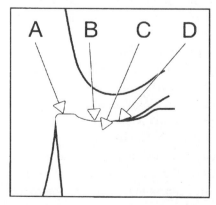

Figure 8.18 Typical wear zones.

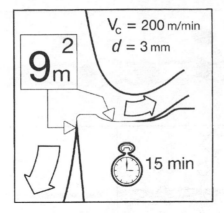

Figure 8.19 Amount of metal passing the edge.

As a result of these complex stresses applied to the tool, five basic wear mechanisms are considered in metal cutting (Figure 8.20).

1. Abrasion wear.
2. Diffusion wear.
3. Oxidation wear.
4. Fatigue wear.
5. Adhesion wear.

The tool's ability to resist loads will determine how it will be affected by these wear mechanisms.

Abrasion wear is very common, and is caused mainly by hard particles in the workpiece material. The cutting edge's ability to resist abrasion is largely connected to the tool hardness. A tool material (densely) packed with the hardest of particles will stand up well to abrasive wear but may not cope with other load factors during machining.

Diffusion wear is more affected by the chemical load during the cutting process. The chemical properties of the tool material and the affinity of the tool material to the workpiece material will decide the development of the diffusion wear mechanism. The metallurgical relationship between the materials will determine the amount of the wear mechanism. Some cutting tool materials are inert against most workpiece materials, while some have high affinity to some.

Tungsten carbide and steel have affinity towards each other leading to the diffusion wear mechanism developing. This results in the formation of a crater on the chip face of the insert. The mechanism is very temperature dependent and is thus greatest at high cutting speeds. Atomic interchange takes place with a two-way transfer of ferrite from the steel into the tool and carbon, being small and ready to move in iron, diffuses into the chip.

High temperatures and the presence of air means *oxidation* for most metals. Tungsten and cobalt form porous oxide films which are more easily rubbed off by the chip. Some oxides like aluminium oxide, however, are much stronger and harder.

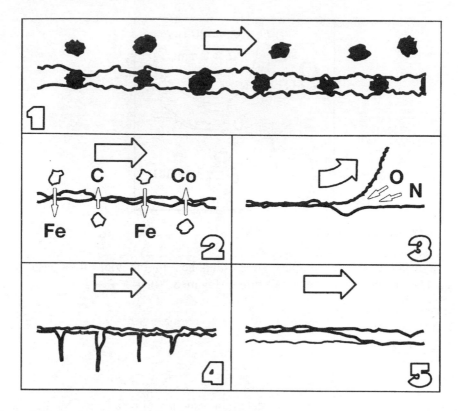

Figure 8.20 Basic wear mechanism in metal cutting.

Some cutting tool materials then are more prone to wear due to oxidation than others, especially at the interface part of the edge, where the chip width finishes (at the depth of cut) and air gains access to the cutting process. Oxidation there leads to typical notches being formed in the edge but is a relatively uncommon phenomenon in today's machining.

Fatigue wear is often a thermo-mechanical combination. Temperature fluctuations and the loading and unloading of cutting forces can lead to cutting edges cracking and breaking. Intermittent cutting action leads to continual generation of heat and cooling as well as shocks of cutting edge engagement. Some tool materials are more sensitive than others to the fatigue mechanism. Pure mechanical fatigue can occur also from the cutting forces being too high for the mechanical strength of the cutting edge. This can be from hard or strong workpiece materials, very high feed rates or when the tool material is not hard enough. However, plastic deformation dominates in such cases.

Adhesion wear (also known as attrition wear) occurs mainly at low machining temperatures on the chip face of the tool. It can take place with long-chipping and short-chipping workpiece materials – steel, aluminium and cast iron. This mechanism often leads to the formation of a built-up edge (BUE), between the chip and the edge. It is a dynamic structure, with successive layers from the chip being welded

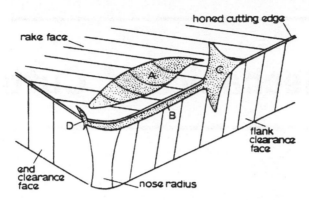

Figure 8.21 Regions of tool wear: crater wear at A; flank wear at B; depth of cut notch wear at C; trailing wear edge notch at D.

and hardened, becoming part of the edge. The BUE can be sheared off and commence build-up again or cause the edge to break away in small pieces or fracture. Regions of tool wear are shown in Figure 8.21.

1. *Flank wear,* as the name indicates, takes place on the flanks of the cutting edge, mainly from the abrasive wear mechanism. The clearance sides: leading, trailing and nose radius or parallel land are subjected to the workpiece moving past during and after chip formation. This is usually the most normal type of wear and to maintain safe progressive flank wear is often the ideal. In the end excessive flank wear will lead to poor surface texture, inaccuracy and increasing friction as the edge changes shape.

2. *Crater wear* on the chip face can be due to abrasive and diffusion wear mechanisms. The crater is formed through tool material being removed from the chip face either by the hard particle grinding action or at the hottest part of the chip face through the diffusive action between the chip and tool material. Hardness, hot hardness and minimum affinity between materials minimises the tendence for crater wear. Excessive crater wear changes the geometry of the edge and can deteriorate chip formation, change cutting force directions and also weaken the edge.

9 Mechanics of Cutting

9.1 Units and measurement

Table 9.1 lists the units of measurement and symbols in common use.

9.2 Cutting force analysis

The range of cutting forces acting upon a cutting tool has already been studied and it is now necessary to evaluate the effect of these upon the machine tool, and to analyse the cutting process in some detail (Figure 9.1). The vertical cutting force, F_c, (sometimes called the tangential force) is the one that does all the work, since it is operating with a high cutting velocity, V_c. The cutting power is the product of $V_c \times F_c$. The other two forces acting are the radial force, F_r, and the axial force, F_a. The radial force is produced by the approach angle of the tool, and is the force needed to hold the tool against the workpiece. It usually has zero velocity and thus zero power. The axial force, F_a, has a very low velocity, and thus a very small power consumption.

Note: This discussion of force directions is based upon the axial turning of a shaft. The directions in which these forces act would be changed if the machine was cutting in a different direction, e.g. by 90° if the tool was facing off the end of a workpiece.

These three forces can be resolved to determine the total resultant force acting, F. This can be very large and can easily cause significant deformation of the workpiece, thus finishing cuts are usually taken with small forces to minimise the forces acting, and to provide accurate cutting conditions.

For most workpiece materials higher cutting speeds mean lower cutting forces, the higher temperatures on the flow zone and reduced contact area contribute towards this effect. The decrease in force varies with the type and condition of the material and the range of cutting speeds in use. For a heat resistant nickel based alloy steel the initial chip forming force can be ten times larger than that required to cut unalloyed aluminium.

Table 9.1

	Symbol	Metric units
Cutting speed	V_c	m/min
Cutting force	F_c	newtons (N)
Depth of cut	d	mm
Feed/rev	f	mm
Metal removal rate	w	mm³/s
Power consumed		watts (W)
Specific power consumption $\left(\dfrac{\text{Power}}{w}\right)$		W/mm³/s
Specific removal rate $\left(\dfrac{w}{\text{Power}}\right)$	K	mm³/s/W
Specific cutting pressure	K_c	N/mm² (or MN/m²)

The main cutting force F can be split up into three components

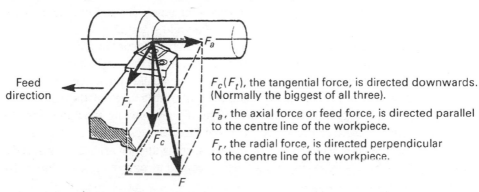

Feed direction

$F_c (F_t)$, the tangential force, is directed downwards. (Normally the biggest of all three).

F_a, the axial force or feed force, is directed parallel to the centre line of the workpiece.

F_r, the radial force, is directed perpendicular to the centre line of the workpiece.

The approximate relationship of these components to each other is: $F_c : F_a : F_r = 4 : 2 : 1$ (for the cutting conditions shown)

Figure 9.1 Cutting force analysis.

Specific cutting pressure

A very convenient way has been developed for estimating the size of the cutting force. A property known as the specific cutting pressure, K_c, has been defined as the vertical cutting force, F_c, divided by the area of the cut being taken, i.e.:

$$K_c = F_c/A = \frac{F_c}{d \times f} \tag{9.1}$$

Figure 9.2 shows that the *specific cutting pressure* (K_c) and the undeformed chip thickness, varies with the type of material (A) stainless steel, (B) alloyed steel and

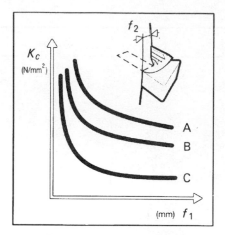

Figure 9.2 Specific cutting pressure variation with materials.

(C) grey cast iron. The pressure depends upon the shear yield strength of the workpiece material and the area of the shear plane. This area varies considerably and with it the cutting pressure. It is thought to be more influential than the yield strength of the material, which, in fact, does not vary that much for the cutting process. Alloying and heat treatment, however, increase the yield strength.

It should be apparent that this is not actually a true stress value, in fact, it is usually three or four times the yield strength of the material, since the actual contact area of the chip on the tool is never known precisely. Since the actual contact area varies, the assumption is made that the chip only contacts the tool over the area of the cut.

The previous discussions of the complex nature of metal cutting show the range of factors interacting as the metal flows over the tool face, and tests show that K_c reduces with increasing feed rates, increasing top rake angles, and with increasing cutting speeds.

Specific cutting pressure is important when it comes to power (P) calculation for any metal cutting process (Figure 9.3).

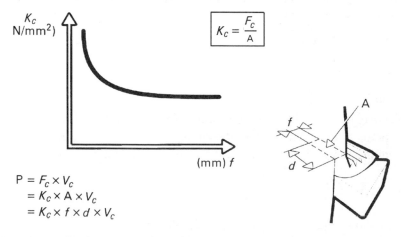

$$P = F_c \times V_c$$
$$= K_c \times A \times V_c$$
$$= K_c \times f \times d \times V_c$$

Figure 9.3 Specific cutting pressure/feed per revolution.

The value of the specific cutting pressure is available for various materials (Table 9.2), enabling the calculation of workpiece material removed per power unit. It is also a measurement of the machinability of materials. The value is valid for a material under certain conditions and cutting data. For instance, the value will vary with the cutting speed: a higher speed, leading to higher cutting temperatures, generally leads to a lower value, (A). Also the geometry of the cutting edge is influential, in that a positive rake angle leads to a smaller value than a negative one.

Table 9.2 Specific cutting force (K_c) values for a range of common materials.

Material	Hardness HB	Condition	$K_c 0.4$ (N/mm^2)
Unalloyed	110	C < 0.25%	2200
steels	150	C < 0.8%	2600
	310	C < 1.4%	3000
Low alloy	124–225	Non-hardened	2500
steels	220–420	Hardened	3300
High alloy	150–300	Annealed	3000
steels	250–350	Hardened tool steel	4500
Extra hard steel	>450	Hardened and tempered	4500
Malleable	110–145	Short chipping	1200
cast iron	200–230	Long chipping	1300
Grey	180	Low tensile	1300
cast iron	260	High tensile, alloyed	1500
Nodular cast iron	160	Ferritic	1200
SG-iron	250	Pearlitic	2100
Steel	150	Unalloyed	2200
castings	150–250	Low alloy	2500
	160–200	High alloy	3000
Stainless steels	150–270	Ferritic, martensitic 13–25% Cr	2800
	150–275	Austenitic Ni >8%, 18–25% Cr	2450
	275–425	Quenched and tempered, martensitic >0.12% C	2800
	150–450	Precipitation hardened steels	3500
Heat resistant	180–230	Annealed or solution treated	3700
super alloys	250–320	Aged or solution treated and aged	3900
Fe based titanium alloys	950 MPa	α, near α and β alloys in annealed condition	1675
	1050 MPa	$\alpha + \beta$ alloys in aged condition, β alloys in annealed or aged condition	1690
Aluminium	38–80	Wrought and cold drawn	800
alloys	75–150	Wrought and solution treated and aged	800
	40–100	Cast	900
	70–125	Cast, solution treated and aged	900
	80	Unalloyed, Al ≥99%	400
Aluminium with		10–14% SI	900
high SI content		14–16% SI	1500

The specific cutting pressure is closely related to the size of the undeformed chip thickness/feed rate. An increase of f leads to a reduction of K_c. This means, that the smaller the chip cross-section used in a process, the higher the specific cutting pressure – and the more unit power needed.

It also leads to the recommendation that feed rates are maximised in the metal cutting process.

It is convenient to resolve the resultant cutting force F acting on a tool into three vectors (Figure 9.4):

1. The radial force F_r.
2. The vertical force F_c.
3. The axial force F_a.

The radial cutting force component (F_r) is directed at right angles to the tangential force from the cutting point.

The axial cutting force (F_a) is directed along the feed of the tool, axially along the direction of machining of the component. It is an important force factor in drilling operations. The cutting ability of the drill geometry will considerably influence the size of the force needed and as a rule the axial feed force requirement rises with the diameter of the drill.

Geometry, especially the entering angle, will determine the size of the two force components. Their relationship becomes especially important when deflection of tool with large overhang or a slender workpiece is a factor as regards accuracy and vibration tendencies. The rake angle also influences the size of the radial cutting force component. Positive rake angles, of course, also mean lower cutting forces in general.

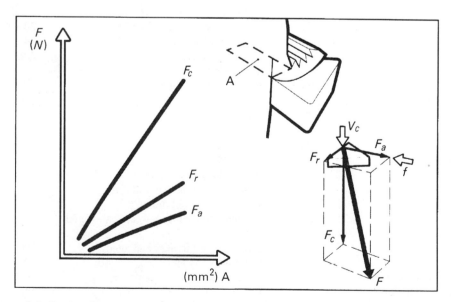

Figure 9.4 Cutting force components.

For most workpiece materials, increasing cutting speed leads to lower cutting forces. The higher temperature in the flow-zone and reduced contact area contribute towards this effect. The decrease in force varies with the type and condition of material and the range of cutting speed in question, (Figure 9.5).

As may be expected, the size relationship between the force components varies considerably with the type of machining operation. The tangential force often dominates in milling and turning operations, especially to do with power requirements. The radial force is of particular interest in boring operations and the axial, feed force in drilling. The size of the radial cutting force is dependent upon the entering angle used and the nose radius. A 90° entering angle and small nose radius will minimise the radial cutting force component, which strives to deflect the tool and gives rise to vibrations.

All three components increase in size with increasing chip cross-section, the tangential one most of all. For rough turning, a typical relationship might be for $F_c : F_r : F_a$, 4:2:1. The tangential cutting force is twice as large as the radial and four times that of the axial force. In drilling the relationship would be quite different and highly dependent upon the feed rate.

Vibration tendency is one consequence of the cutting forces. As well as tool or workpiece deflection, these can be affected by variations in the cutting process such as varying working allowance or material conditions as well as the formation of built-up edges.

Along with the importance of the design of the cutting geometry, to provide smooth chip breaking, and the use of a positive rake angle (Figure 9.6), higher cutting speeds generally have a favourable influence on the cutting forces/ vibrations.

It is very important to achieve stability of the complete system, that is formed by the factors in the machining process. The quality of the toolholder and its ability to securely hold the indexable insert is one of the more important factors.

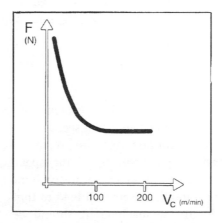

Figure 9.5 Reduction of cutting force with increasing cutting velocity.

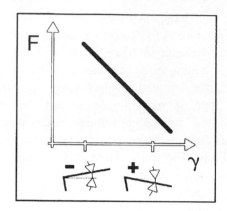

Figure 9.6 Reduction of cutting force with increasing top rake angle.

Basic cutting equations

The cutting speed V_c is the relative velocity of the workpiece and tool at the tool edge. For any point on the tool edge it can be considered as a vector, and the cutting force F_c is measured along the line of this vector. The elementary relationships should be noted:

1. Energy/min consumed in cutting $= F_c V_c$ Nm/min.
2. Power consumed in cutting $= F_c V_c / 60$ watts.

3. Metal removal rate, $w = df \dfrac{V_c}{60}$ mm^3/s.

The power required to remove a unit volume of material per minute is a measure of its resistance to cutting; it also gives an indication of the effectiveness of the cutting conditions. Typical values for common materials enable estimates of cutting power and cutting force to be made for specified metal removal rates. These values are of use to machine tool designers and production planning engineers.

4. $\dfrac{\text{power}}{w} = \dfrac{F_c \times V_c}{60} \dfrac{60}{df V_c \times 10^3}$, i.e. $\propto \dfrac{F_c}{df}$
$\hfill (9.2)$

Hence the specific power consumption is independent of cutting speed, except in so far as the cutting force changes with the cutting speed.

Units of specific power consumption are W/mm^3/s, which is equivalent to J/mm^3. Some authorities quote values of the specific energy required for cutting, i.e. the energy required to remove a unit volume of material. However, at the practical level, machine tool users will tend to think in terms of the power available at the spindle of the machine. It should be recognised that for a specific energy of, say, 2 J/mm^3 it will require 2 W to remove metal at a rate of 1 mm^3/s.

Measurement of the input to the electric motor of a machine tool provides a convenient method of measuring the power consumed in cutting. If the value of the power supplied when the machine is running idle is subtracted from the power reading taken under the cutting load, a reasonable estimate of the power consumed in cutting is obtained. The method is an approximate one, because the efficiency of the drive under varying loads is not taken into account. Watt–meters are commonly used to measure the power, but some machines will still have motors rated in hp (746 watts = 1 hp).

Example 9.1

A lathe is running at 1000 rpm machining a bar of steel which is 100 mm outside diameter. The cutting force applied by the tool to the work is 700 newtons.

(a) What is the cutting velocity when the tool is starting to cut the bar at a diameter of 100 mm?
(b) What is the cutting velocity when the tool has reduced the bar to a diameter of 50 mm?
(c) What power is being used at these two diameters?

Solution

(a) Bar diameter = 0.1 m

$\therefore$ Bar circumference = $\pi \times 0.1 = 0.3412$ m

Spindle runs at 1000 rpm $= \dfrac{1000}{60}$ revs/sec $= 16.66$ revs/sec

$\therefore$ Cutting velocity = 16.66×0.3413 m/sec = 5.236 m/sec

(b) Bar diameter = 0.05 m

$\therefore$ Bar circumference = $\pi \times 0.05 = 0.1706$ m
Same spindle speed, 16.66 revs/sec
This gives a velocity of $16.66 \times 0.1706 = 2.843$ m/sec
i.e. half as much (since diameter of bar is now halved).

(c) Power = force × velocity
   (i)   at 100 mm diameter power = 5.236 m/sec × 700 newtons = 3665 watts
   (ii) at 50 mm diameter power = 2.843 m/sec × 700 newtons = 1832.6 watts

□   □   □

Example 9.2

The power required to turn a medium-carbon steel is approximately 3.8 W/mm^3/s.

If the maximum power available at the machine spindle is 5 hp, find the maximum metal removal rate. Also find, for a cutting speed of 36 m/min and feed rate of 0.25 mm/rev, the depth of cut and the cutting force which will occur when the metal removal rate is at the maximum value.

Solution

$$w_{max} = \frac{5 \times 746}{3.8} = 982 \text{ mm}^3/\text{s}$$

$$F_c = \frac{5 \times 746 \times 60}{36} = 6216 \text{ N}$$

$$d = \frac{60 \times 982}{10^3 \times 36 \times 0.25} = 6.55 \text{ mm}$$

□   □   □

Example 9.3

A lathe running idle consumes 325 W. When cutting an alloy steel at 24.5 m/min the power input rises to 2580 W. Find the cutting force and torque at the spindle when running at 124 rev/min. If the depth of cut is 3.8 mm and the feed 0.2 mm/rev, find the specific power consumption.

Solution

Power consumed in cutting $= 2580 - 325 = 2255 \text{ W}$

$$w = 0.2 \times 3.8 \times 24.5 \times \frac{10^3}{60} = 310 \text{ mm}^3/\text{s}$$

$$\text{Specific power consumption} = \frac{2255}{310} = 7.27 \text{ W/mm}^3/\text{s}$$

$$\text{Torque at the spindle} = \frac{2255 \times 60}{2\pi \times 124} = 174 \text{ Nm}$$

$$\text{Cutting force } (T) = \frac{2255 \times 60}{24.5} = 5522 \text{ N}$$

□   □   □

Example 9.4

The given data relates to the rough turning of an alloy steel (SAE 3140) heat treated to 285 HB and having a UTS of 986 MN/m^2.

Depth of cut 6.4 mm, approach angle (k) 60°, nose radius 6.4 mm, side rake angle 14°, back rake angle 8°.

Cutting speed (V_c) m/min	55.5	35.4	22	14.3	9.5	6.4
Feed (f) mm/rev	0.05	0.10	0.20	0.40	0.80	1.60
kW consumed in cutting	1.27	1.42	1.50	1.64	1.87	2.16

Draw a graph to show the variation of specific cutting pressure with feed. Give the average value of the specific power consumption for cutting this material.

Solution

Feed (*f*)	0.05	0.10	0.20	0.40	0.80	1.60
Metal removal rate (*w*)	296	378	470	610	810	1093
$\dfrac{\text{Power (W)}}{w}$	4.29	3.76	3.19	2.69	2.31	1.98
k_c N/mm²	4290	3760	3190	2690	2310	1980

Note $K_c = \dfrac{10^3 \times \text{power(W)}}{w}$ from equation 9.1

Average value of specific power consumption = 3.04 W/mm³/s.

□   □   □

9.3 Merchant's analysis of metal cutting

Dr Merchant, of the Cincinnati Milling Machine Company, took an idealised concept of chip formation for which a precise geometry may be derived as the basis for his studies of the mechanics of metal cutting.

Figure 9.7 illustrates the concept. Workpiece material advancing at velocity V_c towards the tool edge, is compressed at the tool rake face causing failure to occur by plastic shearing along the plane PQ, called the shear plane.

This is a simplified concept, but one which is a very useful introduction to the cutting mechanism. Examination of the previous illustrations of chip formation (Figures 8.2 and 8.3) shows that there is actually a zone of plastic deformation, and slip line field analysis can be applied to this zone for deeper analysis of the process.

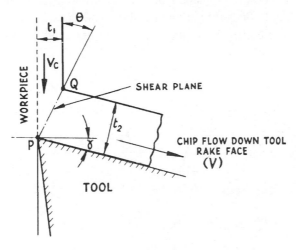

Figure 9.7 Merchant's 'idealised' concept of chip formation.

See, for example, Palmer and Oxley, *Mechanics of Orthogonal Machining*, Proceeding I.MechE. 1959, pp. 173, 623 for more detail. Merchant's pioneering analysis does form a valuable introduction and so is considered in detail, it assumes that under suitable conditions a continuous and steady rate of deformation occurs along the shear plane and the deformed material slides at a uniform velocity V_c down the rake face of the tool in the form of a continuous chip. Before deformation the thickness of the chip is t_1 and $t_1 = \text{feed/rev}$ when $k = 90°$ (i.e. during orthogonal cutting); during deformation the cut chip increases in thickness to t_2. The rake angle (γ) plays an essential part in the action and shearing occurs along some definite plane, the position of which is given by the shear plane angle θ. Since t_1 and γ are known, and t_2 may be measured for a chip cut under steady conditions, the shear plane angle θ can be found from a simple cutting test.

Direct measurement of t_2 is not practical because of the roughness of the upper side of the cut chip. The following method permits reasonably accurate values of t_2 to be found.

A small shallow saw notch is milled along the periphery of the bar to be machined (Figure 9.8). The depth of cut d is obtained by measuring the diameter of the bar before and after cutting and from these measurements the mean diameter of the bar D_m is also obtained. The mean length of the cut chip l_2, measured between two successive notches, is determined by measuring along the inner and outer edges of the chip and finding a mean value. The mean distance before cutting $l_1 = \pi D_m$. Assuming no deformation occurs in the depth d, which is probably nearly true,

$$t_1 . l_1 = t_2 . l_2$$

and

$$t_2 = \frac{t_1 . l_1}{l_2}$$

The ratio t_1/t_2 is a measure of the amount of deformation occurring during cutting; it changes according to the particular material cut and the value of the effective rake angle employed. The higher the value of this ratio obtained from a cutting test, the

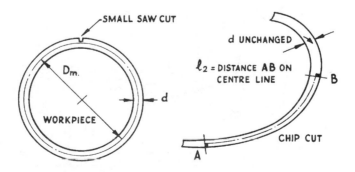

Figure 9.8 Method of obtaining chip length l_2.

lower the energy consumed in shearing along PQ. Efficient cutting lubricants tend to increase the ratio and so reduce the total energy requirements for cutting. For this reason changes in the value of t_2, other factors remaining constant, may be used to test the relative effectiveness of various cutting lubricants.

θ may be obtained from a simple large-scale drawing once the values t_1, t_2 and γ are known. By calculation

$$\tan \theta = \frac{\gamma_c \cos \gamma}{1 - \gamma_c \sin \gamma}$$

where $\gamma_c = t_1 / t_2$.

Example 9.5

A bar 76 mm diameter, is reduced to 71 mm diameter by means of a cutting tool for which $k = 90°$ and for which the cutting edge lies in the plane containing the work axis of rotation. The mean length of the cut chip $l_2 = 73.9$ mm, the rake angle $\gamma = 15°$ and a feed of 0.2 mm/rev is used. Find the cutting ratio t_1 / t_2 and the value of the shear plane angle.

Solution

$$t_2 = \frac{73.5\pi}{73.9} \times 0.2 \times 0.63 \text{ mm}$$

$t_1 / t_2 = 0.2 / 0.63 = 0.32$

θ may then be found by scale drawing as shown in Figure 9.9. Set off TP and PS such that $\angle$TPS $= 90° + 15°$; PS represents the rake face of the tool. To some convenient scale set off a line parallel to TP at distance 0.2 mm, and a line parallel to PS at distance 0.63 mm to intersect at Q. Join PQ and measure the resulting angle TPQ to obtain the value of θ.

By measurement $\theta = 19°$.

□   □   □

By using the concept of chip formation described above and by measuring the forces F_c and F_a with a cutting tool dynamometer, Merchant was able to build up a picture of the forces acting in the region of cutting which give rise to plastic deformation and sliding of the chip down the tool rake-face. The theory assumes that a continuous-type chip without built-up edge is produced during the cutting, and the force system is illustrated in Figure 9.10.

The forces exerted by the workpiece on the chip are:

F_{cm} = compressive force on the shear plane
F_s = shear force on the shear plane

The forces exerted by the tool on the chip are:

N = normal force at the rake face of tool
F_f = frictional force along the rake face of tool

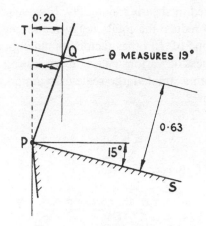

Figure 9.9 Graphical solution of shear plane angle.

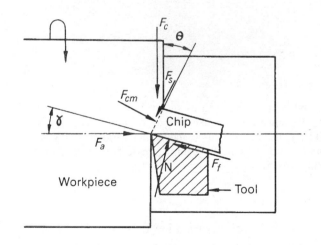

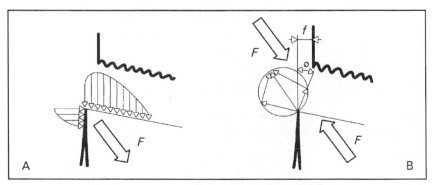

Figure 9.10 Compressive stress and forces at the cutting edge.

The forces acting on the tool, measured by the cutting-tool dynamometer, are the forces F_c, F_r and F_a shown in Figure 9.4.

As explained on p. 256 and illustrated by Example 9.5, the shear plane angle θ can be found from a simple cutting test. Since γ is known and θ can be determined, the directions of all the forces shown in Figure 9.10 can be found.

It is reasonable to suppose that the action of the workpiece on the chip is balanced by the reaction of the tool on the chip, hence the resultant of forces F_{cm} and F_s, and the resultant of forces N and F_f, will be common in magnitude and direction (but opposite in sense). The action of the workpiece on the tool is measured by means of a cutting-tool dynamometer, and since F_c and F_a give rise to the other forces, their resultant is also common in magnitude and direction with the above.

Using the concepts illustrated in Figure 9.10, it is now possible to find graphically the magnitude of the mutually perpendicular pairs of forces F_c, F_s and N, F. Use is made of the geometrical property of the angle in a semicircle, and the required graphical construction is illustrated in Figure 9.11.

The vector diagram of forces is constructed as follows. Draw F_a and F_c to some convenient scale and join AB to obtain their resultant. Bisect AB and draw a circle having the resultant force as its diameter. Set off BE making angle θ with force T, to cut the circle at E. Join EA. The magnitudes of F_s and F_c are now shown.

Set off a line BG, at an angle of $(90° - \gamma)$ from force F_c, to cut the circle at G. Join GA. The magnitude of forces N and F_f are now shown, as also the coefficient of friction at the chip–tool interface (F_f/N). Angle BAG is the angle of friction τ,

$$\tan \tau = \frac{F_f}{N}$$

Friction at the chip–tool interface appears to be of a different order of magnitude from that of normal sliding friction. For this reason the symbols μ and φ, as used for sliding friction, have been avoided.

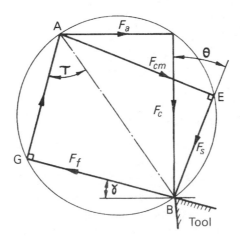

Figure 9.11 Merchant's circle.

The following expressions, derived from the geometry of Figures 9.7 and 9.11, enable the values from Merchant's analysis to be calculated.

Let $r_c = t_1/t_2$, generally called the chip thickness ratio.

Then

$$\tan \theta = \frac{r_c \cos \gamma}{1 - r_c \sin \gamma}$$

Compressive force on the shear plane;

$$F_{cm} = T \cos \theta - F_f \sin \theta$$

Shear force on the shear plane;

$$F_s = F_a \cos \theta + F_c \sin\theta$$

Friction force at the chip–tool interface;

$$F_f = F_a \cos \gamma + F_c \sin \gamma$$

Normal force on the chip–tool interface;

$$N = F_c \cos \gamma - F_a \sin \gamma$$

The coefficient of friction at the interface;

$$\tan \tau = (F_a + F_c \tan \gamma)/(F_c - F_a \tan \gamma).$$

Example 9.6

The following values relate to a cutting test under orthogonal cutting conditions and for a chip type as in Figure 8.2, 5 on page 229.

$$F_c = 1500 \text{ N}, \quad F_a = 1000 \text{ N}, \quad \gamma = 10°, \quad t_1/t_2 = 0.37$$

Determine, using Merchant's theory, the cutting forces F_{cm}, F_s, N and F_f, and also the coefficient of friction at the chip–tool interface.

Solution

Find the shear plane angle θ. Draw the Merchant Circle diagram as explained on p. 259. The following values found by this method are subject to normal graphical errors:

$$\theta = 21\tfrac{1}{2}°, \ F_{cm} = 1504 \text{ N}, \ F_s = 1024 \text{ N}, \ F_f = 1260 \text{ N}, \ N = 1313 \text{ N}$$

$$\text{Coefficient of friction at the chip–tool interface} = \frac{1260}{1313} = 0.96$$

□   □   □

The very high value of the coefficient of friction is typical of the results obtained using Merchant's analysis, sometimes values larger than 1 arise, which is an invalid friction coefficient. Such results suggest that the normal conditions of sliding friction do not apply at the chip–tool interface. The fact that pressure welding may occur between the chip and the tool rake-face, as in built-up edge conditions, gives further confirmation of this.

The theory then does not give a completely satisfactory picture of metal cutting, and discussion of some of its weaknesses can be found in later research papers. The theory is, however, a useful step forward in the attempt to present an adequate picture of the chip-forming process, and it has been the stepping stone to much of the subsequent research work in this field.

9.4 Merchant's analysis, work done in cutting

This can best be illustrated by a worked example.

Example 9.6

When machining at 165 m/min, $F_c = 1080$ N, $F_a = 1000$ N, $\gamma = 10°$, the shear plane angle was found to be 19°. Determine the velocity of shearing along the shear plane and the velocity of sliding at the chip–tool interface, and find the work done per minute in shearing the metal and against friction. Show that the work input is equal to the sum of the work done in shearing and against friction.

Solution

First draw the vector diagram of velocities as shown in Figure 9.12.
    The vector equation is:

Velocity of workpiece relative to the tool (V_c) – Velocity of workpiece relative to the chip (V_s) + Velocity of workpiece relative to tool (V_f)

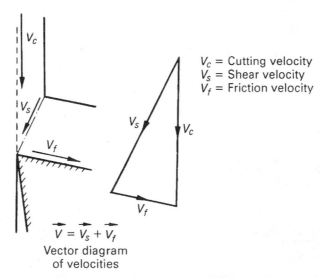

V_c = Cutting velocity
V_s = Shear velocity
V_f = Friction velocity

$$\vec{V} = \vec{V_s} + \vec{V_f}$$
Vector diagram
of velocities

Figure 9.12 Vector diagram of velocities. (V_c = cutting velocity; V_s = shear velocity; V_j = friction velocity).

From the vector diagram:

$$V_s = 164.6 \text{ m/min}, \; V_c = 55 \text{ m/min}$$

By drawing the Merchant circle diagram:

$$F_s = 690 \text{ N}, \; F_f = 1180 \text{ N}$$

Hence:

1. Work done per min in shearing metal $= 690 \times 164.5 = 113\,500$ J
2. Work done per min against friction $= 1180 \times 55 = 64\,900$ J

$$\text{Sum of (1) and (2)} = 178\,400 \text{ J}$$

$$\text{Work input per min} = 1080 \times 165 = 178\,200 \text{ J}$$

It can be seen that the work input is equal to the sum of the work done in shearing the metal and the work done against friction, within the limits of accuracy to be expected from graphical methods.

□   □   □

Relationship of velocities

See Figure 9.13.

$$\frac{V_s}{\sin(\phi - \alpha)} = \frac{V_c}{\sin(90 + \alpha - \phi)} \qquad \therefore \quad \frac{V_s}{\cos \alpha} = \frac{V_c}{\cos(\phi - \alpha)}$$

$$\therefore \quad V_s = \frac{V_c \cos \alpha}{\cos(\phi - \alpha)}$$

Similarly

$$\frac{V_f}{\sin \phi} = \frac{V_c}{\cos(\phi - \alpha)} \qquad \therefore \quad V_f = \frac{V_c \sin \phi}{\cos(\phi - \alpha)} = V_c r_c$$

Strains

See Figure 9.14.

$$\text{Shear strain } \varepsilon = \frac{\Delta s}{\Delta x} = \frac{AB}{CD} = \frac{AD + DB}{CD}$$

$$\therefore \quad \varepsilon = \cot \phi + \tan(\phi - \alpha)$$

$$\therefore \quad \varepsilon = \frac{\cos \phi}{\sin \phi} + \frac{\sin(\phi - \alpha)}{\cos(\phi - \alpha)}$$

$$= \frac{\cos \phi \cos(\phi - \alpha) + \sin \phi \sin(\phi - \alpha)}{\sin \phi \cos(\phi - \alpha)}$$

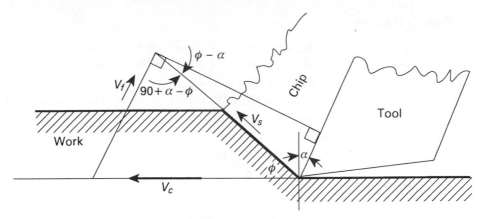

Figure 9.13 Relationship of velocities.

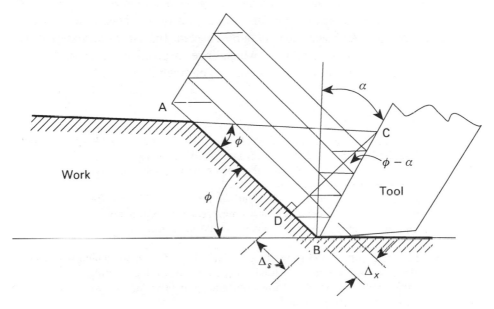

Figure 9.14 Relationships of strains.

but

$$r_c = \frac{\sin \phi}{\cos(\phi - \alpha)}$$

$$\therefore \quad \varepsilon = \frac{r_c \cos \alpha}{\sin \phi \cos(\phi - \alpha)} \times \frac{\cos(\phi - \alpha)}{\sin \phi}$$

$$\therefore \quad \varepsilon = \frac{r_c \cos \alpha}{\sin^2 \phi}$$

Merchant's constant (C)

It seems reasonable to suppose that the angle at which the metal shears when forming a continuous-type chip is related in some manner to the cutting conditions and to the mechanical properties of the material cut.

By making a few assumptions about the relationships involved, Merchant deduced mathematically that the shear plane angle was dependent upon the rake angle, upon the angle of friction at the chip–tool interface and upon an angle (C) which is a constant for the material cut. Constant C represents in some manner the mechanical properties of the material at its point of failure in plastic shear. The relationship given by Merchant is:

$$2\theta + \tau - \gamma = C \tag{9.2}$$

Experimental results show that C is not strictly constant, and for additional reasons more recent researches tend to cast doubt on the validity of equation 9.2. Unfortunately there is no space here for a discussion of the assumptions made in deriving the equation. As however, C is fairly constant for particular materials over the normal range of cutting conditions which arise in practice, equation 9.2 has a certain utility. This is illustrated by Exercises – 9, numbers 14 and 15.

Exercises – 9

1. State the approximate pressure in N/mm^2 of chip cross-sectional area likely to be exerted on a single-point HSS tool when turning: (a) mild steel; (a) brass.

    Use the figures stated to determine the power at the motor necessary to turn these materials under the following conditions:
    (a) Mild steel; $V_c = 24$ m/min, depth of cut $= 6$ mm, feed $= 0.25$ mm/rev.
    (b) Brass; $V_c = 73$ m/min, depth of cut $= 3$ mm, feed $= 0.36$ mm/rev.
    The overall machine efficiency may be taken as 70% in each case.

2. Define orthogonal cutting and show the geometrical conditions for which it will occur. Why are these conditions frequently used in experimental work on cutting?

    Sketch the conditions of chip formation, and give brief notes upon them, for the production of continuous type chips:
    (a) Without built-up edge.
    (b) With built-up edge.
    Why is the first of these chip types the more desirable one for experimental work on cutting?

3. Show from the geometry of chip formation for a continuous type chip without built-up edge, that the shear plane angle θ is given by,

    $$\tan \theta = \frac{r_c \cos \gamma}{1 - r_c \sin \gamma}$$

    where r_c is the chip thickness ratio t_1/t_2.

4. A 100 mm diameter bar is turned by means of a tool for which $\phi = 90°$ and $\gamma_s = 15°$. The depth of cut is 5 mm and the feed is 0.25 mm/rev.

    If the mean length of a cut chip representing one rotation of the workpiece is 92 mm, find the shear plane angle.

5. You are required to test the relative machinability of two different alloy steels.
    Indicate the information which you would require concerning these steels and the particular tests you would carry out in order to reach your decision.

6. (a) Outline the causes of built-up edge and its effects upon:
       (i) tool life and the manner of tool failure;
       (ii) the finish of the surface being machined.
    (b) Describe two different methods by which continuous chips can be broken, for practical convenience, into short lengths.

7. Steel is being cut on a centre lathe under the following conditions: $V_c = 30$ m/min, $\gamma = 15°$, depth of cut $= 3.8$ mm, feed $= 0.25$ mm/rev. The cutting forces acting are:

    Vertical (cutting) load, 196 N;
    Axial (feed) load, 935 N;
    Radial load, 670 N.

    (a) Find the magnitude of the resultant cutting force, the power consumed in cutting and

    the K factor: $\left(\dfrac{w}{\text{power}}\right)$

    (b) Give the direction and approximate magnitude of the changes which would occur in the values obtained under (a):
       (i) if the cutting speed were increased to 46 m/min;
       (ii) if the rake angle were increased to 25°.

8. The following data relates to the rough turning of an alloy steel at a constant feed of 0.4 mm/rev, constant tool life 60 min.

Cutting speed (V_{60})	31	26	23	22	21	20.5	20
Depth of cut (d)	1.6	3.2	6.4	9.5	12.7	16	19
Power at spindle (kW)	0.67	1.1	2.0	2.9	3.9	4.8	5.7

    Use the values to obtain graphs which show the variation in metal-removal rate and in specific cutting pressure with respect to the depth of cut.

9. For certain cutting conditions the cutting force is related to the depth of cut and feed by the empirical equation:

    $$T = 2000 \, d^{0.82} f^{0.68}$$

    A lathe having 4 kW available at the spindle is used under these conditions to turn a 200 mm diameter bar of metal, depth of cut 7.6 mm, feed 0.5 mm/rev.
    Estimate the maximum spindle speed which may then be used.

10. (a) Explain, using diagrams, the general design of a dynamometer for measuring forces on a tool during the cutting process.
    (b) In a test using such a dynamometer the following data were obtained: depth of cut $= 7.5$ mm, cutting speed $= 30$ m/min, feed rate $= 0.12$ mm/rev, tangential load on tool $= 3.5$ kN.
       Determine the efficiency of the machine tool if the input to the driving motor was

3 kW and its efficiency 85 per cent. Find the specific cutting pressure and the rate of metal removal.

11. The given data relates to turning tests carried out on a medium nickel–chrome steel in a heat treated condition, 285 HB. UTS $= 980$ MW/m^2. The cutting speed is that for a tool life of 60 minutes; the power is that consumed in cutting.

Depth of cut, mm (d)	1.6	3.2	6.4	12.7	25.4
Feed, mm/rev (f)	1.6	0.8	0.4	0.2	0.1
Cutting speed m/min	5.5	7.3	10.7	16.2	25
Power (kW)	0.67	0.75	1.12	2.09	3.88

Use this information to plot graphs which show the influence of the chip proportions, (ratio d/f) upon the following:
(a) the metal removal rate,
(b) the power consumed per unit volume of metal removed per second.
(The ratio d/f should be plotted to a logarithmic scale.)
    Of what significance are the facts which may be deduced from these graphs?

12. During a metal-cutting test under orthogonal conditions a lathe knife tool, rake angle 20°, was used to machine the end of a steel tube of wall thickness 3.2 mm, at a feed of 0.38 mm/rev. The following data were obtained from the test:

Vertical (cutting) load, 2340 N;
Axial (thrust) load, 1000 N;
Average chip thickness, 0.9 mm.

Determine by graphical means, or by calculation:
(a) the coefficient of friction at the chip–tool interface;
(b) the angle of inclination of the shear plane;
(c) the shear stress on the shear plane.

13. The following data were obtained from a cutting test: $\gamma = 20°$, $k = 90°$, depth of cut 6.4 mm, feed 0.25 mm/rev, chip length before cutting 29.4 mm, chip length after cutting 12.9 mm. The cutting forces were: axial force 427 N, vertical force 1050 N.
    Use Merchant's analysis to calculate:
(a) the direction and magnitude of the resultant force;
(b) the shear plane angle;
(c) the frictional force;
(d) the friction angle.

14. (a) In orthogonal turning how can the shear plane angle and coefficient of friction along the rake-face of the tool be found?
(b) How would the shear plane angle be affected if the coefficient of friction is decreased?
(c) If Merchant's machinability constant $C = 70$ when the rake angle $\gamma = 20°$ and the friction angle $\tau = 40°$, what will be the magnitude of the shear plane angle?

15. Determine Merchant's constant C for aluminium from the following information: $\gamma = 35°$ $F_c = 200$ N; $F_a = 90$ N; $t_1 - 0.125$ mm; $t_2 = 0.25$ mm; cutting speed $= 30$ m/mm; width of cut $= 2.5$ mm.
    What amount of work per minute is done against friction at the chip–tool interface?

10 Cutting Tool Technology

10.1 Introduction

While a study of the geometry and mechanics of metal cutting is necessary in order to achieve efficiency, metal cutting is primarily an economic activity. When roughing, the aim is to remove a particular volume of metal in minimum time or at minimum cost; when finishing, the area of finished surface produced is the criterion.

Throughout the following discussion the economics of cutting are developed in relation to the volume of metal removed, but the same principles are applicable to finishing cuts, where the area of surface produced is the decisive feature. The economics can be based either upon time or cost; the former is important when maximum output from the available plant is required, the latter is the more common condition where production costs must be kept as low as possible.

10.2 Variables affecting metal-removal rate

Since the metal removal rate $w = df \times V_c$ mm^3/s, any increase in cutting speed, depth of cut or feed will give a directly proportional increase in the metal removal rate. The power available at the machine spindle is one factor limiting the metal removal rate.

If a further condition is imposed and the tool is required to cut effectively for a specified tool life, there is a limit placed upon the cutting speed for each particular combination of depth of cut and feed employed.

For any given set of cutting conditions there is a relationship between cutting speed and tool life which is a vital one for the economics of the process. This relationship may be represented by the empirical law $V_c T^n = C$, where C and n are constants. It must be understood, however, that the law is true only if other possible variables, depth of cut, feed, rake angle, plan approach angle, etc., are kept constant. Any change in these will have a marked effect upon the value of C. The law, introduced by F. W. Taylor as a result of his experimental work on the cutting of metals, is illustrated graphically in Figure 10.1.

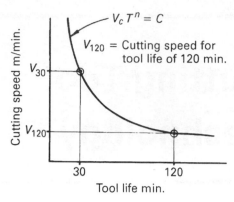

Figure 10.1 Cutting speed *vs* tool life.

Tool life

The tool life of a cutting edge is limited by the breakdown of the edge to the extent that it cannot perform the task it was chosen for. This may be surface texture, dimensional accuracy, ability to control chips satisfactorily or when tool wear has been generated to the extent that the edge should not be relied upon any longer. Ultimately, tool life ends with breakdown or fracture of the edge. But metal cutting correctly applied with modern tooling, need not reach such a drastic stage. *Tool life predictability* is an important factor, especially as most machining today is behind the closed doors of a machine and even unmanned.

The criteria vary, depending on whether a finishing or roughing operation is being carried out. When the end of the tool life is reached, the cutting edge is changed before any unacceptable parts are made or breakdown with damage occurs. It is important in this context that a cutting edge is worn out, not broken down.

Clear definitions need to be established for when cutting edges are to be classified as worn out. With different operations, criteria and tools, this needs some attention for decisions to be made at the machine or when used tools are inspected. Obviously, when an edge cannot generate the required finish or keep within a tolerance, it cannot be used for that particular operation any longer.

Metal cutting is, as already implied, a far from fully analysed science and some of what goes on between the chip and cutting edge is still unknown. Therefore experience, the best possible starting conditions and knowledgeable support are still the way in much of machining.

Tool wear varies and there are several different types and effects. A norm needs to be established, for instance: the height of flank wear below the cutting edge line or the depth of a crater on the tool face. As flank wear is a common wear pattern and one which is often aimed for, this will be used to develop the tool wear/time relationship. Procedures and test for tool life have been standardised internationally.

Tool wear does not necessarily follow a straight line in a wear (VB)/time (T) diagram (see Figure 10.2). The curve often has a typical development for flank wear, first a moderate growth and then rapid escalation. The form varies, especially with

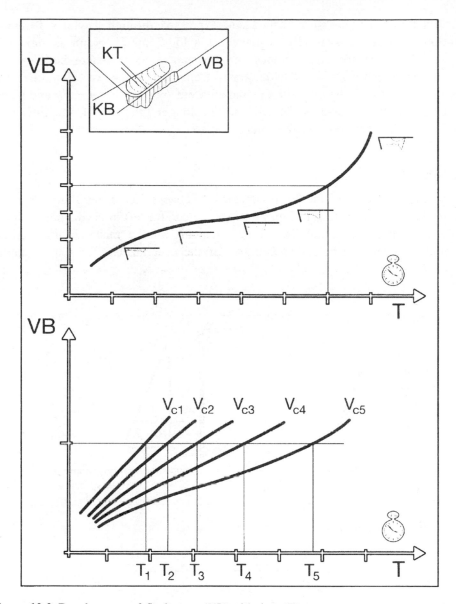

Figure 10.2 Development of flank wear (*VB*) with time (T).

the cutting speed and for each speed a specific wear development curve is plotted. A limit for maximum wear is set. The slower speed to the right has a slower growth while the higher is more of a linear relationship. Higher speed generally gives faster wear.

The cutting time (*T*), is the tool life of a cutting edge before a specific amount of wear is established.

The next step is to establish a direct relationship between time and cutting speed. This can be plotted on a logarithmic diagram for time (*T*) and speed (*V$_c$*).

This relationship is well-established and expressed through Taylor's formula, mentioned above. It is usually displayed as a $\log T/\log V_c$ graph as shown in Figure 10.3. This formula includes two constants (n, C) where both can be established graphically in the diagram from the slope of the line – the relationship between the opposite (x) and adjacent (y) lengths; $n = x/y$ and where the extended line intersects the X-axis. The longer the part of the line used graphically, the more reliable the value.

Cutting data

The position and slope of the line is affected by the workpiece material, tool material and feed rate. A higher feed rate may move the line to the left in the diagram (Figure 10.4). A cutting speed/tool life (V_c/T) diagram can be made up for various materials and can also be supplied as part of the material data for some material suppliers.

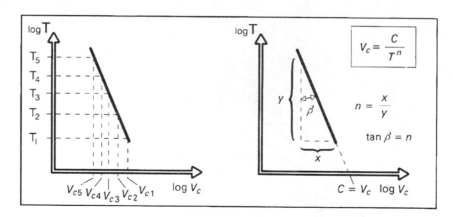

Figure 10.3 Time/cutting speed diagram.

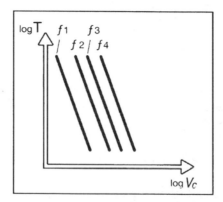

Figure 10.4 Influence of speed.

When tool life criterion and time have been established, the speed/tool life diagram will indicate a suitable cutting speed. The indicated speed should be regarded as a starting value for machining, to be more precisely set according to other factors in metal cutting.

The described procedure is based on tool and data values, including a specific feed rate. If the feed is altered, so will the position of the curves. A higher feed rate basically leads to shorter tool life for a certain cutting speed, but also leads to faster machining and higher removal rates. Several combinations of cutting speed and feed rate can be achieved to give the same tool life, of course. Again there are aspects to be taken into account for various feed levels, such as surface texture in finishing and cutting edge strength in roughing. For this reason, there should be cutting speed/tool life curves for various feed rates. It should also be kept in mind that the curves are related to the tool life criterion set in the wear/time diagram and that diagrams can be drawn up for various wear types. Cutting depth, nose radius, entering angle, etc. are of secondary importance due to their often negligible influence.

To begin to establish economical parameters for *rough machining*, to remove metal as efficiently as possible, feed rates can be raised until workpiece limits are reached, chip control is lost, tool breakdown is reached or until the machine is stalled. Then the tool life diagrams can be established through suitable tests at various cutting speeds and recording the number of components machined during the tool life(see Figure 10.5). Comparisons can then be made with the cutting speed/tool life formula and the optimum tool type and speed and feed can be chosen. Good machine tool utilisation and chip control should always be the aims in metal cutting.

The exponent n depends mainly upon the cutting tool material and has values, when rough turning, of approximately $\frac{1}{8}$ for HSS tools and $\frac{1}{5}$ for TC tools. The constant C is related both to the tool material and to the material cut; it is also affected by other variables of the cutting process, as indicated above, i.e. proportions of the cut, tool geometry and the employment of cutting fluids. Since $V_c T^n = C$, the constant C has the value of V_c for a tool life of $T = 1$ minute.

Values of n and C can be found only from cutting tests. The most comprehensive cutting test data available are those published in the American Society of Mechanical Engineers handbook and the Production Engineering Research Association data sheets.

Example 10.1

A tool life cutting test of HSS tool material, used to cut a special die steel of 363 HB, gave the following values:

Cutting speed,	49.74	49.23	48.67	45.76	42.58
Tool life, T min	2.94	3.90	4.77	9.87	28.27

Use the values to obtain the constants of the tool life equation, $V_c T^n = C$.

Solution

To obtain satisfactory results from the test data a straight-line graph should be drawn (Figure 10.6). This can be obtained by plotting the equation,

$$\log V_c + n \log T = \log C$$

$$\log V_c = -n \log T + \log C$$

$\log V_c$	1.6967	1.6924	1.6873	1.6605	1.6292
$\log T$	0.4683	0.5911	0.6785	0.9943	1.4513

Consider points A and B on the graph:

1. The slope $-n$ $= -0.07$,
2. The value of $\log C = 1.704 + 0.07\ (0.4)$
   $= 1.732$

hence

$$C = 54$$

Tool life equation is, $V_c T^{0.07} = 54$

$\square$ $\square$ $\square$

Example 10.2

When turning 19 mm diameter bar on an automatic lathe employing TC tools the value of n is $\frac{1}{5}$, and the value of V_{c60} is 104 m/min.

At what speed should the spindle run to give a tool life of 6 hours? If a length of 50 mm per component is machined and the feed used is 0.16 mm/rev, what is the cutting time per piece and how many pieces can be produced between tool changes?

Solution

$$V_c \times 360^{1/5} = 104 \times 60^{1/5}$$

$$V_c = 104 \left[\frac{60}{360} \right]^{0.2}$$

$$= 73 \text{ m/min}$$

$$\text{Spindle speed} = \frac{10^3 \times 73}{\pi \times 19} = 1220 \text{ rev/min}$$

$$\text{Cutting time per piece} = \frac{50 \times 60}{1220 \times 0.15} = 16.5\text{s}$$

$$\text{Number of components per tool change} = \frac{6 \times 60 \times 60}{16.5} = 1300$$

$\square$ $\square$ $\square$

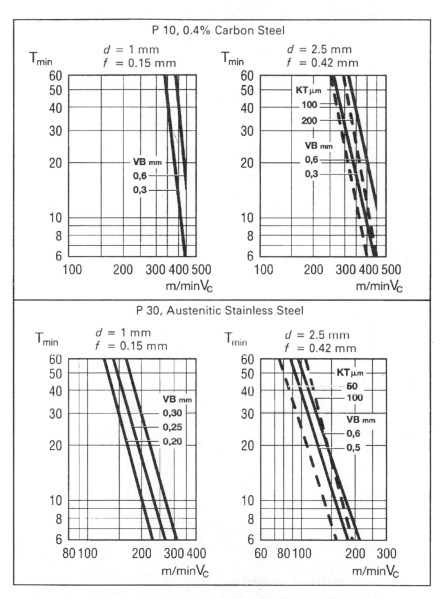

Figure 10.5 Tool life test results on medium-carbon steel and stainless steel.

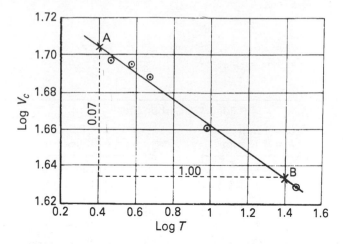

Figure 10.6 Log V against Log T.

Example 10.3

Find the percentage change in cutting speed required to give an 80% reduction in tool life (i.e. reduce tool life to $\frac{1}{5}$ of its former value) when the value of $n = 0.12$.

Solution

$$V_{c1}T_1^{0.12} = V_2 T_2^{0.12}$$

$$\frac{V_{c2}}{V_{c1}} = \left[\frac{T_1}{T_2}\right]^{0.12}$$

$$\frac{V_{c2}}{V_{c1}} = \left[\frac{1}{1/5}\right]^{0.12} = 5^{0.12}$$

$$\frac{V_{c2}}{V_{c1}} = 1.214$$

Increase in cutting speed = 21.4%.

☐   ☐   ☐

10.3 Economic cutting speed

An increase of cutting speed has two main effects upon the economics of cutting; the metal removal rate is increased, the tool life is decreased. An increase in the metal removal rate will lower the direct cost of metal removal; a reduction in the tool life will increase the costs of servicing and replacing worn out tools. The two separate effects, and their combined influence upon the total cost of machining, are best illustrated graphically, as shown in Figure 10.7. The following deductions can be made from the graphs:

1. As V_c increases, the time required to remove the metal (and hence the cost of its removal) will fall. The cost of cutting $\propto 1/V_c$.
2. As V_c increases, the tool life T falls. The costs of tooling $\propto 1/T$ but by Taylor's equation $1/T = [\,V_c/C\,]^{1/n}$ and since C is a constant, the costs of tooling $\propto V_c^{1/n}$.
3. The inclusive costs of machining will be the sum of the separate costs, and by addition of the separate cost values for each value of V_c shown, a third graph showing the change of the inclusive cost with respect to changes in cutting speed is obtained. This graph has a minimum value for inclusive cost, and the ideal (or optimum) cutting speed is the value of V_c at the minimum point.

Since tooling costs can be seen to depend upon the value of n, it would appear that this exponent is important in relation to the economics of metal cutting.

Example 7.4
Under certain machining conditions the tool life equation is $V_c T^{0.2} = 180$. The time taken to change a tool is 10 min. Show that operating at a cutting speed of 90 m/min gives higher output than operating at either 120 m/min or 60 m/min, other cutting conditions remaining constant.

Solution
One way of demonstrating the superiority of a cutting speed of 90 m/min is to determine the *average* cutting speed where the time spent in tool changing (during which no cutting occurs)

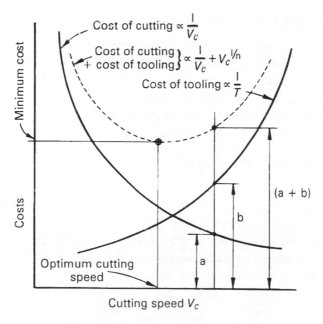

Figure 10.7 Economic cutting speed.

is taken into account. The output will be highest where the average cutting speed is highest because the remainder of the cutting conditions are constant.

$$\text{Tool life } T = \left[\frac{180}{V_c}\right]^5$$

When

$V_c = 60, \; T = 243 \text{ min}$

$V_c = 90, \; T = 32 \text{ min}$

$V_c = 120, \; T = 8 \text{ min}$

When

$V_c = 60$, average cutting speed $= \dfrac{243 \times 60}{243 + 10} = 58 \text{ m/min}$

$V_c = 90$, average cutting speed $= \dfrac{32 \times 90}{32 + 10} = 69 \text{ m/min}$

$V_c = 120$, average cutting speed $= \dfrac{8 \times 120}{8 + 10} = 53 \text{ m/min}$

On a time basis 90 m/min is the best of the three cutting speeds quoted. However, on a cost basis the servicing of the tool every 32 min might be too expensive, and on this basis a speed somewhere between 60 and 90 m/min might be more profitable.

□   □   □

Economic relationships of cutting

Let H = hourly cost of running the machine, i.e. operator's wage and overheads.

J = total cost per tool change; i.e. tool grinding, setting and eventual replacement costs.

P_m = cost of machining metal per unit volume.

P_t = cost of servicing tools per unit volume of metal cut.

P = total costs per unit volume of metal cut, i.e. $P_m + P_t$.

Time to machine a unit volume of metal is

$$\frac{1}{w} = \frac{1}{df \times 10^3 V_c} = \frac{k}{V_c}$$

where k is a constant. (Since we are only concerned with the influence of V_c upon the economics of cutting, d and t are regarded as constants.)

Cost of machining per unit volume of metal cut

$$P_m = \frac{Hk}{60V_c} \tag{10.1}$$

The number of tool changes in k/V_c min $= k/MV_c$, where T is the tool life in min, but from Taylor's equation $T = [C/V_c]^{1/n}$

Cost of tool servicing per unit volume of metal cut

$$P_t = \frac{Jk}{V_c}\left[\frac{V_c}{C}\right]^{1/n} = \frac{JkV_c^{(1-n/n)/n}}{C^{1/n}} \tag{10.2}$$

By addition of equations 10.1 and 10.2

$$P = \frac{Hk}{60V_c} + \frac{JkV_c^{(1-n)/n}}{C^{1/n}}$$

Differentiating and equating to zero

$$\frac{dP}{dV_c} = \frac{Hk}{60V_c^2} + \frac{1-n}{n}\frac{Jk}{C^{1/n}}V_c^{(1-2n)/n} = 0$$

$$\frac{H}{60V_c^2} = \frac{1-n}{n}\frac{J}{C^{1/n}}V_c^{(1-2n)/n} \text{ since } k \neq 0$$

$$\frac{H}{60} = \frac{1-n}{n}\frac{J}{C^{1/n}}V_c^{(1-2n)/n}V_c^2$$

$$= \frac{1-n}{n}J\left[\frac{V_c}{C}\right]^{1/n}, \text{ but } \left[\frac{V_c}{C}\right]^{1/n} = \frac{1}{T}$$

Hence

$$\frac{H}{60}\frac{T}{J} = \frac{1-n}{n} \tag{10.3}$$

represents the economic cutting conditions.

Since $(H/60) \times T$ is the cost of operating the machine between tool changes (tool life of T min), and J is the total cost per tool change, the equation can be written:

$$\frac{\text{Cost of operating between tool changes}}{\text{Total cost of a total change}} = \frac{1-n}{n} \tag{10.4}$$

The expression $(1-n)/n$ shows the significance of the exponent n of Taylor's equation in relation to the economics of cutting. The expression is sometimes called the *costs ratio* for economic cutting.

The relationship proved above has been obtained on the assumption that depth of cut (d) and feed (f) are constants. Since, however, the ratio of the costs depends only upon the value of n, and n is not significantly changed by normal variations of d and f, the costs ratio relationship is valid regardless of the particular values of d and f being used.

As is evident from the equation for economic conditions, the tool life T is a most significant factor because the substitution, $1/T = [V_c/C]^{1/n}$, has been used to eliminate the constant C of Taylor's equation. The variables d and f have a marked influence upon the value of C, and it should be clear that for the economic value of the tool life, V_c will depend upon the particular values of d and f being used. Reference to Table 10.1 (p. 306) will confirm this. To summarise; the economic value of T (*economic tool life*) depends only upon the particular value of n relating to the cutting process and upon the operating and tooling costs involved. The appropriate cutting speed required to give this economic tool life depends upon the material of the tool and workpiece, upon the tool geometry and also upon the depth of cut and feed employed.

Example 10.5

For certain machining conditions the total cost of operating the machine is £6 per hour and the total cost of a tool change is £3.60. If, for the depth of cut and feed employed, $V_{60} = 36$ m/min and $n = 0.14$, find the economic cutting speed.

Solution

$$\frac{6}{60} T \times \frac{1}{3.6} = \frac{0.86}{0.14} \text{ from equation 10.3}$$

$T = 221$ min (economic tool life)

Since

$$V_{c_{60}} = 36 \text{ m/min for the cut employed}$$

$$V_{c_{221}} \times 221^{0.14} = 36 \times 60^{0.14}$$

$$V_{c_{221}} = 36 \left[\frac{60}{221}\right]^{0.14} = 30 \text{ m/min}$$

□   □   □

When maximum output rather than minimum cost is required, the economic relationships have the form

$$\frac{\text{Operating time between tool changes } (T)}{\text{Time to change tool}} = \frac{1 - n}{n} \tag{10.5}$$

This relationship can be proved from first principles in the manner of equation 10.4.

Attention should be given to the significance of n in both equation 10.4 and equation 10.5. For HSS tools n is approximately $\frac{1}{8}$ and the cost (or time) ratio for economic conditions is 7; for TC tools n is approximately $\frac{1}{5}$ and the ratio is 4. These easily remembered figures provide some guidance for judging machining practice in workshops.

When rough turning with HSS tools an operator can regrind and reset a tool in under 3 minutes. The economic tool life is then around 20 minutes. Very few operators run their machines at cutting speeds high enough to achieve this condition. (In some machine shops the rate at which they would consume cutting tools might be questioned if they did so!)

The use of cemented carbide in the form of 'throw-away' tips is now virtually standard practice. A square tip, with suitable clamping to the tool shank, may be indexed round the four edges of both faces as wear occurs. While the limited range of plan shape is a restriction, this arrangement greatly reduces tool servicing costs, leading to an economic tool life at relatively higher metal removal rate than for more conventional tools giving an economic tool life of around 15 minutes.

Servicing costs for the tools rise as tooling becomes more complex, and the economic cutting conditions will then occur at reduced cutting speeds. Given reasonably accurate cost figures, it is easy to check if these are in the ratio of $(1 - n)/n$, as is required for economic cutting conditions.

10.4 Cutting tool materials

Correct selection of a tool material is of course, a key issue for economic machining. Machine tool downtime due to broken and worn tools is one of the main limitations to productivity and the selection of tool material and, subsequently, the right grade is critical. Not one tool material can meet all demands of machining although some grades have broad application ranges that cover many operations.

Figure 10.8 shows a selection of tool materials and Figure 10.9 a selection of carbide turning inserts.

High speed steel is around twice as tough as cemented carbide, which in turn is about three times as tough as ceramic. Polycrystalline diamond is very brittle, but very hard. Ceramic is somewhat harder than cemented carbide and has better thermal and chemical stability than carbide. The mix of tool material properties is considerable and to optimise operations with the right choice needs a basic knowledge of the materials and analysis of the previously mentioned factors before the right direction can be taken. The ideal tool material can be said to:

- Be hard, to resist flank wear and deformation.
- Have high toughness, to resist fracture.
- Be chemically inert to the workpiece.
- Be chemically stable, to resist oxidation and dissolution.
- Have good resistance to thermal shocks.

Sources to aid the correct selection of tool materials (e.g. Figure 10.10) are multiple and can be combined for the best result: standards and classifications are excellent in pointing the way and comparing various materials and grades; tool supplier data indicates what is available and for what application and cutting data; the qualified support by a specialist, trained and experienced, will go a long way towards optimising the operation and, then, to build one's own experience from operations performed; tests will also provide the basis for continued high performance.

Polycrystalline diamond

Coated cemented carbide

Cubic boron nitride

Coated cemented carbide

Pure-ceramic

Uncoated cemented carbide

Mixed ceramic

Coronite

Silicon nitride base ceramic

Cermet

Figure 10.8 Tool materials.

High-speed steel

Tungsten is generally the major alloying element, as in the 18–4–1 type; 18 per cent tungsten, 4 per cent chromium, 1 per cent vanadium. There are, however, HSS alloys in which a lower tungsten content is used, supplemented by larger additions of other alloying elements, generally molybdenum, cobalt and chromium.

Figure 10.9 A selection of carbide turning inserts.

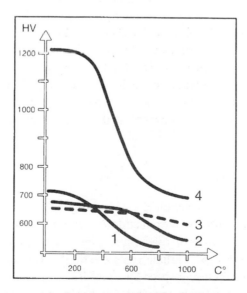

Figure 10.10 Hot hardness of various tool materials. One of the primary reasons for the improvement in performance was due to the better hot hardness of the material. (1) Carbon steel; (2) HSS; (3) cast alloys and (4) WC cemented carbide.

Special (or super) grade HSS

Tungsten may be raised to 20 per cent or more, or other alloying additions made to improve the cutting properties. Better hot-hardness and wear-resistance properties are then obtained.

Cast non-ferrous alloys

These are generally cobalt based, with alloying additions of chromium, tungsten and carbon; a typical analysis is 45 per cent cobalt, 35 per cent chromium, 18 per cent tungsten and 2 per cent carbon. They do not contain iron except as an impurity. The metal cannot be forged or readily cut to shape when cold; it must be cast to the required shape and then dressed by grinding.

Cemented carbide

As the name implies, cemented carbide is a tool-material made up of hard carbide particles, cemented together by a binder. It has an advantageous combination of properties for metal cutting and along with high-speed steel, has dominated metal cutting performed at higher cutting speeds. Development during the past 60 years has been intense with considerable improvements in the productivity achieved from a cutting edge. Coated cemented carbides (described separately) have taken over to the extent that uncoated grades are not among the first recommendations any more. Uncoated cemented carbides are now for aluminium, special purpose and back-up. A basic introduction to this, the step that has so influenced metal cutting, is useful.

Cemented carbide is a powder metallurgical product, made primarily from a number of different carbides in a binder. These carbides are very hard and those of tungsten carbide (WC), titanium carbide (TiC), tantalum carbide (TaC), niobium carbide (NbC) are the main ones. The binder is mostly cobalt (Co). In addition, however, the carbides are soluble in each other and can form a cemented carbide without a separate metal binder. The hard particles vary in size, between 1–10 microns and usually make up between 60 to 95 per cent in volume share of the material.

Coated cemented carbides

One of the big steps in the development of cutting tool materials was taken towards the end of the 1960s with the introduction of cemented carbides with a very thin coating of carbides. The layer of titanium carbides was only a few microns thick but changed the performance of carbide tools overnight. By changing to a CC insert from an uncoated insert, cutting speed and/tool life was dramatically increased. The effect of the coating continues long after it has partly worn off, resulting in the reduction of crater wear when machining steel. Higher temperatures could be tolerated and thus higher speeds and feeds.

Coated carbides (CC) succeeded in breaking the Achilles' heel of cemented carbide, and of cutting tool materials generally: namely, that wear resistance decreases as toughness increases (Figure 10.11), keeping the ideal combination out of reach, forcing users to move along a series of compromise grades. Coated carbides provided a cutting tool material much closer to the ideal, with new combinations being formed. Since then, the range has been moved and extended continually with new generations of coated carbide grades. More than 75 per cent of turning operations and more than 40 per cent of milling are today performed with coated carbides.

Today, nearly all first choice cemented carbide grades for turning are coated. These predominate in turning operations everywhere, representing three-quarters of indexable insert consumption and performance and reliability have been dramatically improved since their introduction. In recent years, coated grades have been developed and found wide acceptance in drilling and milling tools, in cast iron and steel machining. The main coating materials are titanium carbide (TiC), titanium nitride (TiN), aluminium oxide–ceramic (Al_2O_3) and titanium carbonitride (TiCN). Titanium carbide and aluminium oxide are very hard materials, providing wear resistance, and are chemically inert, providing a chemical and heat barrier between tool and chip. TiN is not such a hard material but gives a lower coefficient of friction to the faces of the insert and better cratering resistance.

Modern coating technology has been developed to handle the higher temperatures needed for all types of coatings and for giving the insert the desired properties (Figure 10.12). Various combinations of multiple coatings are continuously being developed to incorporate the best properties from the coating materials. Combinations have been established to provide grades with broad application ranges through having high wear resistance in many respects and to maintain hot hardness and to avoid affinity with workpiece materials.

Hand-in-hand with coating developments have gone the insert substrate and manufacturing process developments. Whereas the first coated grade was a cemented carbide grade (having sufficient toughness) with a layer of titanium carbide

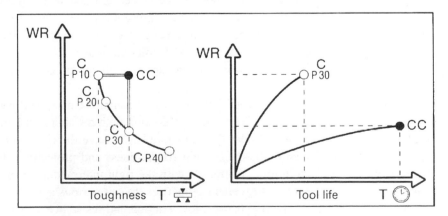

Figure 10.11 Wear resistance against toughness and tool life for coated cemented carbides.

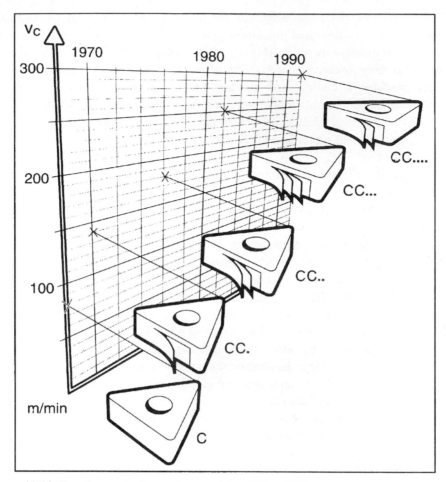

Figure 10.12 Development of coated cemented carbides (CC).

metallurgically bonded to it, today's coated grades have substrates and processes finely adapted to meet the intended properties of the coated grade and to eliminate possible hazards. For instance, free carbon being present in the substrate and especially the surface part, next to the coating, has a negative effect on the properties of the cutting edge. Also, the existence of a brittle composition of cobalt–tungsten–carbon, called an eta-phase, which means not enough carbon content, will be negative for the tool life of the insert.

Titanium carbonitride, although a good bonding layer, contains carbon and needs a heat barrier. For this reason, it makes a good combination with a stable heat barrier such as aluminium oxide, on the outside. The substrate has to have the right amount of ductility and be complemented by the coating for hardness and protection. Both aluminium oxide and titanium nitride, combined with the right intermediate coating, form outer layers with different, but excellent properties, to suit various applications: wear resistance, chemical and thermal barrier, low friction and resistance to built-up edge formation, where applicable.

Coated cemented carbides are first choice grades for a majority of turning, milling and drilling operations in most materials. The grades have very broad application, covering such areas as P05 to P40, M10 to M25 and K01 to K25.

Manufacture of coated cemented carbides

The improvements of the bonding between different coating layers and the insert substrates have led to new generations of coated cemented carbides. These have single, double, triple and even more layers to combine the various properties that each type of coating material has to offer. Thicknesses of coatings on indexable inserts vary between 2 to 12 microns (the average human hair has a diameter of 75 microns). Coated cemented carbides are manufactured by putting layers on inserts, mainly through the chemical vapour deposition (CVD) technique (Figure 10.13). The combination of optimised substrate composition and the developed CVD process combine to manufacture today's generation of coated cemented carbides for turning, milling and drilling.

Basically, CVD coating is done through the chemical reactions of different gases. In the case of coating with titanium carbide: hydrogen, titanium chloride and methane. Inserts are heated to about 1000 °C. Like sintering (Figure 10.14) this is a carefully controlled process where the carbon content, either free or as eta-phase, has to be

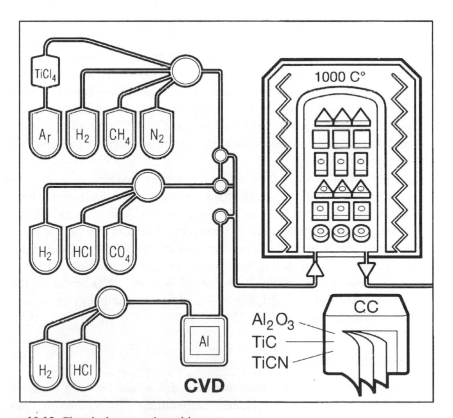

Figure 10.13 Chemical vapour deposition.

Figure 10.14 Sintering.

monitored through an extra carburisation stage, before coating. Aluminium oxide coating is carried out in a similar way as is titanium nitride coating using other gases, aluminium chloride or nitrogen gas, respectively. The CVD process is well adapted to apply multi-layered coatings as the process is relatively easy to regulate as regards various gases. Different types of coating can be performed in the same equipment.

The CVD process today is a mature, automated process that is widely used since almost any insert substrate can be coated, the coating is uniform and homogeneous, and the adhesion between coating and substrate is excellent. Aluminium oxide can be coated on to a tough substrate, providing coated inserts with very high performance and reliability.

A complementary coating process, used to a lesser extent for cemented carbide, is the physical vapour deposition (PVD) method (Figure 10.15). Used widely for high speed steel coating, it can to some extent also be used for coating cemented carbide, although the process needs to be carefully suited to the insert material. Temperatures of about half those used in the CVD process are used (500 °C). The PVD process is good for coating contoured and very sharp tools, such as endmills and drills and to some extent, threading inserts.

The process is based on the principle that the coating material is moved from a material source to the substrate through either vaporisation or sputtering. There are several variants of these processes as they are used widely by high-speed steel suppliers. The PVD process takes place in temperatures around 500 °C. For instance, titanium is ionised with a focused electron beam as energy source, to form a plasma stream. Along with nitrogen this is then coated on the insert. Normally, a PVD coating is thinner than a comparable CVD coating. With the CVD process, a thicker coating means improved wear resistance, especially with aluminium oxide, up to a thickness of twelve microns.

Cermets – cemented carbide

Cermet is the collective name for cemented carbides where the hard particles are based on titanium carbide (TiC), titanium carbonitride (TiCN) and/or titanium nitride (TiN) rather than tungsten carbide (WC). The name comes from CERamic/

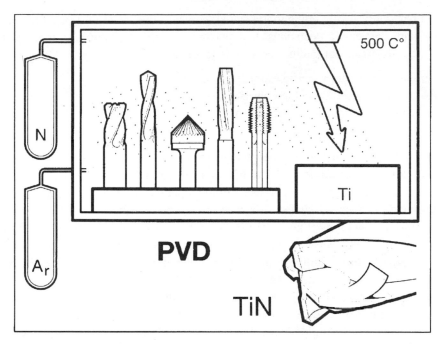

Figure 10.15 Physical vapour deposition.

METal, as ceramic particles in a metal binder. It can be argued that cermets, as a powder metallurgical product, are cemented carbides and that all hard metals are cermets but, in practice, cermets have come to denote cemented carbides based on titanium rather than tungsten carbide.

Cermets are not new, in spite of their growing use in recent years. This is due more to the improvement of the cermet grades, improved stability, speeds and conditions as well as the trend towards near-to-finish workpiece blanks. Titanium based grades were made in 1929 and have undergone considerable development. At first they were difficult to make and were very brittle. With molybdenum (Mo) added to form molybdenum carbide (Mo_2C) and improved manufacturing techniques, reasonable ISO, P01 grades were introduced. Larger amounts of titanium nitride and binder also led to better toughness. (See Figure 10.16.)

Figure 10.16 Cermet milling insert.

From being a relatively brittle tool material, cermets have developed better toughness to cope with demanding operations. It is not only a light steel finishing material but, today, there are grades for milling and stainless turning, etc. Cermets have:

- High flank and crater wear resistance.
- High chemical stability and hot hardness.
- Low tendency for built-up edges to form.
- Low tendency for oxidation wear.

Cermet grades today span application ranges from P01 to P20, M05 to M15 and K01 to K10 in turning. For milling, P01 to P30 and M01 to M25. This means application ranges which include general-purpose milling operations also in tough materials. Improvements in thermal shock resistance means that cermets can be suitable for certain milling operations (Figure 10.17). With their slow rate of wear development, they have long tool lives during which they provide high accuracy and surface texture. Cermets are undergoing more intense development leading to even better grades for the application areas they are most suited for.

Cermets are often advantageous for high cutting speeds, combined with lower feeds and cutting depth and when accuracy and finish are criteria for the operation. Machining conditions should ideally be stable and without severe interrupted cuts. A sharp long-lasting cutting edge is often advantageous for high-volume machining of components when the machining allowance is kept within limits.

Figure 10.17 Cermet milling of die steel.

Compared to tungsten-based cemented carbides and coated carbides, cermets (Figure 10.18) have the following properties:

- About the same edge strength at smaller constant loads.
- Better, longer-lasting ability to produce high finishes.
- Better capacity for high cutting speeds.
- Higher notch wear resistance from oxidation on trailing edge.
- Better ability to produce good finish in ductile and smearing materials with lower built-up edge (BUE) formation.

These are very finish oriented and, to some extent, semi-finish oriented properties, and explain the suitability of cermets in a selective number of applications. Moving on to the more demanding semi-finish and roughing properties of a cutting tool, cermets have:

- Inferior strength at lower and higher feed rates.
- Inferior toughness at varying, medium to heavy loads.
- Inferior abrasive wear (AW) resistance.
- Inferior notch wear resistance on leading edge due to mechanical wear.
- Inferior strength for shock loads.

The complete comparison illustrates why tungsten cemented carbide and coated carbides have dominated the application area and will continue to do so. Cermets are finding applications within their property parameters and, in line with the continued development of near-to-finish blanks, for metal cutting. The adjoining graph illustrates three typical tool life (T) curves, plotted for cutting speed and feed and shows how for a suitable application, a breaking point occurs between coated carbide and cermets, while the uncoated tungsten-based carbide maintains a position under the two (Figure 10.19).

Demanding profiling operations are not suitable applications for cermets. Cermets are advantageous for lighter copying of smaller, well-established working allowance with moderate to high speeds, medium range feeds and where tool life/finish is the criterion in favourable conditions.

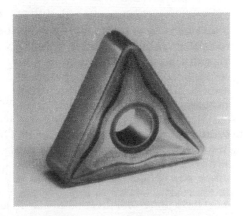

Figure 10.18 Cermet finishing insert.

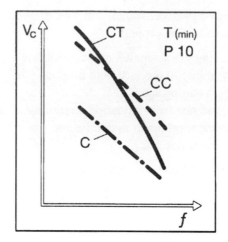

Figure 10.19 V_c against f.

For milling, cermets have a finishing to general purpose role for a wide range of materials. High cutting speeds with moderate feeds and large but even cutting depths can be used but abrasive wear, from casting skin, should be avoided. The higher toughness of cermet grades for milling, means that also stainless steel machining, including austenitic, as well as quite hard steels are suitable operations.

Compared to the wide application possibilities of coated, tungsten-based cemented carbides, cermets uphold a minority area. They remain a good choice for certain finishing operations, involving especially smearing materials and close-limit tool life criterion where the higher hardness at elevated temperatures and lower tendency for diffusion are best utilised. Cermets can, however, be an interesting problem-solver and should be looked upon as an alternative source to improve productivity – in selective operations (Figure 10.20).

Ceramics

Today, ceramics is the collective name for a range of different cutting tool materials (Figure 10.21). In recent times, ceramic cutting tools were first used at the beginning of the twentieth century, along with high-speed steel. These first cemented oxide tools were very brittle leading to grossly inconsistent tool life, due to a mixture of poor manufacturing quality and wrong application.

Ceramics have undergone considerable development and the inserts of today are not comparable to early ones. Also machinery and methods of application have changed to better accommodate the excellent productivity that can be offered by ceramics. However, this versatile material still only represents a few per cent of the cutting tool materials used, being applied mostly to machining cast iron, hard steels and heat resistant alloys.

Ceramic cutting tools are hard, with high hot hardness, and do not react with the workpiece materials. They have long tool lives and can machine at high cutting speeds. Very high metal removal rates are achieved in the right application.

Some of the main property differences between non-metallic ceramics when

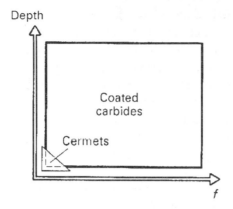

Figure 10.20 Depth against f.

compared to steel are: a density of around a third of steel and very high compressive strength, in relation to its tensile strength, whereas steel is more balanced; no plastic elongation such as that of steel and extremely brittle in comparison; the modulus of elasticity for pure ceramic is almost twice that of steel; ceramics have very low thermal conductivity while steel has high conductivity.

There are two basic types of ceramics:

1. Aluminium oxide-based (Al_2O_3).
2. Silicon nitride-based (Si_3N_4).

Aluminium oxide-based ceramics

Aluminium-oxide (alumina)-based ceramics are divided into:

1. Pure.
2. Mixed.
3. Re-inforced.

Figure 10.21 Ceramics.

The *pure* oxide-based ceramic has relatively low strength, toughness values as well as low thermal conductivity. These are obviously not the best values to have in metal cutting and cause cutting edge fracture if conditions are not right. The addition of small amounts of zirconium oxide to the composition significantly improve the properties of the pure ceramic. The mechanism that the zirconia grades offer means improved toughness. Durability, density and uniformity of grain sizes are important factors as are the various amounts of added zirconia to suit the application area. Any porosity will deteriorate tool performance. The pure ceramic is white if it is manufactured through cold pressing and grey if hot pressed.

The *mixed*, aluminium oxide-based ceramic material has better thermal shock resistance through the addition of a metal phase. This type is less sensitive to cracking through improved thermal conductivity. The improvement is relative and toughness achieved cannot be compared to that of cemented carbides. The metal phase consists of titanium carbide and titanium nitride amounting to 20–40 per cent in content. Other additives do occur and they are produced through hot pressing, leaving inserts black in colour and with a much wider application range to include most of the operations and materials for which ceramics are suitable.

The *re-inforced* ceramic, based on aluminium oxide is a relatively new development. This type is also called whisker re-inforced ceramic, from single crystal fibre called a whisker. These whiskers are only about one micron in diameter with a length of more than twenty microns. They are thus very strong and made of silicon carbide. The effect of this reinforcement is dramatic. The toughness, strength and thermal shock resistance are considerably increased and the grades have high hot hardness and wear resistance. This new tool material is undergoing considerable development as the mechanisms, such as the hindering of crack propagation, in the material are enhanced and put to use. The whiskers make up some 30 per cent of the contents.

Manufactured through hot pressing, which distributes the whiskers advantageously, the inserts are green in colour. The balanced hardness, toughness and thermal shock resistance, mean that these ceramic grades can take on more strength demanding applications involving heat resistant alloys, hardened steel and cast iron as well as interrupted cuts.

Silicon nitride-based ceramics

The silicon nitride-based ceramic is a completely different material and is better than aluminium oxide-based ceramics in standing up to thermal shocks and for toughness. It is the first choice for machining grey cast iron with very high removal rates. Grey cast iron is relatively easy to machine, but does make demands on the tool material when being machined at high removal rates and speeds when high hot hardness, strength, toughness and resistance to thermal shock as well as chemical stability are required.

The silicon nitride-based ceramic is excellent at maintaining hot hardness at temperatures higher than those suitable for cemented carbide, and is tougher than aluminium oxide-based ceramics. Although it does not have the chemical stability of the aluminium oxide-based ceramics when machining steel, it is excellent for machining grey cast iron in dry and wet conditions and at cutting speeds over 450 m/min.

The silicon nitride ceramic material is a two-phase material where the longer silicon nitride crystals lie in a binder and where properties are determined by composition. Production is through cold pressing and sintering or, more advantageously, through hot pressing and grinding into shape. It is also a relatively new material, developed since 1970, and with applications also as a construction material for high-performance components in engines etc.

The main application areas for ceramics are: grey cast iron, heat resistant alloys, hardened steels, nodular cast iron and, to some extent, steel.

For roughing and finishing machining of *grey cast iron*, pure aluminium oxide ceramics offer good performance, especially when there are no casting inclusions or skin. The silicon nitride ceramics stand up well to intermittent cuts and when depths of cut vary. The mixed, aluminium oxide grades are good for finishing when surface texture is the tool life criterion. This is because of the better notch wear resistance, which directly affects the finish.

For *heat resistant alloys*, the development of ceramics has meant considerably improved performance. Having been previously machined by uncoated cemented carbides, ceramics cope with these materials by machining at much higher cutting speeds and lasting several times longer. High strength at high temperatures and great notch wear generators, the heat resistant alloys, such as the nickel-based alloys, are advantageously machined by the mixed and reinforced aluminium oxide-based ceramics. The correct application method is vital because of the depth-of-cut localised notch-wear tendency having to be distributed over the edge.

Hardened steel and *chilled cast iron* are areas of hard part turning being improved by ceramic machining. The mixed and reinforced aluminium oxide-based grades, as well as the silicon nitride-based, are good for these applications in being able to stand up to the demands for thermal shock resistance and hot hardness. Operations vary considerably with the workpieces whether in the form of hard rolls, with various surface conditions, chilled cast iron and welded or sprayed steel components.

Turning hardened steel parts has been very successful in replacing grinding operations. The wear resistance and chemical stability of modern ceramic grades have provided improved productivity in this area. For chilled cast iron, the ability to stand up to abrasive wear resistance is vital because of the hard carbides.

General steel machining is dominated by coated cemented carbides because toughness is an important property, especially at the lower speeds encountered generally. The toughest of the ceramic grades, the silicon nitride-based ceramics, are, unfortunately, not stable chemically in ferrous machining (with the exception of grey cast iron) and generally, the other grades lack sufficient toughness to find wide application possibilities. The reinforced ceramics, also, are too low in chemical stability in steel machining.

The successful application of ceramics depends on the match between the operation types, machining conditions, workpiece material, machine tool performance, general stability, the method by which machining is performed and the cutting edge preparation, especially as regards strengthening chamfers, and presentation to the cut.

Coronite

Coronite is a new cutting material combining the toughness of high-speed steel with the wear resistance of cemented carbide (Figure 10.22). These properties allow Coronite endmills to machine faster than other similar tools in this area, with further improvements in tool life, reliability and surface finish. It is a cutting tool material mainly developed for steel machining, but performs well also in titanium and various light alloys.

Coronite fills an application gap which has existed between cemented carbide and high-speed steel, as well as having the ability to lift the productivity of tools that are typically made in high-speed steel.

Today's range of cutting tools, mostly limited to endmills, are suited for operations involving rough to finish machining and for cutting grooves, pockets and for profiling. They form a new high-performance alternative to high-speed steel tools and are suitable for most workpiece materials.

The key to the new material properties of Coronite lie in the grain size and with an advanced technique for producing and handling extremely small titanium nitride grains (TiN), as small as 0.1 micron. (This can be compared with cemented carbide or high-speed steel where the grain size varies between 1 and 10 microns.) Using special technology, the small TiN grains can be evenly dispersed in a heat treatable steel matrix to form between 35 per cent and 60 per cent of the material's volume. As a result the proportion of hard grains is higher than is possible to produce in high-speed steel, but less than the lowest limit for cemented carbide.

Figure 10.22 Coronite cutter.

This new group of materials, with extremely small grained hard constituents, is a hard material. It has properties which are more closely related to traditional cemented carbide than to high-speed steel, even if the actual production method can be said to be a form of particle metal technology. This is the first material of its kind, containing 50 per cent hard materials.

The properties of Coronite are found within a wide range, between those of cemented carbide and high-speed steel. The properties can be varied by changing the content of alloy material, especially the carbon content, and/or by means of heat treatment. However, the phase transformation diagrams which should apply to the steel matrix are not to be found because of the high content of fine grained TiN, which gives a very large phase boundary area and short mean free path length in the binder.

These advantages of Coronite are developed from the extremely fine grained hard material used to make it. But why are the small grain sizes so much better? Among the reasons are that it is easier to grind a sharp edge, the edge is more wear resistant while it wears, it keeps its sharpness even while being worn. It is self-sharpening, which neither high-speed steel nor cemented carbide is. Also, the finer grain size means that the machined surface integrity is much finer.

The combination of small grains and the high proportion of them means that the grains, and thereby the wear resistance, exist throughout the material with no weak points. There is no cemented carbide which contains as many hard particles as Coronite. One grain in high-speed steel corresponds to more than 1000 grains in Coronite and in addition, Coronite has more than twice the volume share of hard materials as high-speed steel.

TiN, which is the dominating hard material in Coronite is chemically very stable. This means that the crater wear resistance is high and the tendency to smear is low. It also means that a Coronite edge, and thereby the workpiece being machined results in good surface texture.

The manufacturing process uses advanced compound and coating techniques. That is to say the endmills, with the exception of the ballnose endmills, are not made of solid Coronite. Instead they consist of three parts:

1. A steel core.
2. A layer of Coronite which represents about 15 per cent of the diameter.
3. An outer PVD coating of TiCN or TiN, approximately two microns thick.

The core of steel, which in the drilling endmill consists of high-speed steel and in the non-drilling endmills of spring steel, means that the endmill is relatively tough (Figure 10.23).

The application of a PVD coating of TiCN or TiN to a cutting tool is today a relatively conventional method, but Coronite has been developed, with the intention of being coated. Since Coronite contains as much as 50 per cent TiN, an unusually strong bond is achieved between the titanium nitride in the substrate and in the coating, which reduces the risk of flaking. Contributing to this is the fact that titanium nitride consists of grains which are evenly distributed. Furthermore, the

Figure 10.23 Coronite blank and end mill.

thermal coefficients of expansion for the coating and substrate are approximately the same.

The layer of TiCN gives increased wear resistance on the clearance side of the cutting edge: high flank wear resistance. Coronite, from the material point of view, has a high resistance to crater wear. For this reason, re-grinding the Coronite rake surfaces, produces a new tool, without needing to be coated again.

The manufacture of Coronite is carried out through a unique process: the basic powder is produced through the addition of nitrogen in a double-chambered furnace. Carried out at relatively low temperatures, the powder does not melt.

The core of high-speed steel or spring steel, depending upon the type of tool, is covered with Coronite powder whereupon it is pressed to a brittle but unified body. This is compacted to the desired density through hot extrusion at a temperature well above one thousand degrees to maintain material stability. The bar thus produced is the raw material for Coronite cutting tools. After manufacture, it is coated with titanium carbonitride or titanium nitride.

The properties of Coronite in relation to high-speed steel and even cemented carbide for its application area are advantageous: with toughness and bending strength values similar to that of high-speed steel and better than cemented carbide, stiffness as represented by the modulus of elasticity being less than the high value of cemented carbide but better than high-speed steel, hot hardness is considerably better than high-speed steel, smearing of workpiece material on the cutting edge is relatively low compared to both, crater and flank wear resistance is better than high-speed steel and the ability to produce improved surface texture is better than both. Balanced wear and the ability to keep edge sharpness are better.

Cubic boron nitride

CBN – cubic boron nitride – is one of the really hard cutting tool materials, second only to diamond. It is an excellent cutting tool material in that it combines extreme hardness, high hot hardness up to very high temperatures (2000 °C), excellent abrasive wear resistance and generally good chemical stability during machining. It is a relatively brittle cutting tool material but is tougher than ceramics.

Compared to ceramics generally, it is harder but does not have as good thermal and chemical resistance. It is also a relatively young material, introduced during the 1950s but more widely during the 1970s. In spite of its high cost, it has found wide application in turning hard components that were previously ground.

Steel forgings, hardened steel and cast iron, surface hardened components, cobalt- and iron-based powder metals, forming rolls pearlitic cast iron and heat resistant alloys are among the main types of applications for CBN.

Applications in this area should always be analysed to reach a basis for whether a ceramic or CBN grade is the most suitable for optimum results and economics – the application areas for these two cutting tool materials are interlinked.

CBN is manufactured through high temperature and pressure to bond the cubic boron crystals together with a ceramic or metal binder. The randomly oriented particles form a very dense polycrystalline structure. The actual CBN crystal is similar to that of synthetic diamond. The properties of the CBN cutting tool material can be varied by altering the crystal size, content and the type of binder to make various grades. *Low content* CBN, in combination with a ceramic binder, has better wear resistance and chemical stability, and are more suited to hard steel components, but also in cast iron. *Higher content* CBN, which has more toughness, is more suitable for hard cast iron and steel as well as heat resistant alloys.

When the CBN cutting tool material is made with a ceramic binder, better chemical stability and wear resistance, but somewhat poorer toughness, is achieved. By bonding the CBN material onto a cemented carbide substrate, a tough, shock resistant support is created for the relatively brittle cutting edge. Titanium nitride is also added to the composition.

CBN should be applied to hard workpiece materials, over 48 HRC. If components are too soft, excessive tool wear is generated – the harder the material, the less the tool wears. Excellent surface textures can be achieved with CBN edges, making turning a very attractive alternative to grinding.

Cutting forces tend to be high, partly because of the negative cutting geometry used with many CBN tools and partly because of the demanding workpiece materials and high friction during machining. General high stability and machine power are crucial elements.

Tool and machine rigidity are essential and a large enough tool radius is also important. Interrupted cuts should be carefully assessed to ensure that tool and set-up are the most suitable (Figure 10.24).

Edge preparation with strengthening chamfers and correct tool application are essential. The grains in CBN are very small and hard and to avoid microchipping, the edge must have suitable chamfers and honing characteristics for the type of operation and material it is to machine. Correctly applied, CBN inserts will provide extremely good wear resistance for a hard, sharp cutting edge.

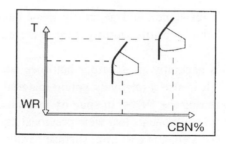

Figure 10.24 Toughness versus per cent CBN.

CBN inserts (Figure 10.25) are excellent for finishing to close tolerances in operations with hard steel. Surface texture with Ra 0.3 and tolerances of +/−0.01 mm are turned with CBN. Much longer tool life than cemented carbide and ceramics is achieved. When correctly applied the component being machined remains relatively cool as the heat is taken up mainly by the chip.

Moderately high cutting speeds and relatively low feed rates are recommended for CBN machining and, if coolants are to be used it must be copiously around the cutting edges, otherwise dry machining is always recommended to avoid thermal cracking.

Polycrystalline diamond

The hardest material known is the natural monocrystalline diamond and almost as hard is the synthetic polycrystalline diamond (PCD) (Figure 10.26). Its considerable hardness enables it to stand up to very abrasive wear – it is used to dress grinding wheels, for instance. Fine diamond crystals are bonded during sintering, under high temperature and pressure. The crystals are randomly oriented to eliminate any direction for crack propagation. This results in hardness and wear resistance uniformly high in all directions.

The small PCD cutting edges are bonded to cemented carbide inserts which add strength and shock resistance. Tool life can be many times longer than cemented

Figure 10.25 Cubic boron nitride insert.

Figure 10.26 Polycrystalline diamond.

carbide – up to one hundred times (Figure 10.27). However, the drawbacks for this seemingly perfect cutting tool material are:

1. Cutting zone temperatures must not exceed 600 °C.
2. It cannot be used for ferrous applications due to affinity between carbon in tool and work.
3. It cannot be used for tough, high-tensile workpiece materials.

In practice this excludes PCD for the majority of metalworking operations. Although limited in range, PCD is an excellent tool material in the right application

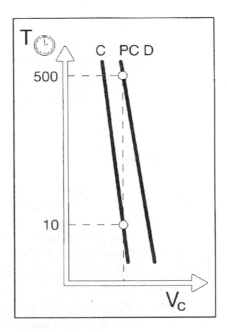

Figure 10.27 T against V_c.

of machining that does not include the above limitations – especially in abrasive non-ferrous and non-metallic materials requiring accuracy and high finish.

PCD is also a relatively new material introduced at the beginning of the 1970s. It is today used for turning and milling especially the abrasive silicon–aluminium alloys when surface finish and accuracy are criteria. In fact, uncoated fine-grain cemented carbide and PCD are the two main cutting tool materials for aluminium. Sharp cutting edges and positive rakes are essential. But also other abrasive non-metallic materials such as composites, resins, rubber plastics, carbon, pre-sintered ceramics and carbide and sintered carbide, as well as metals like copper, Babbit metal, bronze, brass, magnesium alloys, zinc alloys and lead can be machined with PCD.

Smearing of workpiece material is usually not a problem for PCD cutting edges thanks to the high chemical stability. Burr generation on workpieces is often eliminated with PCD and tool life is normally extended very many times.

Because of the very brittle nature of PCD, very stable conditions, rigid tools and machines and high speeds are necessary for machining with PCD. Cutting fluids can be used, generally for cooling. Finishing and semi-finishing in turning and boring are typical operations. For facemilling, PCD tipped inserts can be used as ordinary or wiper inserts in special seats. Lighter feeds, lower depth of cut and avoidance of interrupted cuts and shocks are important.

For turning, the largest possible tool shank should be used with minimum overhang. For milling, axial and radial run-out should be kept to a minimum, each insert is adjusted individually for height.

10.5 Cutting fluids

Cutting oils have good lubricating properties and provide good protection against corrosion but do not provide such good cooling as water-based cutting fluids. Neat cutting oils, that is oils which are not mixed with water, can be divided into the following main groups:

- Mineral oils.
- Fatty oils.
- Mixtures based on mineral oil and fatty oil.
- EP (extreme pressure) oils.

Fatty oils are based on animal or vegetable fats, for instance, colza oil. They are fatty and give very good lubrication but offer poor resistance to welding. Today they have been replaced by mineral-based oils, partially because fatty oils are expensive and difficult to obtain but, above all, because the development of additives which are mixed with mineral oil provides a substantially more efficient cutting fluid.

Mineral oil is used either neat or mixed (Figure 10.28). Neat mineral oil has very good lubricating properties and provides very good protection against corrosion. Because its cooling properties are not so effective, mineral oil is primarily used for lighter machining, for instance, in brass, cast iron and light alloys.

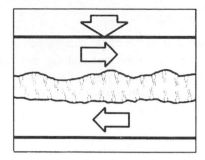

Figure 10.28 Lubrication with mineral oil.

In order to cope with machining operations where the loading between the workpiece and tool sets greater demands on the cutting medium's bearing strength, mineral oil can be mixed with additives of fatty oil. Additives of fatty oil provide a thin coating with high bearing strength and low shearing resistance. This coating provides lubrication and prevents the friction between the tool, chips and workpiece, even when the film of oil has broken down. However, with difficult machining, the additives of fatty oil are not sufficient to provide lubrication. Fatty mineral oil is primarily used in applications where it is wished to improve the surface finish by means of increased lubrication, for instance, when machining harder brass and copper and where more active additives cause corrosion attack.

In applications where the cutting forces are high, the cutting fluid must provide lubrication even when there is great pressure between the sliding surfaces. In order to cope with this, cutting oils with EP additives are used for difficult machining (Figure 10.29).

Such additives form compounds with the metal of the sliding surfaces. The effect takes place at those points where unevenness in the surfaces breaks through the film of oil and the compound forms a firm lubricant which prevents welding between the opposing peaks on the surfaces. The additives consist of sulphur, chlorine and phosphor compounds which react at high temperatures to form metallic sulphides, chlorides and phosphides.

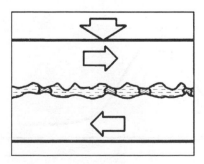

Figure 10.29 Lubrication with EP additives.

Primary functions of cutting oils

High temperatures are generated in the metal cutting zones and high friction forces arise at the point of contact between the tool and the workpiece. In many cases this would lead to unacceptable results if the machining was carried out without cutting fluid, the primary functions of which are to lubricate, cool the tool and transport chips away from the cutting area.

When two surfaces slide past one another without lubrication, uneven points on the surfaces will rub against each other (Figure 10.30 – A). This means that either the material will be heated by friction or that pieces will be worn free from the surfaces. The lubricating properties of the cutting fluid reduce the wear between the workpiece and the tool by separating the surfaces from one another (Figure 10.30 – B).

The energy which must be supplied to deform the metal when forming chips causes high temperatures in the cutting zone. Tool wear is very much affected by the temperature, and efficient cooling is important in order to extend the life of the tool edge.

In Figure 10.31, the specific surface texture (R) is illustrated as a function of the edge temperature (T). The worst finish is obtained with extensive built-up edge (2). At lower temperatures there is what is known as a reaming zone (1) and the surface is more even.

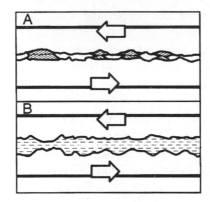

Figure 10.30 Surface friction: (A) without lubrication; (B) with lubrication.

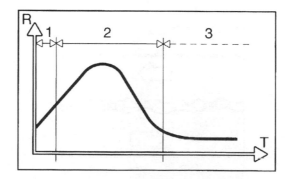

Figure 10.31 Graph of surface finish against cutting edge temperature.

The best surface finish is obtained in the free cutting zone (3). When there is a risk of built-up edge various measures can be taken to change the edge temperature. These depend upon the circumstances: if the edge temperature is close to the free cutting zone, the risk of built-up edge can be reduced by increasing the cutting speed, that is, the edge temperature is increased. At lower edge temperatures one possible solution is to use a cooling cutting fluid to reduce the temperature and so avoid the built-up edge zone. Although a cutting fluid with a strong cooling effect can be used when the edge temperature lies in the free cutting zone, if the temperature is close to the built-up edge zone, such a cutting fluid can increase the risk of built-up edge.

The machined surface must not be damaged by chips and other particles which break free during the machining process. Therefore the cutting fluid is used to remove this material in an efficient way.

The cutting fluid then, performs the following functions:

- Lubricates and increases the service life of the tool cutting edges.
- Cools the workpiece and tool thereby increasing its resistance to wear.
- Removes the chips.
- Facilitates chip breaking in certain materials.
- Prevents the formation of built-up edge.

Choice of cutting fluid

The choice of cutting fluid is dictated by the machining operation, the workpiece material, the tool material and the cutting data (Figure 10.32).

Generally, better *lubrication* should be sought with:

- Low speeds.
- Difficult-to-machine materials.
- Difficult operations.
- Demands for a better surface texture.

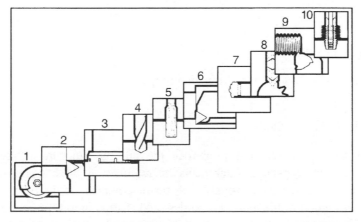

Figure 10.32 Cutting operations shown in increasing order of severity of use and need for cutting oil.

Improved *cooling* should be chosen for:

- High speeds.
- Easy-to-machine materials.
- Simple operations.
- Problems with built-up edge.

The following list shows various machining operations (illustrated in Figure 10.32) in order of metal cutting demands, from relatively less-demanding operations (such as grinding) to highly demanding operations (such as thread turning and threading with a screw tap):

1. Grinding.
2. Turning.
3. Milling.
4. Drilling.
5. Reaming.
6. Boring.
7. Deep-hole drilling.
8. Gear making.
9. Thread turning.
10. Threading with a screw tap.

However, the degree and number of demanding factors depend on conditions, cutting data, tool geometry and material.

In addition to the lubrication and cooling properties, it must also be borne in mind when choosing a cutting fluid that the environment should not be affected. The cutting fluid must not cause rust or corrosion to the workpiece but should rather protect the material. The machine must not be damaged and the cutting fluid must not constitute a health risk for the operator. Water-soluble cutting fluids should be compatible with the water that is used. The pH value of the water is important in this respect.

Types of cutting fluid

Cutting fluids can be divided into two main groups:

1. Neat cutting oils.
2. Water-soluble cutting fluids.

In order to have good cooling properties the cutting fluid must have a high temperature diffusion capacity. Water meets this requirement but has poor lubricating properties and it also reacts corrosively with ferrous metals. On the other hand, neat cutting oil has good lubricating properties and forms good protection against corrosion. However, the oil's low heat conductivity and specific heat means that its cooling properties are poor.

The good cooling properties of water have been utilised by the development of various water-soluble cutting fluids. These can be divided into:

1. Oil emulsions.
2. Synthetic or chemical cutting fluids.

The use of oil emulsions is the traditional way of combining the cooling properties of water with the lubricating and corrosion-protecting properties of oil. To call a cutting fluid water-soluble is a modification of the truth since oil is not soluble in water. The oil is dispersed as small drops in the fluid (Figure 10.33 – A) and is stabilised by means of additives known as emulsifiers.

Synthetics contain no oil but consist of glycols, for example, which are emulsified or dissolved in water. These are transparent and provide a good opportunity to see the operation compared with oil emulsions, which have a milky appearance. By mixing a small quantity of emulsified oil with a synthetic cutting fluid, a semi-synthetic fluid is obtained, where the advantages of synthetics are combined with the substantially better lubricating properties of oil emulsion.

In certain cases gas is also used as a cooling and lubricating medium. The gas is applied at high pressure and evacuates chips and other particles. Air is the most common gas used but carbon dioxide and liquid argon and nitrogen are also to be found.

Water-soluble cutting oils

Water-soluble cutting oils are supplied as concentrates and the user must prepare the oil emulsion by mixing the concentrate with water. When using oil emulsions, the lubricating and corrosion protecting properties of the oil are combined with the cooling property of water. The water-soluble concentrate also contains additives to improve various properties, for example:

- Emulsifiers to prevent separation.
- Lubrication to prevent corrosive attack.
- A conservation medium to prevent the growth of bacteria.
- Grease to improve the lubricating properties.
- EP additives to improve the bearing strength.

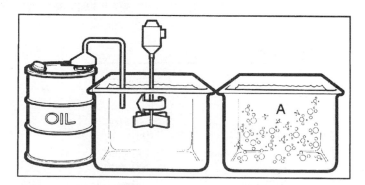

Figure 10.33 Mixing of an oil/water emulsion.

It is extremely important that the preparation of the oil emulsion is done correctly so that the cutting fluid is stable:

- The oil concentrate should be added to the water and not the other way round.
- The mixture should be stirred continually when preparing the emulsion.
- The mixing ratio of oil to water should be measured accurately.
- Clean vessels and tank should be used.
- The water should be of a suitable pH value and hardness.
- The oil concentrate should not be added more quickly than it can be converted to emulsion.

The quantity of undissolved salts, mainly calcium, magnesium and iron, affect the hardness of the water. In order to obtain softer water, soda is added. The hardness of the water is normally expressed in the number of parts per million (ppm) of water-free sodium carbonate which are needed to make the water completely soft. Undissolved salts can react with emulsifiers and, in addition to the emulsion breaking down quickly, a foam is formed which can block the filter, piping and pumps in the cutting fluid system. In addition, hard water which contains more than 200 ppm can reduce the corrosion protecting properties. Table 10.1 shows an approximate classification of the hardness value of water expressed in ppm is given. Too great an addition of soda increases the pH value of the water. Low pH values increase the tendency to corrode and the growth of bacteria but, on the other hand, high pH values involve the risk of skin irritation. The pH value for oil emulsions should be between 8.5 and 9.3.

Synthetic and semi-synthetic cutting fluids

Synthetic cutting fluids contain no mineral oil and were used previously solely for grinding operations. However, the development of synthetics with improved lubrication properties and improved rust protection has increased their field of application considerably in recent years.

The use of synthetic cutting fluids has economic advantages over oil-based fluids and offers quick heat dissipation, cleaning properties, simple preparation and provides good protection against rust. In addition they are transparent, which helps the

Table 10.1

	Hardness in ppm
Soft	0–50
Moderately soft	50–100
Slightly hard	100–150
Moderately hard	150–200
Hard	200–300
Very hard	300–

operator to monitor the job. With difficult operations the lubrication properties may possibly be insufficient, which can cause sticking and/or wear to the sliding surfaces.

Since totally synthetic cutting fluids form a true solution with water the concentration rather increases with use, because the water evaporates. Since the synthetics have strong cleaning and alkaline properties (pH 9–9.5), the concentration should be checked more often so that an increased pH value does not cause allergies for the operator in the form of skin irritation.

Since oil is dissolved in synthetics, what are known as semi-synthetics are obtained. Semi-synthetics have, on the whole, the same properties as totally synthetic cutting fluids but offer substantially better lubrication properties. Semi-synthetics are more suitable than oil emulsions for grinding operations since they contain a smaller quantity of oil. Oil often tends to cause overloading of the grinding wheel.

In semi-synthetics the oil particles are smaller than in oil emulsions. The high content of emulsifiers provides extra capacity to emulsify leakage oil (Figure 10.34 – A) from the machine. However, this way of absorbing leakage oil cannot continue indefinitely. When excess emulsifiers in the cutting fluid have been used up in order to emulsify leakage oil, the oil floats up and forms a coating on the surface (Figure 10.34 – B) while the oil particles in the fluid increase in size (Figure 10.34 – C). This oil coating on the surface can then increase the growth of bacteria.

Cutting fluid and the environment

Cutting fluid should primarily provide good lubrication and cooling but there are a number of other requirements which must be taken into account when choosing a cutting fluid (see Figure 10.35):

1. The cutting fluid must not produce unpleasant side effects like smells or allergic reactions, (A).
2. It should be able to cope with high pressure equipment, such as centrifuges, without foaming (B).
3. The cutting fluid must not dissolve paint thereby affecting the coating on the machine (C). Neither should it corrode seals.

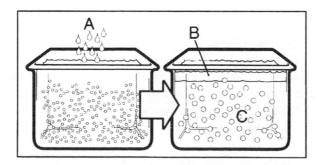

Figure 10.34 Contamination of synthetic cutting fluids by machine oil.

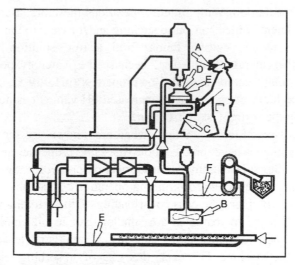

Figure 10.35 Cutting fluid in the environment.

4. The cutting fluid must not be the cause of corrosion attacking the workpiece (D). Since a variety of materials are usually machined the cutting fluid should be suitable for these, or most of them, without the need to change cutting fluids. Of particular importance is the risk of corrosion attack with non-ferrous materials, for instance, copper, brass and aluminium.
5. The cutting fluid must not cling or stick causing the chips and particles to become attached, thereby making cleaning of the tank more difficult or spoiling the surface of the workpiece (E).
6. Most machines leak oil and in modern machines this can be about one litre per day. Therefore it is preferable if the cutting fluid can dissolve leakage oil without its performance deteriorating (F).

Storage, maintenance and waste disposal of cutting fluids

Water-based cutting fluids should not be exposed to extreme temperatures in either direction. High temperatures can result in water evaporating and low temperatures can result in the separation of certain additives. Even concentrates normally contain a small quantity of water and, should this freeze, the oil will be very difficult to mix when the temperature rises again. The barrels of oil should be stored under cover so that they are not exposed to wetness. When the oil is stored outdoors temperature variations can cause the barrel to let in water which has collected on top of the cover. In order to avoid this the oil barrels should be stored on their sides.

Water-soluble cutting fluids require more maintenance than cutting oils. This applies above all in limiting the number of microorganisms in the water since these shorten the service life of the cutting fluid and can cause corrosion problems, clogging of supply lines and valves as well as causing an unpleasant smell. There are three types of microorganism: bacteria, fungi and algae. When using water-based cutting fluids where accumulations of bacteria often occur in difficult to reach spaces, a bacteria killing agent, known as biocides, must be added when cleaning. If pollutants and contaminated cutting fluid remain, fresh cutting fluid which is added will soon become contaminated.

The concentration and composition of cutting fluid changes with use. This occurs through water evaporation, cutting fluid losses, reaction when leakage oil mixes with the cutting fluid, etc. Therefore it is important that the concentration of the mix is checked regularly. When the concentration needs to be changed, mixed cutting fluids are added. The water added must not be too hard since the fluid may then separate, with the result that a film of oil is obtained on the surface. On the other hand water which is too soft causes frothing. Modern systems for the reclamation of cutting fluids should have automatic equipment for measuring, filling up and mixing supplementary cutting fluid.

Rejected cutting oil and oil emulsion must not be released into the public sewage system. Emulsions are treated so that oil and water are separated. In order to break down emulsion, ferrous sulphate, salts and, in certain cases, strong acids are used. If acids are used the water is neutralised before it is released into the sewage system. The oil which is separated is strongly contaminated by the substances which are used in the treatment. The oil can, like the rejected neat cutting oils, be burned in special furnaces. For most workshops it is not economically viable to handle rejected cutting fluids themselves. Therefore most companies engage contractors who specialise in this type of waste disposal.

Sedimentation of undissolved particles using the force of gravity is the easiest way to remove pollutants from the cutting fluid. However, since sedimentation takes a relatively long time, bacteria growth can create a problem and, furthermore, leakage oil is not removed. If neat cutting oil is used it is not certain that very fine particles will drop to the bottom of the container. Instead, due to the oil's high viscosity, particles can be enveloped and accompany the oil. This could damage the circulation pump and/or impair the surface finish when machining. The cutting fluid should always pass a filter (Figure 10.36 – 4) before passed on with a pump (Figure 10.36 – 5).

In order to obtain quicker separation of the pollutants, sedimentation tanks are used whereby the cutting fluid is separated into layers. In this way the distance the particles must fall is reduced. At the same time leakage oil has a shorter distance to the surface where it is continually skimmed off. Heat treatment of the cutting fluid can speed up the separation. It may be difficult to use this method with water-soluble cutting fluids, where leakage oil can be emulsified. It is best suited to the cleaning of synthetics where the oil floats up to the surface in free form. To remove emulsified leakage oil effectively, high speed centrifuges are used.

In central systems the cutting fluid circulates continually and the sedimentation method does not work. Instead, various types of filters are used. With fluids of high viscosity, filtering is a slow process since the flow speed through the filter is reduced. The filtering of oil of high viscosity is facilitated since the oil is heated up and therefore becomes thinner. In order to remove the smallest particles with a filter the mesh size must be small which can mean that certain additives in the cutting fluid are also filtered away unintentionally. Another disadvantage is that leakage oil can clog up the filter. Therefore the use of centrifuges is also common in central systems.

The cutting fluid system should ensure that the cutting fluid maintains a working temperature of approximately 20 °C. If the volume of the tank does not allow sufficient cooling, various types of heat exchangers or cooling units can be installed.

Re-use of cutting fluid

The technique of recirculating cutting fluid for re-use which is practiced today, requires continual maintenance and inspection. Pollution, microorganisms and changes in the concentration are continually monitored in order to keep the breaking down of the cutting fluid under control and extend its service life.

When a central system is used, a number of machines use the same cutting fluid. The advantage of this is that maintenance and inspection can be carried out at one single container. However, central systems demand that the same type of cutting fluid is suitable for all the machines. Any contamination which causes the cutting fluid to be replaced long before its expected service life is over is very expensive since very large quantities of cutting fluid are involved when central systems are used.

Special plants for the reclamation of cutting fluids from chips are not economically viable when water-based cutting fluids are used. However, when using cutting oil, up to 300 litres of cutting oil/ton of chips can be reclaimed (Figure 10.36). In order to enable the bulky chips to be handled more easily chip crushers are used (1), after which the crushed metal mass is placed in a centrifuge (2) so that the oil is thoroughly removed. The oil which is reclaimed is passed on for cleaning together with the used cutting fluid from the machine.

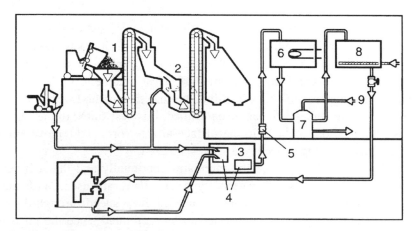

Figure 10.36 Re-use of cutting fluid.

Cleaning systems vary from simple sedimentation tanks to advanced plants which can consist of: .

- Chip crusher (1).
- Centrifuge (2).
- Sedimentation (3).
- Filter (4).
- Pump (5).

- Heat treatment (6).
- Filtering/centrifugation (7).
- Temperature regulation (8).
- Automatic inspection and filling up (9).

10.6 Other variables influencing the economics of cutting

Taylor's equation, upon which the economics of cutting primarily depend, is not independent of such variables as depth of cut feed, rake angle and plan approach angle. Particular values of these may permit a higher metal removal rate at the optimum tool life than will other values. In its simplest form this may be noted by reference to Figure 10.37, which shows the effect of varying the rake angle. The graph shows that, all else remaining constant, there is an optimum value for effective rake angle. The graph would not necessarily have the maximum cutting speed at the same value of γ if the feed were changed. However, the curve is fairly flat in the region of the maximum cutting speed, and the value of V_{20} is not likely to depart much from the maximum value on account of small changes in the other cutting conditions.

The general impact of other cutting variables can best be shown in sketch graphs:

1. Figure 10.38 shows the effect of varying the depth of cut and feed upon the cutting speed, tool life constant.
2. Figure 10.39 shows the effect of varying the depth of cut and feed upon metal removal rate, tool life constant.

Notice that in Figure 10.38 it can be seen that V_{20} falls as d or f is increased, but from Figure 10.39 it can also be seen that the metal removal rate (w) *rises* despite this fall in V_{20}. The use, when roughing, of full depths of cut and maximum feeds at cutting speeds which give the economic tool life for the selected conditions, is in accordance with the economics of the process.

3. The effect of the plan shape of the tool upon cutting speed, tool life constant, is shown in Figure 10.40.

As angle k falls below 90° the cut is spread over a longer cutting edge. The tool is less likely to fail if a small nose radius is used instead of a sharp point, and a small trailing angle gives good support to the cutting edge and increases the section through which heat can pass from the tool edge. All these features of tool plan shape tend to increase the value of V_{20}. The general impact of the two main variables of cutting, cutting speed and feed, on the economics of cutting is illustrated in Figure 10.41.

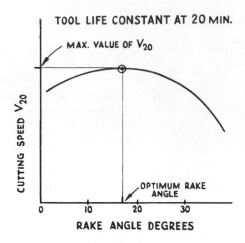

Figure 10.37 Cutting speed versus top rake angle.

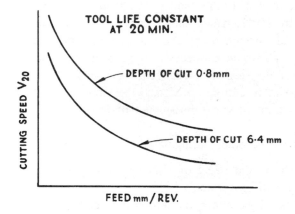

Figure 10.38 Influence of feed rate upon cutting speed, tool life constant.

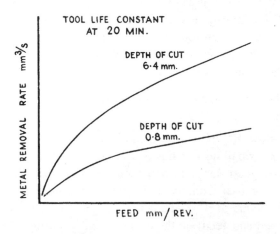

Figure 10.39 Influence of feed rate upon metal removal rate, tool life constant.

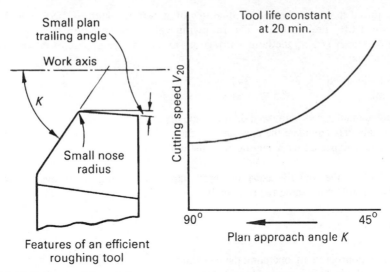

Features of an efficient
roughing tool

Figure 10.40 Influence of tool shape upon cutting speed, tool life constant.

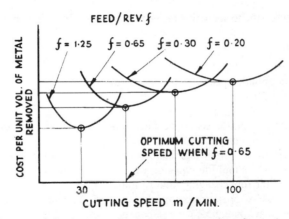

Figure 10.41 Influence of feed and speed upon the economics of metal removal.

Exercises – 10

1. (a) What is the relationship between tool life and cutting speed?
   (b) If chip thickness is increased, how would the value of C be affected?
   (c) If $n = 0.2$ and $C = 30$, what cutting speed would give a tool life of 1 hour?

2. When machining on a lathe taking 2.5 mm depth of cut and 0.25 mm/rev feed, the vertical cutting force on the tool is 588N.
   (a) What volume of metal can be removed per minute per kW?
   The speed/life relationship of a tool is given by $V_c T^n = C$ for a given set of conditions. What would be the effect on C:
   (b) If the chip thickness is increased?
   (c) If the depth of cut is increased?
      (Assume no nose radius.)
   (*Hint*. Solution to (*b*) and (*c*) should include sketch graphs. Solution can vary according to the value of ϕ taken; $\phi = 90°$ is suggested.)

3. It is required to assess the cutting qualities of a certain batch of HSS tools by means of a 'short life' practical test. The following values of spindle speed (rev/min) and length of travel (L) of the lathe carriage up to the failure of each tool are applicable to such a test:

Spindle speed	395	343	256
L mm	26.5	64.8	398

Workpiece diameter 90 mm, feed 0.5 mm/rev. depth of cut 1.4 mm.

From this data calculate the constants of the tool life equation $VT^n = C$. Determine the tool life to be expected when cutting at 45 m/min.

4. If $V_c T^n = C$ is the tool life equation representing the cutting conditions, show that the inclusive cost of machining (S) is given by:

$$S = kT^n \frac{H}{60} + kT^{(n-1)} J$$

where H = hourly rate of operating the machine;
    J = the inclusive costs of a tool change;
    k = a constant for the cutting conditions.

Use the above expression to show that the economic tool life is given by:

$$T = \frac{1-n}{n} J \frac{60}{H}$$

Find the optimum spindle speed for turning a 150 mm diameter shaft if the ratio $H/J = 2.25$ and $V_c T^{0.12} = 50$.

5. Illustrate in a sketch graph the economic factors involved in metal cutting and show that cutting speed has an optimum value.

When finish turning a 50 mm diameter × 130 mm long bore in high-tensile steel at a feed of 0.13 mm/rev the tool life equation was $V_c T^{0.12} = 146$. The cost of regrinding the tool was 8 units and the tool changing time 10 min.

If the inclusive cost of operating the machine is 13.5 units/hr, determine:
(a) The economic machining time per component.
(b) The number of components to be produced between tool changes.

6. It is required to reduce 50 mm diameter MS bar to 32 mm diameter for 152 mm of length by means of a single operation on a lathe. The conditions are as follows:

Operator's hourly rate = 4.2 units
Machine hourly rate = 11.4 units
Time to remove, replace and reset the tool = 0.1 hr
Cost of regrinding the tool = 0.3 units
Initial cost of the tool = 10 units
Number of possible regrinds of the tool = 50
Number of parts to be turned = 10 000
Feed/rev of tool = 0.25 mm
*Operator constant = 1.15
Tool-life law, $V_c T^{0.2}$

Evaluate the economic spindle speed for the lathe.
(*Assumed to mean that the operator is earning 15% bonus.)

7. When turning under certain conditions the relationship between cutting speed, tool life and chip thickness could be expressed as

$$V_c T^{0.1} = C f^{-0.5}$$

where f = chip thickness in mm. For a tool where $\phi = 90°$ cutting at 30 m/min, the tool life was 1 min. Estimate the cutting speed which would give the same tool life if $\phi = 30°$, the other cutting conditions being unchanged.

8. (a) Discuss the criteria by which the 'durability' of cutting tools can be assessed. Compare the relative performance of HSS and TC tools.
   (b) In what manner do modern cutting fluids assist the performance of cutting tools?
   (c) Compare the effectiveness of a cutting fluid used for milling with one used for turning.

11 Turning and Milling

11.1 Introduction

Turning, basically, generates cylindrical forms with a single point tool and in most cases the tool is stationary with the workpiece rotating (Figure 11.1). In many respects it is the most straight-forward metal cutting method with relatively uncomplicated definitions. On the other hand, being the most widely used process and easily lending itself to development, turning has led the field and is a highly optimised process, requiring thorough appraisal of the various factors in applications.

In spite of generally being a single cutting edge operation, the turning process is varied in that the workpiece shape and material, type of operation, conditions, requirements, costs, etc. determine a number of cutting tool factors. Today's turning tool is carefully designed, based on decades of experience, research and development.

From the microgeometry and materials at its point of engagement, to the basic shape and clamping of the indexable insert through to the toolholder, shank type or modular, the tool handles the dynamics of metal cutting today in a way which would have been unthinkable a couple of decades ago. Many of the principles that apply to single point machining apply also to other metal cutting methods, even multi-point, rotating tool machining such as milling.

Turning is an efficient flexible machining method of machining round workpieces, in a large variety of sizes and materials, with a single point tool. There are several basic types of operation, requiring specific types of tools for the operation to be performed in the most efficient way.

Turning can be broken down into a number of basic cuts for selecting tool types, cutting data and also programming for a certain operation. To make tool application more straight forward, there can be said to be three basic turning operations (Figure 11.2):

1. Longitudinal turning.
2. Facing.
3. Turning radii and tapers.

There are also numerous combination cuts such as machining shoulders, diameter transitions and chamfers but they are essentially variants of the three basic operations.

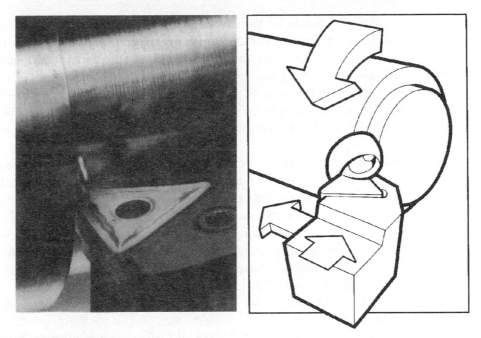

Figure 11.1 Turning.

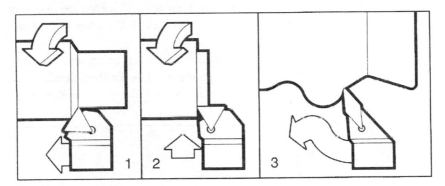

Figure 11.2 Three basic turning operations.

Turning is the combination of two movements: rotation of the workpiece and feed movement of the tool. In some applications, the workpiece can be stationary with the tool revolving around it to make the cut, but basically the principle is the same.

The feed movement of the tool can be along the axis of the workpiece, which means the diameter of the part will be turned down to a smaller size. Alternatively, the tool can be fed towards the centre, at the end of the part, which means the length of the part will be faced down. Often feeds are combinations of these two directions, resulting in tapered or curved surfaces. Today's lathe control units will cope with many possibilities to turn parts straight and round. CNC means very efficient control of the cutting edge, replacing previous concepts of copying according to templates and the use of form tools.

This chapter discusses turning of the outside of workpieces for simplicity, but most of the processes can be performed inside a workpiece using boring tools to reach inside the workpiece. The forces and power that can be used are often limited by the lower strength of the boring bar.

The workpiece rotates in the lathe, with a certain spindle speed (n), at a certain number of revolutions per minute (rpm). In relation to the diameter of the workpiece, at the point it is being machined, this will give rise to a cutting speed, or surface speed (V_c in m/min).

This is the speed at which the cutting edge machines the surface of the workpiece. It is the speed at which the periphery of the cut diameter passes the cutting edge. It should be noted that the cutting speed is only constant for as long as the spindle speed and/or part diameter remains the same. In a facing operation, where the tool is fed in towards the centre, the cutting speed will change progressively if the workpiece rotates at a fixed spindle speed (Figure 11.3). On many modern lathes, the spindle speed is increased as the tool moves in towards the centre, making up for the decreasing diameter (see Figures 11.4 and 11.5).

The *feed speed* (V_f – in m/min) is the machine feed which moves the tool along in various directions.

The feed per rev (f – in mm/rev) is the movement of the tool in relation to the revolving workpiece. This is a key value in determining the quality of the surface being machined and for ensuring that the chip formation is within the scope of the tool geometry. This value influences, not only how thick the chip is, but also the quality of the chip breaking.

The *cutting depth* (d – in mm) is the difference between uncut and cut surface – half the difference in the uncut and cut diameter. The cutting depth is always measured at right angles to the feed direction of the tool, not to the cutting edge. The

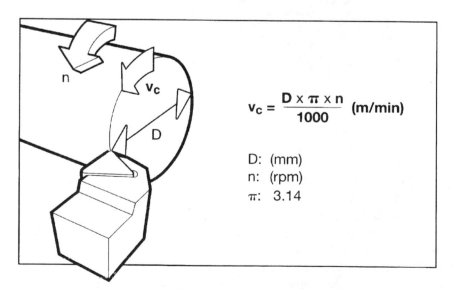

$$V_c = \frac{D \times \pi \times n}{1000} \text{ (m/min)}$$

D: (mm)
n: (rpm)
π: 3.14

Figure 11.3 The cutting speed is the surface speed.

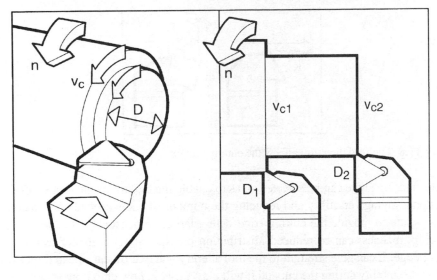

Figure 11.4 Three variables: cutting speed, diameter and spindle speed.

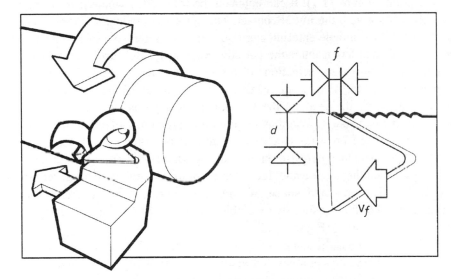

Figure 11.5 The correct depth and feed are critical for chip control.

way in which the cutting edge approaches the workpiece is expressed through the *entering angle* (κ). This is the angle between the cutting edge and the direction of feed (Figure 11.6).

It is an important angle in the basic selection of a turning tool for an operation. In addition to influencing the chip formation, it affects factors such as the direction of forces involved, the length of cutting edge engaged in cut, the way in which the cutting edge makes contact with the workpiece and the variation of cuts that can be taken with the tool in question. The entering angle usually varies between 45° and 90°.

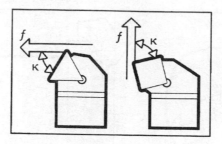

Figure 11.6 The feed direction affects the entering angle.

The entering angle can be selected so as to enable the tool to machine in several feed directions, giving versatility and reducing the number of tools needed. Alternatively it can be made to provide the cutting edge with a larger corner and thus added strength. The chip thickness can be reduced, distributing pressure along a greater length of the cutting edge. It can give strength to the tool at entry and exit of cut, it can direct forces to provide stability during the cut and it will play a role in how chip flow is created.

The entering angle can permit the cutting edge to enter into cut some distance from the point (Figure 11.7). It can enter on the main edge where it is stronger and more able to cope with the initial contact. The cut starts gradually without excessive impact, and with a suitable entering angle, the edge cuts abrasive scale off rather than scraping it off, and lets it enter the cut advantageously. Because of the destructive nature of scale and skin, selection of the correct entering angle has a profound influence on tool life. Also, when it leaves the cut, the entering angle will affect the pressure created by the feed on the last remaining material and on the change of pressure on the edge itself. When an edge with a large entering angle finishes a cut, the release of feed pressure can cause a momentary increase in feed and chip thickness. This extra load can also put the cutting edge at risk from fracturing.

For a given cutting depth and feed per revolution there will be a certain *chip cross-section*. The area and shape of this, however, will be determined by the entering angle. For this reason, two variables are needed to define the change of the resulting chip formation (Figure 11.8).

The *chip width* (l_a) is the same as the effective length of the main cutting edge.

The *chip thickness* (*h*) is measured across the cutting edge, perpendicular to the cut along the main cutting edge. The chip width and thickness are the dimensions defining the theoretical cut of the edge into the workpiece material.

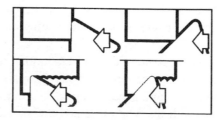

Figure 11.7 Entering angles affect start and exit of cuts.

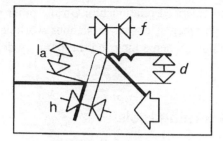

Figure 11.8 Chip width (l_a) and chip thickness (h).

At a 90° entering angle, the cutting depth is the same as the effective cutting edge or chip width, with the chip thickness the same as the feed per revolution.

At a 45° angle, with the same feed rate and cutting depth, the effective cutting edge length will show a dramatic increase – growing to nearly half as long again – as compared to the 90° length. The chip thickness will be considerably smaller.

The tool entering angle (κ) is the angle between the cutting edge and the centreline of the workpiece. The entering angle influences chip thickness, direction of chip flow and cutting forces. The *free cutting angle* (κ_N) (also known as the plan trail clearance angle) is the angle between the secondary cutting edge and the centreline of the workpiece (Figure 11.9).

At small entering angles, the total cutting force is distributed over a longer portion of the cutting edge. This means that the tool is better able to withstand heavy and intermittent machining. The tool is, however, pressed very hard against the workpiece, which can give rise to vibrations and deflections of slender parts.

Large entering angles give lower forces in the radial direction. The tool is instead pressed harder in the longitudinal direction of the workpiece. One disadvantage with a large entering angle is the sudden loading of the cutting edge at the start of machining and the sudden unloading at the end.

The entering angle influences the direction of cutting forces in the horizontal plane. The feed pressure is directed on to the main cutting edge and as such can be seen as force components in the axial and radial direction. A large entering angle gives a large axial force and a smaller radial force, while a smaller entering angle

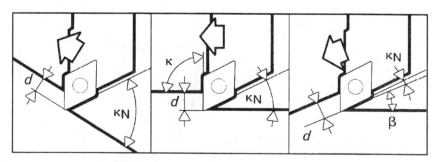

Figure 11.9 Free cutting angles.

will give rise to a more balanced relationship. On the other hand, a dominant axial force is often preferable when it comes to slender workpieces that deflect easily. Also for internal machining, when a long, boring bar may deflect because of higher radial forces.

11.2 Selection of turning tools

The selection of turning tools is a process of logical steps in response to the task of machining a workpiece according to a manufacturing drawing (Figure 11.10). Mostly it is a routine process through a programme of cutting tools affected by the parameters of the operation in question. In addition, rational creative thinking is required to achieve the best result or to solve specific problems.

There are three main variables when it comes to a modern turning tool (Figure 11.11):

1. Insert clamping method.
2. Indexable insert type and geometry.
3. Cutting tool material.

Essentially the application process revolves round these variables. Today's turning tool is, in the majority of cases, a steel toolholder with a mechanism for holding an

Figure 11.10 Application of turning tools.

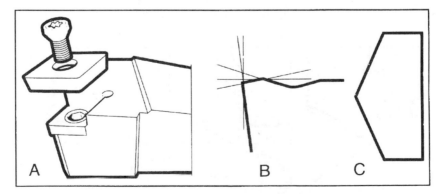

Figure 11.11 Insert clamping, cutting geometry and tool material.

insert. The insert is made from a harder cutting material and has several cutting edges. Indexing takes place when an edge has reached a certain amount of wear, which has made the edge incapable of maintaining the specified dimensional tolerance or surface finish or even making it susceptible to rapid breakdown. The logical defining of the tool variables is taken step-by-step with the tool inventory, machine tool specifications, manufacturing drawing and possibilities for new ideas.

The end-result (R) and performance (P) during the operation are the two overriding criteria for any machining (see Figure 11.12). The end-result has to be a component produced according to specifications and limits as well as to satisfactory machining economics. Performance is also down to economics – the total cost of the operation, as well as how good other factors are such as chip breaking, reliability, cutting data, tool handling and inventory.

The main factors that influence the application of tools for the turning operation are:

1. Workpiece material – machinability, condition, properties, etc.
2. Workpiece design – shape, dimensions and working allowance.
3. Limitations – accuracy, surface texture, etc.
4. Machine – type, power, condition and specifications.
5. Stability – from cutting edge to foundation.
6. Set-up – accessibility, holding, changing.
7. Tool programme – the right tool.
8. Performance – cutting data, tool life and economics.
9. Quality – tool delivery and service.

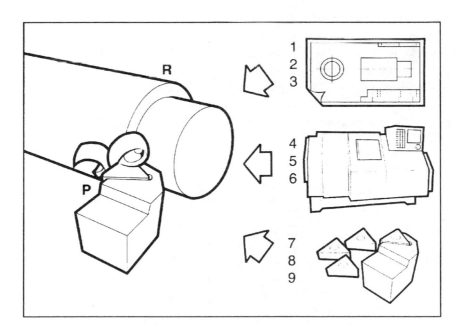

Figure 11.12 Tool application factors in turning.

11.3 The selection process

The application of turning tools can be carried out according to the following step-by-step approach.

1. Edge clamping system.
2. Toolholder size and type.
3. Insert shape.
4. Insert size.
5. Nose radius.
6. Insert type.
7. Tool material.
8. Cutting data.

1. Edge clamping system

The clamping system of the insert in the toolholder should be selected first (Figure 11.13). Toolholders have been designed to provide optimum performance in different applications and usually over a broad area. The type of operation and, to some extent, size of workpiece determines the selection of toolholding system. Roughing operations on large workpieces make considerably different demands to that of finishing of small components.

2. Toolholder size and type

When the edge clamping system is established, the size and type of toolholder has to be resolved (Figure 11.14). The selection is influenced by feed directions, size of cuts, workpiece and toolholding in machines as well as accessibility required. The shape of the workpiece is decisive if contour turning is involved.

The guiding rule is to select the largest toolholder size (h) possible for the machine (Figure 11.15) in order to reduce the tool overhang ratio and to provide the most rigid base for the edge. The toolholder size should then also be co-ordinated

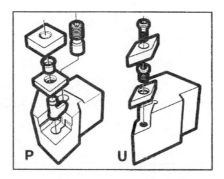

Figure 11.13 The two main, modern edge clamping methods.

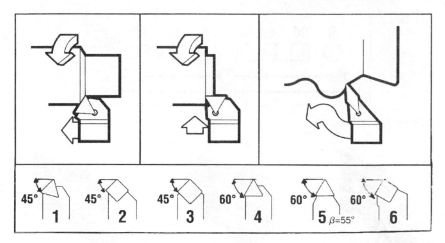

Figure 11.14 A sample of toolholder shapes and different workpiece/tool shape functions.

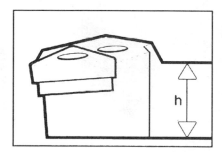

Figure 11.15 Select the largest toolholder size (h) possible for the machine.

with the subsequent selection of insert size, where the effective cutting edge length is determined. Generally, the smallest entering angle that the operations will allow should be selected.

3. Insert shape

Also at this stage, the insert shape should be selected relative to the entering angle required and the accessibility or versatility required of the tool. The largest suitable point angle on the insert should be selected for strength and economy. However, when variations in cut are involved in the operations, strength versus versatility through the use of smaller point angles should always be considered.

Figure 11.16 shows the most common insert point angles, from the round to the pointed 35° insert. Scale 1 indicates that as regards cutting edge strength (S), the larger the point angle to the left, the higher the strength. While as regards versatility and accessibility (A), the inserts to the right are superior.

Scale 2 indicates that the vibration tendency (V) rises to the left while power (P) requirement is lower to the right.

Typical turning tools are illustrated in Figure 11.17.

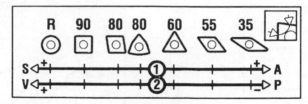

Figure 11.16 Insert shapes.

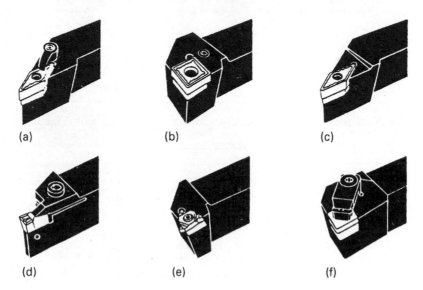

Figure 11.17 Typical turning tools. (a) Profiling tool with top clamping; (b) roughing tool with lever clamping; (c) profiling tool for finishing cuts with screw clamping; (d) grooving tool with top clamping; (e) threading tool with screw clamping; (f) roughing tool for ceramic tips with top clamping.

4. Insert size

The indexable insert is directly related to the toolholder selected for the operation. Toolholder size, entering angle and insert shape have already been established. Through the holder being designed with a seat to take specific insert shape and size, the insert size is to some extent predetermined.

However, the largest cutting depth that has to be taken with the selected tool also influences the toolholder size selection, especially for roughing operations.

5. Nose radius

The nose radius of the insert is a key factor as regards:

● Strength in roughing.
● Surface texture in finishing.

The size of the radius also affects vibration tendencies and at times, feed rates.

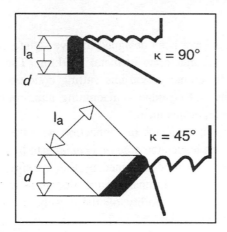

Figure 11.18 Effective area of cut.

Generally for roughing, the largest possible nose radius should be selected so as to obtain the strongest point. A larger radius permits higher feeds but must be checked against any vibration tendencies. An insert is normally available in several nose radius alternatives, which vary considerably. The larger alternatives are intended for roughing operations, the most common being 1.2–1.6 mm.

When establishing the feed rate for roughing operations, it is essential that maximum levels of feed are not exceeded relative to the nose radius. Table 11.1 indicates recommended feed rate ranges for common nose radii. Generally, it can be said that the feed rate for roughing operations should be targeted at around half the size of the nose radius.

The higher feeds in a recommended range apply to inserts that have stronger cutting edges, at least a 60° point angle, single sided, used with a smaller entering angle than 90° and in materials having good machinability characteristics.

In rough turning operations, power and stability of the machine and even chip forming ability can often be limiting factors. The maximum metal removal rate is obtained with a combination of high feed and moderate cutting speed with limiting factors taken into consideration. The power (P) available in the machine is sometimes the limiting factor and in such cases the cutting speed should be lowered to suit.

$$P = k_c \times f \times d \times V_c \,(\text{kW})$$

Table 11.1

r_ε	0.4	0.8	1.2	1.6	2.4
f	0.12–0.25	0.25–0.5	0.36–0.7	0.5–1.0	0.7–1.6

6. Insert type

A number of insert types have been designed to give satisfactory cutting conditions in the various applications that arise in metal working. The insert type is largely determined by the insert geometry. Various cutting conditions and materials make different demands on the cutting edge – machining aluminium is quite different to machining hardened steel, for instance.

Having established the insert shape in connection with the entering angle and the size of the nose radius then leaves the *type of geometry* to be established.

The selection of insert type is determined by the *working area* of the operation and the workpiece material. Additional factors that may influence the choice are machine condition, power, the stability of the set-up, continuous or intermittent machining and vibration tendencies.

Turning can be divided into a number of working areas based upon removing material, generating accurate dimensions with specific surface textures or a combination of the two.

Figures 11.19 and 11.20 show the six main working areas with the most common feed (f) and cutting depth (d) ranges for each. When establishing an insert type, the feed and cutting depth should be identified with one of these working ranges as the various insert types relate to these. However, the suitable working area for an insert varies with the combination of factors such as size, shape and size of nose radius.

7. Tool material

The other major factor, as regards selecting the type of insert geometry, is the workpiece material. There is a huge number of different materials according to several national standards. Insert geometries have been developed to best machine various materials and are rated as regards suitability (Figure 11.21).

Most turning involves six basic workpiece materials:

1. Long chipping, such as most steels.
2. Stainless steels.
3. Short chipping, such as cast iron.
4. Heat resistant materials, such as nickel-based alloys.
5. Soft materials, such as aluminium alloys.
6. Hard materials, such as those harder than 400 HB.

This basic listing is usually sufficient for rating most standard insert geometry suitability. It can also be related to the ISO application system.

Other factors that should affect the selection of insert are related to the operation (see Figures 11.22 and 11.23):

1. Intermittent machining.
2. Vibration tendencies.
3. Limited machine power.

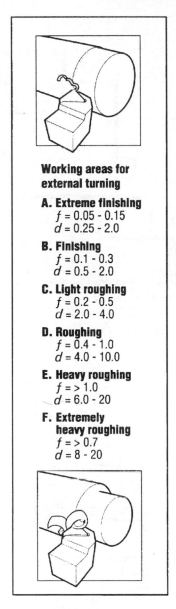

Figure 11.19 Working areas for external turning.

The International Organisation for Standardisation (ISO) gives the following classification of machinability definitions for use as a guide to selecting the correct tool material for a wide range of workpieces. The ISO classification is divided into three areas:

1. Blue P – representing machining of long chipping materials such as steel, cast iron, stainless steel and malleable iron.

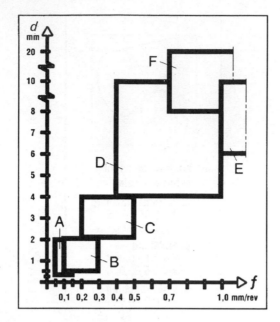

Figure 11.20 Feed (f) and cutting depth (d) ranges.

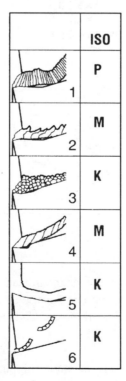

Figure 11.21 Workpiece material types (1)–(6).

Figure 11.22 Operations (1)–(3).

2. Yellow M – representing machining of more demanding materials such as austenitic stainless steel, heat resistant materials, manganese steel, alloyed cast iron, etc.
3. Red K – representing machining of short chipping materials such as cast iron, hardened steel and non-ferrous materials such as aluminium, bronze, plastics, etc.

Within each main area there are numbers indicating the varying demands of machining, from roughing to finishing. Starting at group 01 (Figure 11.24) which represents finish-turning and finish-boring with no shocks and with high cutting speed, low feed and small cutting depth, through a semi-roughing, semi-finishing area to medium-duty, general purpose at 25 and then down to group 50 for roughing at low cutting speeds and very heavy chip loads. Demands for wear resistance (WR) and toughness (T) vary with the type of operation and increase upwards and downwards, respectively.

ISO operations and working conditions

P

P01. Finish turning and finish boring, high cutting speed, small chip cross-section, high quality surface finish, close tolerance, freedom from vibration.
P10. Turning, copying, threading, milling, high cutting speed, small to medium chip cross-section.
P20. Turning, copying, milling, medium cutting speed, medium chip cross-section, facing with small chip cross-section. Mildly unfavourable conditions.
P30. Turning, milling, planing medium-to-low cutting speed, medium-to-large chip cross-section, including operations under unfavourable conditions.
P40. Turning, planing, milling, slotting, parting-off, low cutting speed, large chip cross-section, high top rake possible, very unfavourable working conditions.

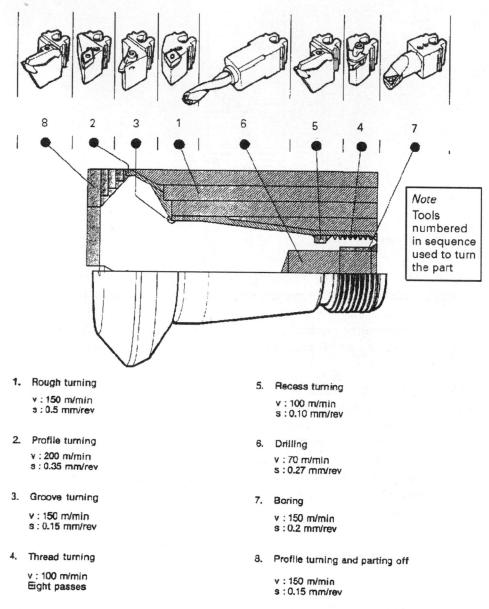

Figure 11.23 Tool selection for a typical turning operation.

1. **Rough turning**

   v : 150 m/min
   s : 0.5 mm/rev

2. **Profile turning**

   v : 200 m/min
   s : 0.35 mm/rev

3. **Groove turning**

   v : 150 m/min
   s : 0.15 mm/rev

4. **Thread turning**

   v : 100 m/min
   Eight passes

5. **Recess turning**

   v : 100 m/min
   s : 0.10 mm/rev

6. **Drilling**

   v : 70 m/min
   s : 0.27 mm/rev

7. **Boring**

   v : 150 m/min
   s : 0.2 mm/rev

8. **Profile turning and parting off**

   v : 150 m/min
   s : 0.15 mm/rev

P50. Where very great toughness is required from the tool in turning, planing, slotting, parting-off, low cutting speed, large chip cross-section, high top rake possible, extremely unfavourable operating conditions.

M

M10. Turning, medium to high cutting speed, small to medium chip cross-section.
M20. Turning, milling, medium cutting speed, medium chip cross-section.

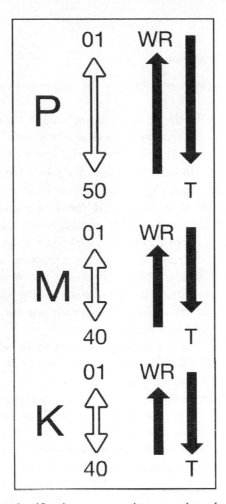

Figure 11.24 ISO tooling classifications: wear resistance and toughness comparison.

M30. Turning, milling, planing medium cutting speed, medium to large chip cross-section.

M40. Turning, profile turning, parting-off, especially in automatic machines.

K

K01. Turning, finish turning and finish boring, finish milling, scraping.

K10. Turning, milling, drilling, counter-boring, etc.

K20. Turning, milling, planing, counter-boring, broaching, operations requiring a very tough tool.

K30. Turning, milling, planing, parting-off, slotting, unfavourable conditions and possibilities of high top rakes.

K40. Turning, milling, planing, parting-off, very unfavourable conditions and very high top-rake possibilities.

8. Cutting data

Tool inclination and rake angles

The cutting ability of the edge is defined by the angles and radii that go to make up the form of the edge.

On the macrogeometry level, the insert is inclined in relation to the horizontal plane. A negative *angle of inclination* (λ) is shown in Figure 11.25A. This angle can also be positive as in Figure 11.25B. This is the angle of the insert seat in the toolholder. It is the angle, seen from the side of the tool – front to back, between the plane of the toolholder and the angle that the insert is inclined.

The inclination angle has to be negative if the *wedge angle* (β) of the cutting edge is 90°, as it most often is on inserts for strength reasons. If the 90° wedge angle insert was not inclined, there would not be any clearance underneath the cutting point against the workpiece. A *clearance angle* (α) is necessary for the cutting edge to work freely without unnecessary rubbing. However, for some types of machining, such as boring and cutting aluminium, where sharp cutting edges are necessary, the wedge angle is smaller and the angle of inclination positive. The top face of the insert is then larger than the bottom face.

It is important to note that the turning tool has to be viewed in at least two directions. A plane parallel to the main cutting edge contains the *angle of inclination* (λ), describing the front-to-back approach.

Perpendicular to the main cutting edge of the tool, in Figure 11.26, the *rake angle* (γ) is a measure of the edge in relation to the cut itself. These two angles only coincide when the tool is fed along a line at 90° to the axis of workpiece rotation, as in some facing, grooving or cutting off when the entering angle is 0° (Figure 11.27A).

The planes in which these angles are measured are at right angles to each other when the cutting action is *orthogonal* – the main cutting edge being at a 90° entering

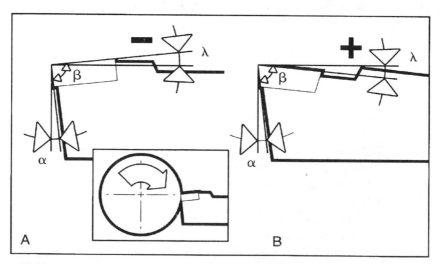

Figure 11.25 Negative and positive angle of inclination.

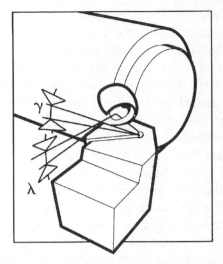

Figure 11.26 Rake and inclination angle.

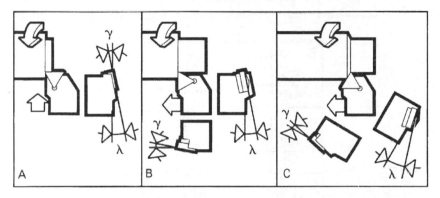

Figure 11.27 Entering angles of 90° and 60°.

angle to the axis of rotation (Figure 11.27B). Often though, the entering angle is neither 0 nor 90° degrees and the cutting action is *oblique* (Figure 11.27C). This means that the plane measuring the angle of inclination and that of the rake angle are less than 90 degrees to each other. The rake angle is also known as the *side rake angle*, as the turning tool normally cuts with the side as it is fed along the workpiece. These angles combine to present the cutting edge to the workpiece in a way that influences chip flow, forces, power, etc.

   Although the angle of inclination in the holder is usually negative, the rake angle of the cutting edge in relation to the cut, is often positive. This, however, is achieved through the insert geometry.

   On the level of actual insert geometry, the modern indexable insert has carefully designed combinations of grooves, flats, angles and curves to provide the actual cutting ability of the edge. This ability is the forming and breaking of the chip during the cut in the workpiece. The technology becomes part of the insert at manufacturing when the insert is pressed or ground.

Integrated into the edge is the rake angle. The size of the rake angle depends to a large extent upon the application area of the insert, such as the material, cutting depth and feed. The rake might function far back along the top face of the insert, as for roughing inserts (Figure 11.28B) while on finishing inserts (Figure 11.28A), the rake is available at the nose itself. Some cutting geometries give combinations (Figure 11.28C) providing the insert with a relatively wide application range as regards cutting depth and feed rate. The toolholder then, has a rake angle (γA), often negative but also neutral or positive, and the insert has a rake angle (γB), usually positive. Together this forms an effective rake angle (γE), adapted to function within the application area of the cutting edge (Figure 11.29).

The basic cross-section of the insert with a 90° wedge angle, has a decisive influence on the strength and stability of the cutting edge but does not decide the cutting ability of the edge because of the modern cutting geometry being pressed into the insert. Providing a cutting edge with a land, before the actual rake angle takes effect, is common practice to strengthen cutting edges. The land is applied in relation to the application area of the cutting edge (Figure 11.30). A finishing insert

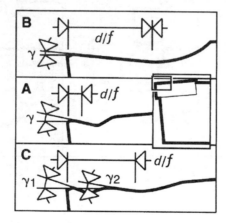

Figure 11.28 Extent of rake and chip breaker on various insert types.

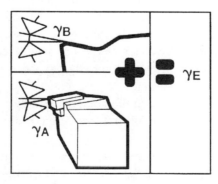

Figure 11.29 Effective rake.

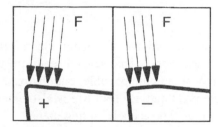

Figure 11.30 Cutting force directions change with the application of a negative land.

working with very small feeds and cutting depths may have no land at all and only some small edge rounding. A heavy roughing insert will have a large, negatively inclined land to provide the best strength. The land will redirect the cutting pressure into the cutting edge where it is well supported instead of down, and out into the more fragile cross-section of the edge.

The arguments for primary land is in some respects similar to that of rake angles. As the cutting forces act at right angles to the top face of the edge, the direction of these forces can be directed to either create tensile or compressive stresses on the edge. Compressive stresses which can be easily handled by modern cutting materials are to be preferred.

Various manufacturers offer a great deal of technical support to engineers involved in selecting the most suitable tools from the great variety that are available. One example of a computerised tool selection system is shown in Figure 11.31.

Figure 11.31 Computerised turning tool selection system.

11.4 Milling

The milling process uses a cutter with several teeth which rotates at high speed, and which moves slowly past the work. Since several teeth are usually engaged in the workpiece high metal removal rates are possible, but obviously the forces are several times larger than in single point cutting. Tool life is also extended, since each tooth spends only a small portion of time actually cutting metal.

There are many types of milling cutter (Figure 11.32) and many ways in which they can be used; we will examine some of the main classifications to illustrate the range of processes.

End milling

An end mill typically has four cutting edges, and is normally manufactured from high speed steel, although end mills utilising brazed on helical carbide teeth have been developed, and small diameter cutters are available in solid tungsten carbide. A recent innovation is milling cutters with extremely fine grained TiN particles bonded to a tough steel core, so that the cutting edges are composed of these extremely fine particles, approximately one thousandth of the size of the HSS grain. These can be

Figure 11.32 A heavy duty milling operation on a horizontal milling machine.

ground to an extremely sharp edge, which they keep over a long cutter life. They are also frequently given a further coating of TiN or TiCN about two microns thick to extend the cutter life. The use of TiN as a PVD coating on a matrix consisting of around 50 per cent TiN produces excellent bonding of the coating, and allows for the regrinding of the rake face of the cutters without damaging the coating on the cutting face.

Cutting operation

End mills do not have cutting teeth across all of the end, but have a 'dead' area in the centre, and thus can only be used for milling using the side of the cutter and the outer periphery of the ending end (Figure 11.33).

While end mills have traditionally had four cutting flutes, recently cutters with three flutes and higher helix angles have been introduced. These have one cutting edge ground right to the centre, and thus can be used to drill directly into the workpiece when creating a recess. Cutters designed for roughing work may have chip-breaking grooves in the flutes, while cutters used for finishing work will have continuous flutes. A high helix angle provides lower loads per cutting edge and a higher degree of accuracy and finish on deep cuts.

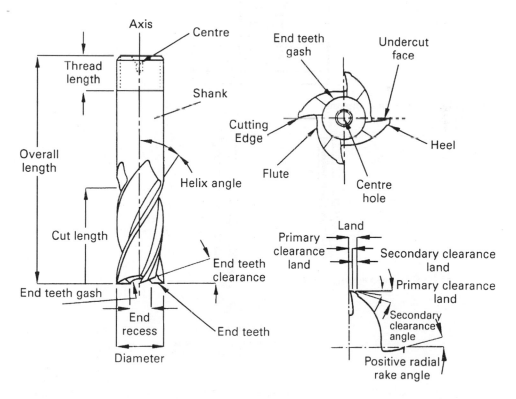

Figure 11.33 End mill cutter geometry.

A special class of end mill is termed the *slot drill*. This is a cutter intended for slotting and keyway cutting. It is provided with end cutting edges which are offset about the centre, so that the longer cutting edge passes slightly over the centre, allowing plunge cutting into the work. It lacks the self-centering action of a true drill, and thus tends to give rise to sideways deflection of cutter or workpiece if there is any lack of rigidity in the set-up (Figure 11.34). For this reason slot drills are kept short.

A useful derivative of the slot drill is the *ball nose cutter*, again with two flutes and with end cutting faces carried slightly over centre. Various contours on workpieces are required in copy-milling (Figure 11.35) and CNC machining. This profiling ability is essential in tool making, and although groove and pocket making cutters can machine some contours, true profiling ability comes from the use of round cutting edge end mills such as ball nose and, to a lesser extent, round insert end mills. The ability not only to mill, but also to drill and follow, and to cut round intricate forms – at efficient machining rates – demands a lot of the tool.

For continuous machining along concave and convex forms in tool steels, the end cutting edge has to be round, with an effective all-round machining capacity and long life. This has been achieved both with a brazed carbide cutting edge and with

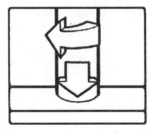

Figure 11.34 Slot milling motion.

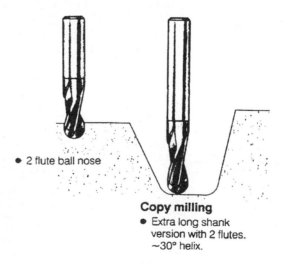

- 2 flute ball nose

Copy milling
- Extra long shank version with 2 flutes. ~30° helix.

Figure 11.35 Copy milling: (a) two flute ball nose; (b) extra long shank version with two flutes (~30° helix).

indexable inserts. The insert type of cutter is intended for roughing and semi-finishing operations, while the brazed type is precision ground for finishing.

This same ball nose indexable insert concept has been broadened to include large size cutters where the cutting edge is made up of several inserts accurately staggered to provide a smooth cutting action. The centre point of the ball nose endmill has an efficient geometry for drilling, it should be noted, however, that a round-form deviation may occur at the centre point and some profile cutting accuracy may be lost (see Figure 11.36).

End mills (Figure 11.37) are obviously subject to severe torsional and bending stresses in use. These limit the size of cut that can be taken. With a cut equal to the full width of the cutter, the maximum recommended depth of cut (a_p) is 0.6 D. If, however, the cutting action is the cleaning up of the edge of a component, with the cut only 10 per cent of the diameter, the depth can be increased to 1.5 D.

The effect of these deflections is that it is not possible to cut an accurate slot with a sized cutter. To accurately cut a, say, 12 mm wide slot with a 12 mm cutter is not possible; a 10 mm cutter should be used, followed by a 1 mm finishing cut (Figure 11.38) around the periphery to finish the slot to exactly 12 mm.

Larger cutters

Cutters above 25 mm diameter offer the possibility of using carbide inserts bolted onto a steel cutter body (see Figures 11.39, 11.40, 11.41, 11.42, 11.43). These allow high metal removal rates by the use of high feeds and speeds.

Face milling

Face milling usually means using a large diameter cutter with several cutting teeth

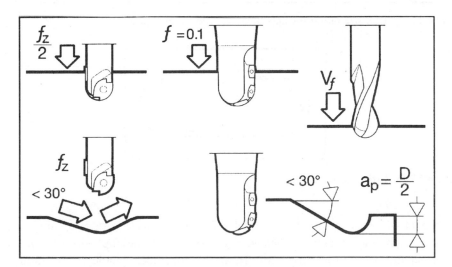

Figure 11.36 Possible milling operations with round insert cutters.

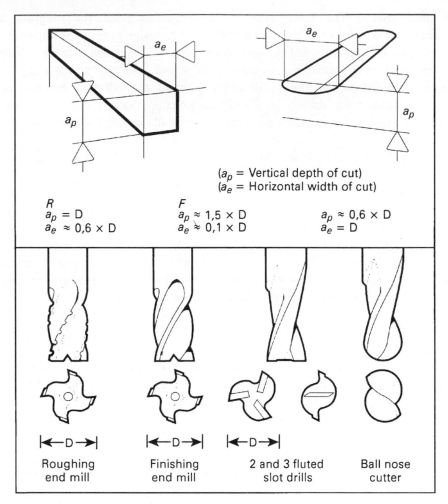

Figure 11.37 Top: size of milling cuts; bottom: types of end mills.

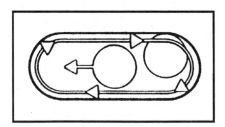

Figure 11.38 1 mm finishing cut.

(Figure 11.44). It is used to produce flat surfaces in a wide variety of applications, e.g. the machining of cylinder heads. High powers can be applied by the use of a cutter having very little projection from the machine spindle, thus minimising deflections (Figure 11.44(a)).

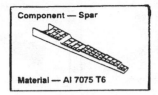 

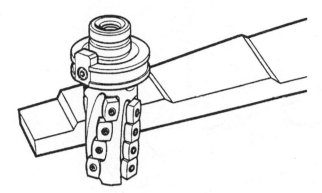

Figure 11.39 Edge finishing cutter.

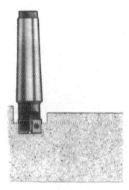

Figure 11.40 Carbide tipped end mill.

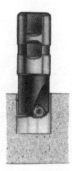

Figure 11.41 'Drilling' end mill.

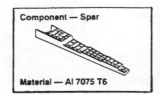

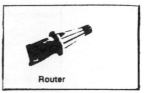

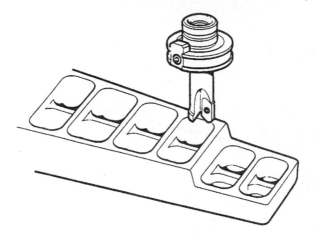

Figure 11.42 Pocketing end mill.

(a)

(b)

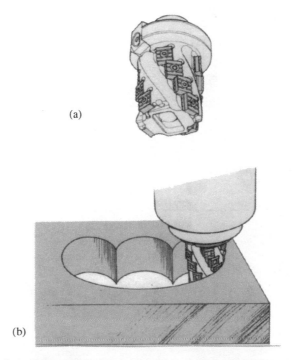

Figure 11.43 Long edge cutter: (a) cutter; (b) typical application.

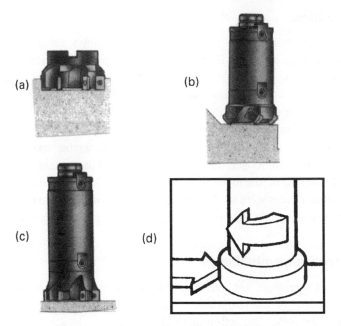

Figure 11.44 (a) Face mill; (b) face mill with 45° angled edge cutters; (c) extended shark face mill; (d) basic face milling motions.

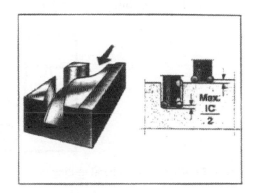

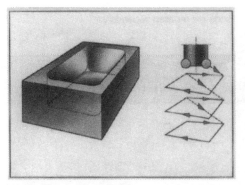
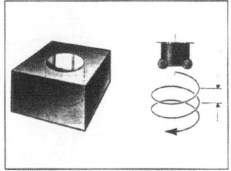

Figure 11.45 Pocketing and slotting operation using a face mill.

Peripheral milling

Peripheral milling uses a 'side and face' cutter mounter on a shaft (the arbor) supported by bearings on each side of the cutter, providing resistance to deflection due to cutting forces (Figures 11.46, 11.47, 11.48). The 'side and face' cutter (Figure 11.49) has strong teeth designed to cut both on the side face and the outer diameter, and may be used to cut on the face of the tooth, on one or other side of the tooth, or on any combination of these.

The 'slotting' cutter (Figure 11.47) is a variant on this theme, specially designed for cutting slots. The teeth are longer, and thus less rigid, and the cutting is done only on the periphery, with the side of the tooth only cleaning up the side of the slot to accurate size and finish. If such a cutter were used to cut on one face only it would distort and possibly fracture. Slotting cutters may have alternate teeth offset to provide chip clearance.

Milling operation terminology

Since several cutting teeth are available on each cutter, the terminology used for single point cutting is inappropriate when describing milling operations. In the succeeding text the following definitions will be used:

- Feed per revolution (f). This is defined as the distance the tool moves during one revolution, normally in mm/rev. It is used in feed calculations and in determining the finishing capability of a face mill.

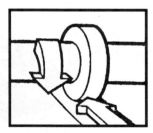

Figure 11.46 Basic peripheral milling motions.

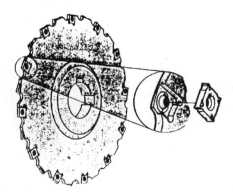

Figure 11.47 Slotting cutter.

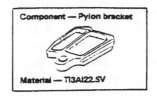

Figure 11.48 Special slotting operation.

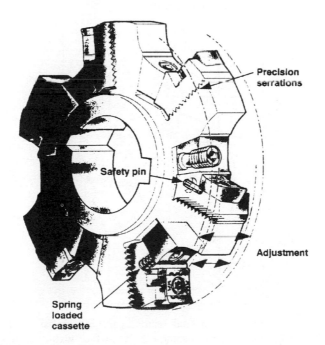

Figure 11.49 Side and face milling cutter.

- Feed per tooth (f_z). This is defined as the linear distance moved by the tool while one particular tooth is engaged in the cut. It is a key value in milling. As the milling cutter is a multi-edged tool, a value is needed to ensure that each edge machines under satisfactory conditions. It is the capability of each tooth which sets the limit for the tool. It is effectively the distance covered by the table feed between the engagement of two successive cutting edges, and is determined by the number of cutting faces on the tool, the speed of rotation and the speed of traverse of the table. (See Figures 11.50, 11.51, 11.52, 11.53, 11.54.)

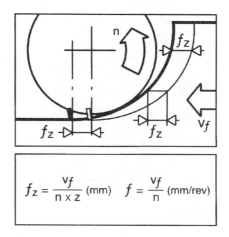

$$f_z = \frac{v_f}{n \times z} \text{ (mm)} \quad f = \frac{v_f}{n} \text{ (mm/rev)}$$

Figure 11.50 Feed per revolution and feed per tooth.

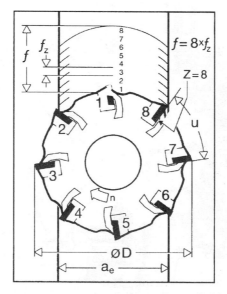

Figure 11.51 Feed per tooth (f_z) is a vital factor in milling, decisive for: metal removal per edge, load per edge, tool life and to some extent, surface texture (u is the pitch of the cutter).

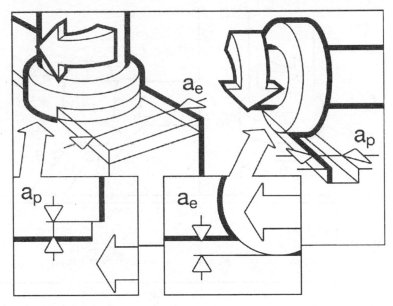

Figure 11.52 Cutting depth in face and peripheral milling.

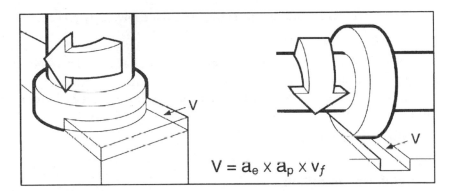

$$V = a_e \times a_p \times v_f$$

Figure 11.53 Metal removal rate.

- Depth of cut. Axial (a_p) in face milling, radial (a_e) in peripheral milling. This is the amount the tool removes from the workpiece, i.e. the distance the tool is set below the unmachined surface.
- Cutting width. Radial (a_e) in face milling, axial (a_p) in peripheral milling. This is the amount by which the tool covers the workpiece surface.
- Spindle speed (n) in rpm is the number of revolutions the milling tool on the spindle makes in unit time. This is a machine-oriented value, and does not say much about what is happening at the periphery, where the cutting is taking place.
- Cutting speed (V_c) in m/min on the other hand denotes the surface speed at which the cutting edge machines the workpiece. This is an important tool-oriented value, and that part of the cutting data which ensures that the operation is carried out with maximum efficiency by the tool in question.

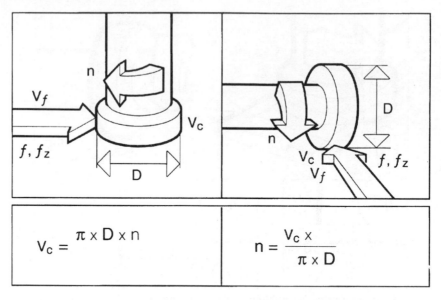

Figure 11.54 Milling definitions.

- The spindle speed, tool diameter, and cutting speed are obviously related:

$$V_c = \pi D_n \qquad n = V_c / \pi D$$

- Feed speed or table speed (V_f) is the feed of the tool against the workpiece in distance per time unit.
- The volume of metal (V) removed in unit time can be established from these definitions:

$$V = a_p \times a_e \times f \times N \text{ mm}^3/\text{min}$$

Up-cut and down-cut milling

In principle, the milling cutter rotates while the workpiece is fed against it. There are two different ways in which this may be achieved, depending on the rotation of the tool in relation to the workpiece. The workpiece may be fed either with or against the direction of rotation of the cutter, and this affects the character of the cut, especially the start and finish. During *up milling* (U) (also called conventional milling) (Figure 11.55), the feed direction of the workpiece is opposite to that of the cutter rotation at the area of cut. *The chip thickness starts at zero and increases to the end of the cut.*

In up milling, with the insert starting its cut at no chip thickness, there are high cutting forces which tend to push the cutter and workpiece away from each other. The insert has to be forced into the cut, creating a rubbing or burnishing effect with excessive friction, high temperatures and often contact with a work-hardened surface, caused by the preceding insert. Forces (F) will also tend to lift the workpiece up from the table, which means precautions need to be taken in the fixturing.

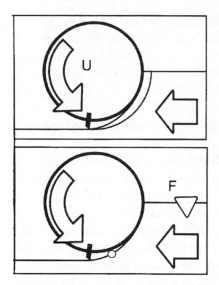

Figure 11.55 Up-cut milling.

During *down milling* (*D*) (also called climb milling) (Figure 11.56), the workpiece feed direction is the same as that of the cutter rotation at the area of cut. *The chip thickness will decrease from the start of the cut until it is zero at the end of the cut in peripheral milling.*

The cutter thus starts with a positive 'bite' into the workpiece, and avoids the rubbing action in up-cut milling. The forces also tend to hold the workpiece down onto the machine table. They also, however, tend to drag the work into the cutter, and the machine

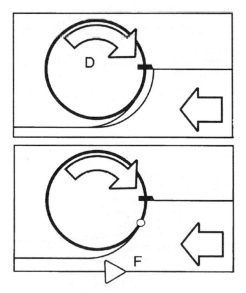

Figure 11.56 Down-cut milling/climb milling.

must have a backlash elimination system fitted to the leadscrew to avoid damage to the workpiece or the machine (and possibly to the operator). All CNC machines will be suitable for down-cut milling, but care must be taken to ensure that older, conventional machines have a backlash eliminator fitted before being used for down milling.

11.5 Peripheral milling – geometry of chip formation

The geometry of chip formation depends upon the path of the cutter edge across the workpiece. The cutter rotates with uniform angular velocity about a fixed axis; the work moves towards the cutter at a uniform feed rate and the path of the cutter relative to the workpiece is determined by a combination of rotary and feed motions. If the workpiece is regarded as stationary, and a motion equal and opposite to the feed motion is given to the cutter axis, the path of the tooth across the workpiece can be set out.

Figure 11.57 illustrates the cutting conditions: the cutter axis is imagined to lie at the centre of a disc of radius r (such that $r = f/2\pi$, where f = feed/rev) which rolls without slip along the straight edge ST. A point Q on the cutter periphery, at radius R from the axis of the disc, will trace out the path of the cutting edge relative to the workpiece. The curve described by Q is a *trochoid*. The chip cut lies between the trochoid AB, traced by the cutting edge, and the trochoid AC cut by the previous tooth. For *up-cut* milling the chip is cut from A to B; the undeformed chip thickness then increases from zero at A to a maximum (t_{max}) just before B is reached.

For a trochoid, the centre of curvature of Q is at U, the point where the rolling disc touches the straight edge, and the velocity of Q relative to the workpiece is perpendicular to UQ and not to OQ.

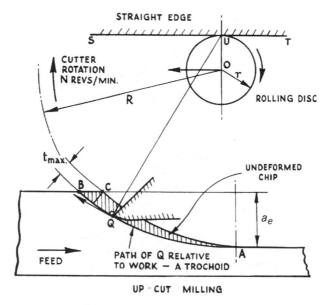

Figure 11.57 Geometry of chip formation, up-cut milling.

If the direction of the feed is now reversed, as shown in Figure 11.58, *down-cut* milling, the chip will be cut from B to A, and t_{max} will occur almost at the start of the cut. The curve BA is a different portion of the trochoid traced by the cutting edge Q, and the path length from B to A is shorter than the equivalent length A to B for up-cut milling.

Because the path of the cutting tooth across the workpiece when down-cut milling comes from a different portion of the trochoid than for up-cut milling, there are small differences between the magnitudes of t_{max}, of chip length and of effective cutting clearance angle between these methods, the cutting conditions being otherwise identical. The mathematics of the trochoid is involved and does not give exact expressions in a convenient form, but approximate solutions, which provide a sufficient demonstration of the above differences, may be found as follows.

Undeformed chip length (AB)

Consider a small element of the trochoid AB (Figure 11.59) generated by rotation of the cutter through a small angle $\delta\theta$. The length of the element PQ depends upon two displacements:

1. $PT = R\delta\theta$, due to rotation of the cutter.
2. $TQ = r\delta\theta$, due to the feed motion of the work.

The vector sum of these displacements is given by

$$\delta\theta[(R \cos \theta + r)^2 + R^2 \sin^2 \theta]^{1/2}$$

and does not lead to a solution in a convenient form.

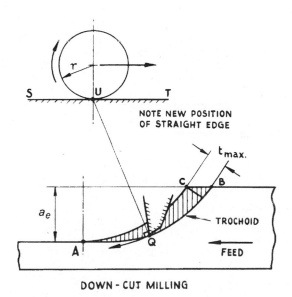

DOWN - CUT MILLING

Figure 11.58 Geometry of chip formation, down-cut milling.

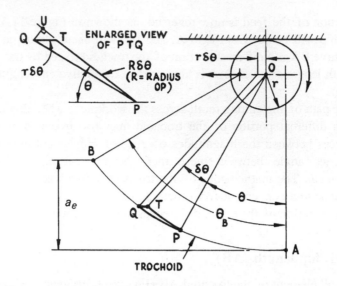

Figure 11.59 Determination of undeformed chip length.

Extend PT (see enlarged view in Figure 11.59) to the point U such that TUQ is a right angle. Now, since r is very small in relation to R, length PQ is very nearly the same as length PU.

$$PQ \approx PU \quad \therefore \quad PQ \approx R\delta\theta + r\delta\theta \cos \theta$$
$$= (R + r \cos \theta) \, \delta\theta$$

Hence

$$AB \approx \int_0^{\theta_B} (R + r \cos \theta) \, d\theta$$

$$= R\theta_B + r \sin \theta_B \quad \text{(up-cut milling)} \tag{11.1}$$

Similar reasoning gives:

$$BA \approx R\theta_B - r \sin \theta_B \quad \text{(down-cut milling)} \tag{11.2}$$

From Figure 11.60 it can be seen that

$$CD = r \sin \theta_F \tag{11.3}$$

which is very nearly the same as $r \sin \theta_B$
    Also, by similar triangles

$$CD = \frac{r}{R} \, (EF) \tag{11.4}$$

and by Pythagoras' theorem

$$EF = [R^2 - (R - a_e)^2]^{1/2} = (2Ra_e - a_e^2)^{1/2} \tag{11.5}$$

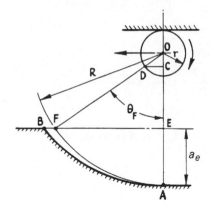

Figure 11.60

Let

$$f = \text{feed/rev of cuter, i.e. } f = 2\pi r$$

hence

$$r = f/2\pi \tag{11.6}$$

By substitution of (3), (4), (5) and (6) in (1) and (2):

$$AB \simeq R\theta_B + \frac{f}{2\pi R}(2Ra_e - a_e^2)^{1/2} \qquad \text{(up-cut milling)} \tag{11.7}$$

$$BA \simeq R\theta_B - \frac{f}{2\pi R}(2Ra_e - a_e^2)^{1/2} \qquad \text{(down-cut milling)} \tag{11.8}$$

Equations 11.7 and 11.8 are given in the Cincinnati reference book, *A Treatise on Milling and Milling Machines*.

The difference in the undeformed chip length as between up-cut and down-cut milling is now seen to be $(f/\pi R)(2Ra_e - a_e^2)^{1/2}$, the length for down-cut milling being the shorter: This difference increases in direct proportion to the feed.

Although this difference is very small, it is a difference in the amount of *sliding* which occurs between the cutter edge and the material cut, and it leads to a slight reduction in the amount of cutter wear achieved by changing from the up-cut to the down-cut method. As will be shown later, it is not the most significant reason for the improvement in tool life normally resulting from this change.

Example 11.1

A 100 mm diameter cutter, having 8 teeth, cuts at 24 m/min. The depth of cut is 4 mm and the table feed 150 mm/min. Find the percentage reduction in sliding between the cutter edges and the material cut, which results from a change from up-cut to down-cut milling.

Solution

$$\text{Spindle speed} = \frac{10^3 \times 24}{\pi \times 100} = 76 \text{ rev/min}$$

$$\text{Feed/rev of cutter} = \frac{150}{76} = 1.974 \text{ mm}$$

Let

$\quad\quad \theta = $ angle of engagement between cutter and work

$$\cos \theta = \frac{50 - 4}{50} = 0.920 \quad \theta = 23°$$

$$= 0.403 \text{ rads.}$$

$$r = \frac{1.974}{2\pi} = 0.314 \text{ mm}$$

Path length AB $= 50 \times 0.403 + 0.314 \sin 23°$
$$= 20.15 + 0.123 = 20.273 \text{ mm}$$

Reduction of path length resulting from change of method

$$= 2 \times 0.123 = 0.246 \text{ mm}$$

$$\% \text{ reduction in sliding} = \frac{0.246}{20.273 \times 100} = 1.2\%$$

□   □   □

The increase of cutter life on account of reduced sliding will be small, and the improvement usually achieved by a change from up-cut to down-cut milling must depend upon additional factors.

Feed rate and cutter wear

The life of a cutter is influenced by the amount of sliding which occurs between the teeth and the work. If a workpiece of length l mm is cut at a feed of f mm/rev the number of rotations of the cutter to remove the metal is l/f (approach distance neglected), and each tooth travels approximately $R\theta_B l/f$ mm through the material. It is obvious that high values of f reduce the amount of sliding per unit volume of material removed. For this reason milling should be done at moderate cutting speeds and high feed rates. The normal upper limit of feed rate will depend either upon the mechanical strength of the cutter or workpiece, or upon the power available at the machine spindle. Vibration may be a limiting factor.

Depth of cut and cutter wear

The depth of cut (a_e) is also significant in relation to cutter wear. The volume of metal removed per pass $\propto a_e$; the cutter wear $\propto R\theta_B$. It can be seen from Figure 11.60 that θ

increases as a_e increases according to the relationship $\cos \theta = (R - a_e)/R$. For typical values of R and a_e an increase in a_e is always proportionally greater than the accompanying increase in θ. The volume of metal removed per pass is thus seen to increase faster than the amount of sliding between the cutter and workpiece, for an increase in a_e. The best relationship between cutter life and volume of metal removed is achieved by taking one pass only at the full depth of cut required. A further advantage is demonstrated on p. 369, namely the torque at the arbor gets smoother as a_e is increased.

Cutter life should also be considered in relation to the principles given on pp. 267–275, Figures 10.6 and 10.7. The main differences between cutting with single-point tools and milling are differences in the geometry of chip formation, which for milling involves intermittent engagement of the cutting edges with the workpiece.

Chip thickness

As may be seen from Figures 11.61 and 11.62, the chip thickness varies during cutting.

When the tool feed axis does not intersect the tool engagement in the workpiece, the chip thickness should be checked as it is less than the feed per tooth. The maximum chip thickness (h_x) should be checked along with the entering angle to ensure a satisfactory entry into cut and to avoid overloading the edge. In face milling, the maximum chip thickness is where up milling changes to down milling (Figure 11.61).

The chip thickness is variable in milling and often complicated to determine. Therefore it is more practical, and more representative in many respects to work with the *average chip thickness* (h_m). This has been established as a key value, especially

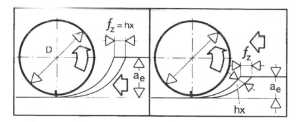

Figure 11.61 Chip thickness in up-cut milling.

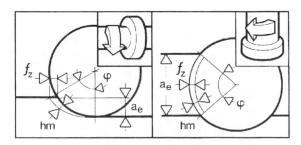

Figure 11.62 Chip thickness in down-cut and face milling.

in peripheral milling where the described relationship between a_e and D occurs. This can be seen in the form of the arc (φ) generated by the tool. In addition to being a measure of the chip load, it is also a factor in power requirements (Figure 11.62).

The average chip thickness is calculated through a relatively simple formula for peripheral and face milling. The feed per tooth is obviously one value as well as cutting depth and cutting width, in the respective milling methods. The diameter of the cutter is also an influential factor as it affects the arc of cut. Also the entering angle affects the chip thickness.

The average chip thickness value should reach certain minimums. For peripheral milling, this is generally in the region of 0.04–0.2 mm. For face milling, 0.1 mm is often regarded as a minimum depending on factors such as material hardness and cutting geometry. Suitable values are provided for each cutter type, enabling the calculation of the feed per tooth.

In face milling, it is often of sufficient accuracy to let the feed per tooth equal the value of the average chip thickness. There may be an exception with a small entering angle. For an entering angle of 45°, the multiplying factor for correction is 0.71. A minimum of 0.1 mm and 0.4 mm as the maximum are values recommended for feed per tooth. A small feed value gives poor usage of inserts and of power (Figure 11.63).

The maximum thickness, t_{max}, occurs almost at the end of the cut for up-cut milling and at the start of the cut for down-cut milling.

A close approximation for the value of t_{max} can be obtained by reference to Figure 11.64.

By similar triangles,

$$\frac{f_t}{t_{max}} = \frac{L}{(2Ra_e - a_e^2)^{1/2}} \tag{11.9}$$

where f_t = the feed/tooth.

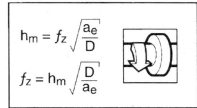

$$h_m = f_z \sqrt{\frac{a_e}{D}}$$

$$f_z = h_m \sqrt{\frac{D}{a_e}}$$

$$h_m = \frac{\sin \kappa \times 180 \times a_e \times f_z}{\pi \times D \times \arcsin\left(\frac{a_e}{D}\right)}$$

ae/D	fz (mm)										
	0,04	0,08	0,10	0,16	0,20	0,30	0,40	0,50	0,60	0,80	1,0
	hm (mm)										
1/50					0,03	0,04	0,06	0,07	0,08	0,11	0,14
1/40				0,03	0,03	0,05	0,06	0,08	0,09	0,13	0,16
1/25				0,03	0,04	0,06	0,08	0,10	0,12	0,16	0,20
1/20				0,03	0,04	0,07	0,09	0,11	0,13	0,18	0,22
1/10			0,03	0,05	0,06	0,09	0,12	0,16	0,19	0,25	
2/10		0,03	0,04	0,07	0,09	0,13	0,17	0,22	0,26		
3/10		0,04	0,05	0,08	0,10	0,16	0,21	0,26			
4/10		0,05	0,06	0,09	0,12	0,18	0,23				
5/10	0,03	0,05	0,06	0,10	0,13	0,19	0,25				

Figure 11.63 Recommended chip thickness values.

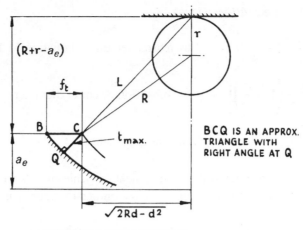

MAXIMUM CHIP THICKNESS

Figure 11.64 Maximum chip thickness.

Also

$$L^2 = (R + r - a_e)^2 + 2Ra_e - a_e^2$$
$$= (R + r)^2 - 2ra_e \qquad\qquad (11.10)$$

by substitution of equation 11.10 in 11.9

$$t_{max} = f_t \left[\frac{2Ra_e - a_e^2}{(R + r)^2 - 2ra_e} \right]^{1/2} \qquad \text{for up-cut milling}$$

Similarly, it can be shown that:

$$t_{max} = f_t \left[\frac{2Ra_e - a_e^2}{(R - r)^2 - 2ra_e} \right]^{1/2} \qquad \text{for down-cut milling}$$

These expressions are not very convenient, and by further approximations more suitable expressions for general use are obtainable.

1. r is generally small and can be neglected, to give,

$$t_{max} = f_t \left[\frac{2Ra_e - a_e^2}{R^2} \right]^{1/2} \qquad\qquad (11.11)$$

Any distinction between the values of t_{max} for up-cut or down-cut milling has now disappeared.

2. If a_e is small relative to R (shallow cuts), a_e^2 can be neglected to give

$$t_{max} = f_t \left(\frac{2Ra_e}{R^2} \right)^{1/2} = 2f_t \left(\frac{a_e}{D} \right)^{1/2} \qquad\qquad (11.12)$$

where D = diameter of the cutter.

3. If a_e cannot be neglected (deep cuts),

$$t_{max} = f_t \left[\frac{a_e}{R} \left(2 - \frac{a_e}{R} \right) \right]^{1/2} \qquad \text{by rearrangement of equation 11.11}$$

$$= 2f_t \left[\frac{a_e}{D} \left(1 - \frac{a_e}{D} \right) \right]^{1/2} \qquad \text{by substition of cutter diameter}$$

$$= \frac{2}{D} f_t [a_e(D - a_e)]^{1/2} \qquad \text{by rearrangement} \qquad (11.13)$$

Equation 11.12 can be obtained by the development of Schlesinger's formula for mean chip thickness. As shown in Figure 11.65, t_{mean} is assumed to occur at angular position $\dfrac{\theta}{2}$,

From the diagram,

$$t_{mean} = f_t \sin \frac{\theta}{2}, \qquad (11.14)$$

but

$$\sin \frac{\theta}{2} = \left[\frac{1 - \cos \theta}{2} \right]^{1/2}$$

and

$$\cos \theta = \frac{R - a_e}{R}$$

hence

$$\sin \frac{\theta}{2} = \left[\frac{a_e}{2R} \right]^{1/2} = \sqrt{\frac{a_e}{D}} \qquad (11.15)$$

where D = cutter diameter.

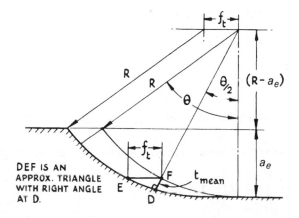

DEF IS AN APPROX. TRIANGLE WITH RIGHT ANGLE AT D.

Figure 11.65 Geometry of Schlesinger's formula for mean chip thickness.

Substituting equation 11.15 in 11.14,

$$t_{mean} = f_t \sqrt{\frac{a_e}{D}} \text{ (Schlesinger's formula)}$$

Since $t_{max} = 2t_{mean}$, Schlesinger's formula agrees with equation 11.12. It should be appreciated that Schlesinger's method is not as accurate for deep cuts as equation 11.13.

Example 11.2

A side and face cutter 125 mm diameter has 10 teeth. It operates at a cutting speed of 14 m/min and table feed of 100 mm/min. Find the maximum chip thickness: (i) for a cutting depth of 5 mm; (ii) for a cutting depth of 25 mm.

Solution

$$\text{Spindle speed} = \frac{10^3 \times 14}{\pi \times 125} = 35 \text{ rev/min}$$

$$f_t = \frac{100}{35} \times \frac{1}{10} = 0.286 \text{ mm}$$

From equation given for (i)

$$t_{max} = 2 \times 0.286 \sqrt{\frac{5}{125}} = 0.114 \text{ mm}$$

From equation given for (ii)

$$t_{max} = \frac{2 \times 0.286}{125} \sqrt{(25 \times 100)} = 0.229 \text{ mm}$$

□   ⊓   □

It has been assumed so far that each tooth of a cutter cuts an equal chip, a condition unlikely to occur in practice due to the eccentric running of cutters and arbors. Where t_{max} is very small, some of the teeth may not cut a chip while others may cut chips of twice the estimated thickness. Eccentric running may cause cutter breakage by the overloading of some teeth; it may also cause unnecessary cutter wear because some of the teeth rub over the surface without cutting a chip. This rubbing is a serious disadvantage when cutting metals which work-harden rapidly.

11.6 Cutting forces and power

Figure 11.66 shows those components of the force exerted by the work on a cutting tooth, which act in a plane perpendicular to the cutter axis. The axial load on the cutter will be treated separately.

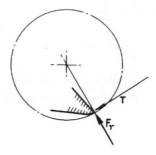

Figure 11.66 Forces acting on cutter tooth.

T, the tangential force, determines the torque on the cutter; F_r, the radial force, may be regarded as the rubbing force between the workpiece and the tooth. The value of T will depend upon the chip area being cut and on the specific cutting pressure. Work done in cutting a chip $= \int_0^\epsilon T \, dl$, where l is the undeformed chip length.

Direct measurement of forces T and F_r is difficult because they are oscillating rapidly during the cutting; a suitable dynamometer must be one capable of recording the fluctuations on a time base. From the general principles deduced from cutting data obtained when turning metals, it is possible to deduce the approximate trend of the forces which occur in up-cut and down-cut milling.

Figure 11.67 shows the trend of these forces, in relation to the undeformed chip section for up-cut milling. Note the upward surge of force F_r near the start of the cut. This is caused by cutting from zero chip thickness at A so that high elastic forces must be exceeded before the tooth edge penetrates the metal to be cut. Obviously at this point T has a value which includes the frictional resistance of the tooth to sliding over the metal. For this reason the values of T, and of the work done in cutting the chip, tend to be higher for up-cut than for down-cut milling. Comparison of Figure 11.67 with Figure 11.68 should make this clear.

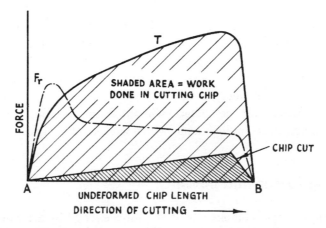

Figure 11.67 Relationship between forces acting on a cutter tooth and undeformed chip thickness, up-cut milling.

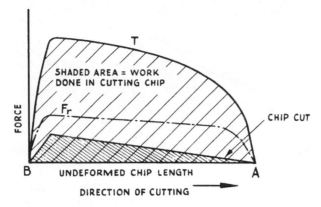

Figure 11.68 Relationship between forces acting on a cutter tooth and undeformed chip thickness, down-cut milling.

The high initial value of F_r causes rapid wear of cutter edges, and this has the effect of work-hardening the surface which the following teeth must penetrate. In the milling of materials which work-harden readily the effect, as shown in Figure 11.69, is quite serious; the cutting edges on the clearance side of the teeth rapidly develop a polished land. If dull cutters are kept in service F_r rises to a very high value and pieces of metal may fracture from the rake face, as shown in Figure 11.70. It is obvious that such conditions are unfavourable for the application of carbide tooth cutters because of their brittleness.

Since down-cut milling avoids the conditions described in the above paragraph, an improved cutter life should generally result from a change to this method. It is normally employed when peripheral milling with carbide tooth cutters (e.g. deep slot milling).

Total forces acting on a cutter

Suppose three teeth are in engagement with the workpiece and the down-cut method of milling is employed. Figure 11.71 illustrates these conditions and shows how the

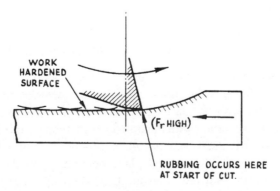

Figure 11.69 Rubbing action caused by radial force.

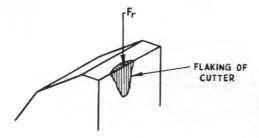

Figure 11.70 Compression fracture at cutting edge caused by very high radial force.

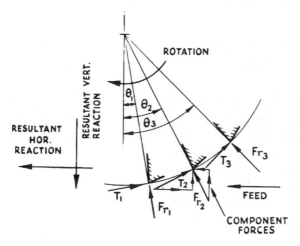

Figure 11.71 Total reactions of cutter on workpiece, down-cut milling.

forces acting may be resolved to give the magnitude of the reactions of the cutter on the workpiece.

$$\textit{The vertical reaction} = T_1 \sin \theta_1 + T_2 \sin \theta_2 + T_3 \sin \theta_3$$
$$+ F_{r1} \cos \theta_1 + F_{r2} \cos \theta_2 + F_{r3} \cos \theta_3$$

For all possible values of θ the vertical reaction is downwards forcing the workpiece onto the machine table.

$$\textit{The horizontal reaction} = T_1 \cos \theta_1 + T_2 \cos \theta_2 + T_3 \cos \theta_3$$
$$- F_{r1} \sin \theta_1 - F_{r2} \sin \theta_2 + F_{r3} \sin \theta_3$$

The horizontal reaction will be in the *same direction* as the table feed for small values of θ and will fall in value (may even become reversed in direction) as θ is increased. A feed drive with backlash eliminator, or a suitable hydraulic feed drive, must be employed in association with the down-cut method of milling.

A similar analysis of the reactions when up-cut milling will show that the direction of the reaction in the vertical plane depends upon the value of θ; it is possible for the work to be lifted from the table over unclamped sections, due to flexure of the workpiece, when deep slots are being milled.

Figure 11.72 shows the axial force acting on a cutter due to the spiral angle of the cutting edge. The value of T may be taken as the sum of the separate values T_1, T_2, etc. Helical cutters should be mounted so that force A pushes the arbor into the spindle nose. Milling machines have bearings designed to carry thrust loads in either direction. The arbor, however, is secured in a non-stick taper by means of a long slender drawbar. Elastic extension of the drawbar may become a source of vibration if force A, which varies during cutting, greatly increases the tensile load in the bar.

Work done in cutting

Figure 11.73 shows the area being cut at rotation angle θ when a straight-toothed cutter is employed. The value of T for this position is given by $T = K_c bt$, where K_c is the specific cutting pressure. Figure 11.74 shows the variation of K_c which occurs with variation of chip thickness, but since t is a function of θ, K_c also is a function of θ.

The work done in cutting a chip (W_c) = torque × angle turned

$$W_c = Rb \int_0^{\theta_B} K_c t a_e \theta, \text{ where } R \text{ is the cutter radius.}$$

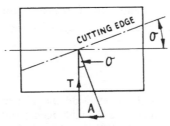

A = AXIAL FORCE ON CUTTER

Figure 11.72 Axial force on helical cutter.

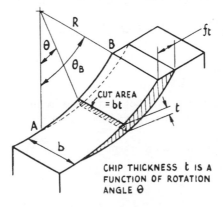

CHIP THICKNESS t IS A FUNCTION OF ROTATION ANGLE θ

Figure 11.73 Changes in chip thickness as the cutter rotates.

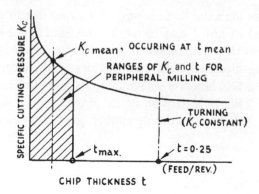

Figure 11.74 Range of specific cutting pressure involved in peripheral milling, compared with turning.

The relationships between t and θ and between K_c and θ are complex, so that the above expression does not lead to practical solutions. An approximate solution is possible in terms of mean values of t and K_c.

Schlesinger's formula gives $t_{\text{mean}} = f_t \sqrt{(a_e/D)}$, and values of K_c for the mean chip thickness can be reasonably estimated from test data, hence,

$$W_c = R\theta_B bf_t \sqrt{(a_e/D)} K_{c\text{mean}} \tag{11.16}$$

If a cutter has N teeth and rotates at S rev/min the total work done/min in cutting $= W_c NS$, and from this expression a value for the power required at the cutter can be estimated.

Example 11.3

A milling cutter is to produce a slot 20 mm wide and 12 mm deep. The cutter is 100 mm diameter, has 10 teeth, cuts at 25 m/min, and a table feed of 125 mm/min is to be employed. If $K_{c\text{mean}} = 4200$ N/mm^2, estimate the power required at the cutter.

Solution

An expression for power can be deduced from the formulae developed on p. 365.

$$\text{Power (watts)} = \frac{W_c NS}{60} \qquad W_c = R\theta_B bf_t \sqrt{\frac{a_e}{D}} K_c$$

$$\text{Power} = \frac{R\theta_B bf_t NS}{60} \sqrt{\frac{a_e}{D}} \times K_c$$

Examination of the units employed will show that R must be in metres.
   Let F = table feed, mm/min. Then $F + f_t NS$ hence

$$\text{Power} = \frac{R\theta_B bF}{60 \times 10^3} \sqrt{\frac{a_e}{D}} \times K_c \quad \text{for } R \text{ in mm}$$

For the given conditions,

$$\cos \theta_B = \frac{38}{50}, \; \theta_B = 0.707 \text{ rads}$$

$$\text{Power} = \frac{50}{10^3} \times 0.707 \times 20 \times 125 \times \sqrt{\frac{12}{100} \times \frac{4200}{60}} \times 10^{-3} \text{ kW}$$

$$= 2.15 \text{ kW}$$

☐ ☐ ☐

The method may appear to give the power at the cutter without taking the cutting speed into account. However, if the table feed remained constant, any increase in cutting speed would reduce the feed/tooth; the specific cutting pressure would then rise because the chip thickness would be reduced, and more power would be required. So long as the table feed F is unchanged the metal removal rate is unchanged and the power required for the cut will remain fairly constant for conventional values of cutting speed.

If the depth of cut (a_e) is considered as a variable, it can be shown that the power criterion watts/w (where $w = \text{mm}^3/\text{s}$ of metal removed) will be reduced by cutting at half the depth and at twice the feed. Proof is left as an exercise for the reader. The power saved is, however, of negligible importance, and as the total amount of sliding between the cutting teeth and workpiece is increased, the tool life will be reduced.

The chief uncertainty in making such estimates as Example 11.3 lies in the value of K_c because this varies with the sharpness of the cutter. K_c may rise to as much as twice the initial value during the period between cutter regrinds. As may be seen by reference to Figure 11.74, values of K_c for milling are higher than for rough turning, hence the power criterion

$$\left(\frac{\text{watts}}{w} \right)$$

will be higher.

A convenient method of estimating the power required for a milling operation is provided by the specific power consumption

$$\frac{\text{watts}}{w}$$

which may be tabulated as a constant for various materials. A comparison between milling and turning of SAE 1020 steel, based on this method, is as follows:

1. Milling SAE 1020, HSS slab mill, cut 125 mm wide ×6.4 mm deep ×76 mm/ min feed.

$$\frac{\text{watts}}{w} = 4.64 \text{ (based on Cincinnati data)}$$

2. Turning SAE 1020, depth of cut 6.4 mm feed 0.38 mm

$$\frac{\text{watts}}{w} = 2.6 \text{ (based on ASME data)}$$

K values for power estimates when milling

Both Cincinnati and the Kearney and Trecker Corporation have published values of an approximate constant (K) in relation to the power required for milling various materials (Table 11.2).

$K = w/$watts and is the volume of metal which can be removed per second by 1 W at the cutter. The values quoted are for cutters near the end of their tool life so that maximum power demands may be estimated.

Example 11.4

Estimate the power required to take a cut, 100 mm wide × 3 mm deep, at 80 mm/min feed, in a 900 N/mm² alloy steel for which $K = 0.17$. If the cutter diameter is 90 mm, and a cutting speed of 14 m/min is employed, find the mean torque at the arbor and estimate the force required to drive the machine table (up-cut milling). If the cutter has a spiral angle of 40° estimated the axial thrust.

Solution

Power at the spindle,

$$\text{Power} = \frac{w}{K} = \frac{100 \times 3 \times 80}{60 \times 0.17} = 2353 \text{ W}$$

Mean force at the periphery of the cutter,

$$T = 2353 \times \frac{60}{14} = 10080 \text{ N}$$

Mean torque at the arbor,

$$\text{Torque} = \frac{90 \times 10080}{10^3 \times 2} = 454 \text{ Nm}$$

Estimated force to drive the machine table: reference to Figure 11.71 shows that the contact angle θ, and magnitude of forces F_r and T, are involved. Since for the cut being considered,

Table 11.2 Typical values of K (mm³/s/watt).

Material cut	Brinell hardness	K
Steel	100	0.27
Steel	200	0.22
Steel	300	0.19
Steel	400	0.17
Aluminium		0.74
Brass		0.61
Bronze		0.47
CI (average)		0.42
CI (hard)		0.28

(Converted from published values in Imperial units)

θ_B is small, the force to drive the table (neglecting friction at the slide) is approximately equal to T (10.08 kN).

Estimated axial thrust:

$$A = 10.08 \tan 40° = 8.46 \text{ kN}$$

Note that the feed force of 10.08 kN estimated above is about 8 times the feed force required to achieve an equivalent metal removal rate on a lathe.

□   □   □

Smoothness of arbor torque

For a given breadth of cut (b) and depth (a_e) the torque at the arbor will fluctuate according to:

1. The number of teeth in the cutter (N_T).
2. The spiral angle of the teeth (σ).

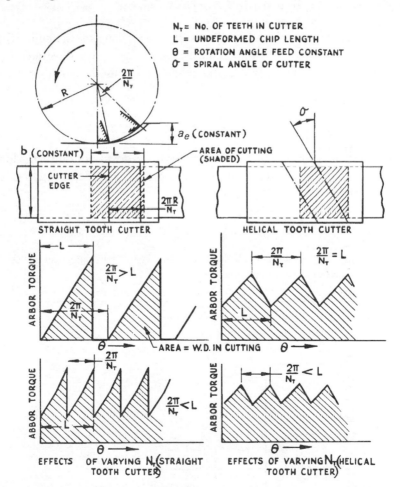

Figure 11.75 Influence of number of teeth in cutter, and of the spiral angle of teeth on arbor torque.

Figure 11.75 illustrates the effect of increasing N_T for a straight-tooth cutter and for a spiral-tooth cutter. The progressive reduction in torque fluctuation should be noted. Since torque × angle turned = work done, the shaded areas of the graphs show the relative amounts of work necessary to achieve a common metal removal rate. The smoother the torque, the *lower* the maximum value required for equal amounts of work done in cutting.

Of the variables which may be employed to obtain a smooth arbor torque, N_T is the least effective due to eccentric rotation of the cutter and the rise in K_c associated with reduced chip thickness (Figure 11.74).

The larger the spiral angle, the smoother will be the torque, and for cutters operating at low values of a_e there is a distinct advantage in employing high values of σ. Increase of the cutting depth (a_e) also results in a smoother torque and is an additional reason why cutters should be set to cut the full depth in one pass.

11.7 Character of the milled surface

The cross-section of a peripherally milled surface is a copy of the cutter form, the longitudinal section is a result of the chip-formation geometry illustrated in Figure 11.76. If the cutter ran perfectly true the height of the tooth marks on the surface would be given by:

$$h \simeq \frac{f_t^2}{8[(R \pm f_t)N_T/\pi]}$$

where the +ve sign is for up-cut and the −ve sign for down-cut milling. More often the surface markings are the result of eccentric running and depend upon the feed/rev of the cutter rather than the feed/tooth.

Exercises – 11

1. Compare up-cut and down-cut milling processes with particular reference to chip formation and forces induced in component and cutter.

2. Illustrate the tooth path relative to the work for up-cut milling, and show:
   (a) increase of the chip thickness t as the cut progresses;
   (b) decrease of effective clearance angle as t increases.

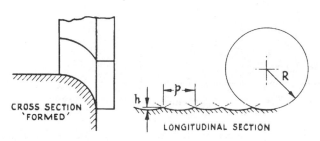

Figure 11.76 Surfaces produced by milling.

A 100 mm diameter side and face cutter has 10 teeth and cuts at 24 m/min, table feed 80 mm/min. If the depth of cut is 4 mm, find the maximum chip thickness.

3. Show for down-cut milling that the tooth path across the workpiece is a trochoid.

   A 180 mm diameter side-and-face cutter has 14 teeth; it is used at a spindle speed of 48 rev/min and table feed of 70 mm/min, to produce a slot 32 mm deep.

   Find the length, and maximum thickness, of an undeformed chip. On what assumption are such values calculated?

4. Show that the power required to drive a milling cutter depends upon the mean chip thickness, the mean specific cutting pressure for the material cut, the angle of engagement between the cutter and the workpiece the width of cut and the table feed.

   If the cutting speed only is increased, what are the likely effects upon the forces and power required for cutting?

5. Illustrate by means of sketches:
   (a) the direction of chip flow;
   (b) the effective rake angle;
   (c) how the magnitude direction and sense of the force acting on the tool may be measured for the following cutters:
      (i) a side turning tool;
      (ii) a cylindrical milling cutter 80 mm diameter, 100 mm long, with cutting edges set at an angle of 20° to the axis of the cutter.

6. Two surfaces are to be milled (about 2 mm metal removal) at production rates.
   (a) Steel forging, surface 40 mm × 70 mm.
   (b) Grey iron casting, surface 400 mm × 600 mm.
   Specify the method of milling and type of cutter you would use in each instance. Give brief reasons for your choice.

   A face milling operation employs a 150 mm diameter cutter having 5 inserted carbide teeth. The surface machined is 115 mm × 380 mm. Illustrate the positioning of the cutter for an angle of entry of 35°. Estimate a cutting time/piece given cutting speed = 180 m/min, feed = 0.25 mm/tooth. Sketch a suitable profile for the cutter teeth.

7. Explain, with reference to chip formation, why the power consumed per unit volume per min of metal removed is lower for face milling than for peripheral milling. (An analytical treatment is expected.)

   Given the specific cutting pressure as 3100 N/mm², estimate the max load/tooth for a slotting cutter 90 mm diameter, of 12 teeth and 10 mm width, cutting a slot 8 mm deep at a speed of 24 m/min and table feed of 10 mm/min.

8. A 180 mm diameter face mill is to cut steel of 350 HB. A cut of 5 mm depth across a slab 95 mm in width is to be taken at a cutting speed of 76 m/min. If the feed/tooth is not to be less than 0.2 mm, and the power of the machine is limited to 7.5 kW at the cutter, find the maximum number of teeth which could be safely employed. (Use Table 11.2.)

9. Discuss, with the aid of diagrams, the following aspects of the milling process.
   (a) The reason why carbide-tooth slotting cutters are generally used with the down-cut milling technique.
   (b) The reason why a slab mill should be mounted so that end thrust of the arbor is towards the spindle nose.
   (c) The reason why the corner edges of an inserted tooth carbide mill are arranged to cut under oblique rather than under orthogonal conditions.

12 Abrasion

12.1 Introduction

The machining of materials using abrasive material is of increasing importance, not only in the engineering industry, but in many other industries. The processes used produce well controlled surfaces of high finish, and tight tolerances are possible. In addition very hard materials can be machined.

Unlike most cutting processes, in abrasive machining the individual cutting edges have a random distribution and orientation. Since the particles providing the cutting edge are small, depth of engagement must similarly be small, and these factors result in variable chip formation, since any individual grain may encounter the work with positive, zero, or in most cases negative rake angle. Thus particles may merely deform the surface, may plough into the surface, or may form chips. Since only a proportion of the grains actually cut, and the others absorb energy without cutting, the energy required is many times higher than cutting for a similar material removal, and there is considerable heat generation.

12.2 Sanding and finishing

The oldest form of abrasive machining is that of rubbing the abrasive along a surface, the abrasive grains being attached to a backing of paper or cloth. The development of modern backings and adhesives have resulted in mechanised forms of this process, using coated belts or disks operating at high speed. Electrostatic techniques have been developed to align the grains, giving less random orientation and making chip formation the dominant mode, allowing high material removal rates with minimum heating. The processes have also been applied to disk backing, and there are few industries which do not make use of sanding or finishing in belt or disk form.

12.3 Grinding

By bonding with an appropriate bonding agent, the abrasive material may be formed into an axially-symmetric wheel, balanced for high-speed rotation. The grinding processes using such wheels have a geometry of chip formation similar to that of milling, with the grit edges which project from the surfaces of the wheel acting as small cutting teeth, but with random cutting angles. When cutting on the periphery of the wheel, Figure 12.1, chip formation occurs under similar geometrical conditions to up-cut or down-cut milling, depending upon the direction of work speed relative to the direction of wheel rotation. When grinding on the face of the wheel, Figure 12.2, the geometrical conditions are similar to face milling.

The principal differences between the cutting action of grinding, and of milling, lie in the following:

1. Grinding grits are sufficiently hard to cut fully hardened steels of the order of 850 HV.
2. The cutting angles of the grits have a random geometry.
3. The pitch of the grit cutting edges is much smaller than the pitch of milling-cutter teeth.
4. The size of the chips cut is very small for grinding, compared with milling.

Any serious study of the grinding process is helped by examination of wheel structure, and of grinding swarf, under a low-power microscope. If a measuring microscope is available the approximate sizes of the grits, and of the chips cut, can be determined. A wide strip of transparent adhesive tape, wrapped sticky side outward round a steel rule and then held in the stream of sparks produced by grinding, may be used to collect a representative sample of the chips.

Figures 12.3 and 12.4, based upon microscopic examination of grinding swarf, show the form of the chip produced when surface grinding on a horizontal-spindle machine as compared with the longer chip formed when grinding on a vertical-spindle machine. The small spheres are chips that have coiled up and fused together.

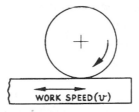

Figure 12.1 Surface grinding on wheel periphery.

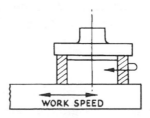

Figure 12.2 Surface grinding on wheel face.

Figure 12.3 Grinding chips (× 30 magnification), surface grinding mild steel on wheel periphery, wheel – A 46 L4 V.

Figure 12.4 Grinding chips (× 30 magnification), surface grinding mild steel on wheel face, wheel – A36 G 10 V.

Composition of grinding wheels

Figure 12.5 illustrates grinding grits cemented together by means of a bond material to make an abrasive wheel.

General information regarding standard methods of wheel specification is widely available, see BS 4481. The following points should be noted.

Abrasive material

1. Silicon carbide (SiC), specification 'C'. Hardness on Moh's scale 9.5 (diamond = 10). A rather brittle grit which tends to fracture under the forces which occur when grinding steels.
2. Aluminium oxide (Al_2O_3), specification 'A'. Hardness on Moh's scale 9. Tougher than silicon carbide and for this reason used to grind steel.

Grit size

Artificially produced abrasive is crushed to produce sharp-edged grits which are graded for size by sieving. A 46 grit indicates that the abrasive particles will just pass through a sieve having 46 openings per inch (25.4 mm) in either direction.

Bond material

Vitrified clay, specification 'V', is the material used to cement the grits of wheels employed for most precision grinding operations. A thin cut-off wheel, or a wheel subjected to shock loads or side forces as in fettling operations, requires a less brittle bond material. Bonds such as rubber or synthetic resin are then employed to reduce the chance of wheel fracture under the operating loads.

Bond strength (grade)

The amount of bond material surrounding the grits largely determines the force required to break a grit from the wheel, a property indicated by the term *grade*. Grade is specified by letter symbol.

Soft ───────────→ **Medium** ───────────→ **Hard**
E F G H I J K L M N O P Q R S T U V W

The workshop terms, *soft* and *hard*, are used to indicate that a relatively small or large force is required to break a grit from the wheel.

Structure

This describes the relative spacing of the grits, e.g. closely packed together or a more 'open' arrangement. For the same size of grit the pitch of the grits round the

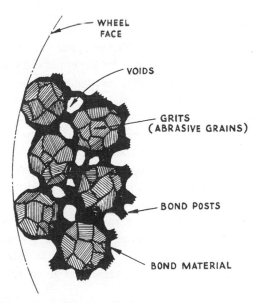

Figure 12.5 Composition of an abrasive wheel.

circumference of the wheel is greater for 'open' than for 'close' structure. The structure symbol is a number.

$$\text{Close} \longrightarrow \text{Open}$$
Structure symbol 1 2 3 4 5 ... 9 10 11 12

Figure 12.5 draws attention to open spaces in the bond of the wheel (*voids*). As the structure is made more open the voids tend to increase in number and size. The cutting life of an open-structure wheel is less than that of a close-structure wheel of the same dimensions, because there are less grits to become worn and be discarded. The increased pitch of the grits improves chip clearance. For certain grinding applications the heating effects of cutting are reduced by the employment of open-structure wheels.

Porosity

This term does not enter into the standard specification of a wheel but is used to indicate the combined effect of grit-size and structure. When the grits are large or the structure 'open', the space between adjacent grits exposed at the wheel face into which a chip can recede during the passage of the grits over the work material, will be larger than when the grits are small or the structure 'close'. The higher the porosity, the greater the chip clearance space between the grits and the more freely will coolant penetrate the wheel. It is sometimes arranged for coolant to enter a porous wheel at the central mounting, and be expelled by centrifugal force at the periphery. Such application of the coolant effectively dislodges swarf from between the cutting grits.

Force and power in grinding

Cutting speeds for grinding are always high, the upper limit being imposed either by the safe centrifugal stresses of the wheel or by the upper limit of rotational speed of the spindle employed. Vitrified wheels may be used up to about 1800 m/min; elastic bonds may be used at much higher speeds, up to 4800 m/min for abrasive cut-off wheels. For the internal grinding of small bores cutting speeds may not exceed 760 m/min due to the speed limitations of belt drives and bearings. A 10 mm diameter wheel must rotate at approximately 25 000 rpm to achieve this cutting speed. Bearings which operate up to speeds of 100 000 rpm have been developed, but the bearing loads must be very light and special driving arrangements are necessary; compressed air-turbine drives achieve this.

Cutting forces in grinding are difficult to measure and at present it is possible only to estimate the forces which act on the individual grits. The same basic mechanics may be applied to the grinding process as are applied to single point cutting. The force acting on a grit in the direction of its velocity is given by:

$$T = K_c \times \text{cut area}$$

It is difficult to obtain accurate values of the cutting force on a grit, or of the cross-sectional area of the chip cut. The sections of the grooves cut by individual grits will vary in shape due to the random geometry of the grits, and any attempt to find the value of the maximum chip thickness, and hence the maximum force on a grit, is necessarily approximate.

The following expression for chip thickness when surface grinding has been derived (Backer *et al.* 1952):

$$\text{chip thickness } (t) = \left[\frac{4v}{VCr} \sqrt{\frac{d}{D}} \right]^{1/2}$$

where v = work speed, m/min.
   V = cutting speed, m/min.
   C = the number of effective grits per mm^2 of grinding wheel surface.
   r = ratio of width to depth of a groove cut by a grit.
   d = depth of cut, mm.
   D = wheel diameter, mm.
   C is measured by rolling the grinding wheel on a piece of smoked glass and counting, under a microscope, the marks left where grit points pierce the smoke film.
   r is measured by taper sectioning of the ground workpiece so that the ratios of width to depth of the grooves cut by the grits can be estimated with reasonable accuracy.

It is possible to verify this expression for chip thickness as follows.

Consider a section through a grinding wheel in a plane normal to the axis of rotation, in which grits are imagined to lie with points projecting as indicated, Figure 12.6, spaced $1/N$ mm apart, where N is the number of grits per mm of the section circumference. By analogy with milling, equation 11.12, p. 359, the maximum chip thickness is given by, $t = 2f_g \sqrt{(d/D)}$, where f_g is the feed of the work per grit and t a uniform thickness across the full width of cut. It can be shown very simply that f_g is given by v/NV, so that for surface grinding,

$$t = \frac{2v}{NV} \sqrt{\frac{d}{D}} \tag{12.1}$$

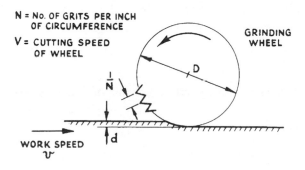

Figure 12.6

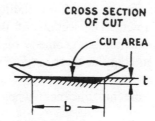

CROSS SECTION
OF CUT

CUT AREA

Figure 12.7 Typical cross-section cut by a grit.

Consider the area of metal removed across a wheel of face width s at the position of maximum chip thickness.

$$\text{Area removed} = \frac{2vs}{NV}\sqrt{\frac{d}{D}} \tag{12.2}$$

This area is, in effect, removed by a succession of grits and Figure 12.7 represents a possible condition for the metal removed by a single grit. The area of the cut at maximum chip thickness is $tb/2$.

Although the grits have a random geometry there are laboratory methods which enable the average ratio (r) of width b to depth t of the cut to be estimated from a grinding test.

By substitution of $b = tr$ the cut area shown in Figure 12.7 becomes $t^2 r/2$.

The total area of metal removed across the wheel face depends upon N and s; the total area removed is therefore

$$\frac{Nst^2 r}{2} \tag{12.3}$$

Equating (12.2) and (12.3),

$$t^2 = \frac{4v}{N^2 Vr}\sqrt{\frac{d}{D}}$$

but $N^2 = C$, by definition; hence

$$t = \left[\frac{4v}{VCr}\sqrt{\frac{d}{D}}\right]^{1/2} \tag{12.4}$$

This expression is the same as that derived by analogy with micro-milling by Backer, Marshall and Shaw (see Bibliography).

Example 12.1

The following conditions relate to a surface grinding operation carried out with a 46-grit wheel:

$V = 1700\ \text{m/min}$ $v = 9\ \text{m/min}$ $C = 2.95$

$r = 17$ $d = 0.025\ \text{mm}$ $D = 200\ \text{mm}$

estimate the maximum chip thickness.

Solution

$$t = \left[\frac{4 \times 9}{1700 \times 2.95 \times 17} \times \sqrt{\frac{0.025}{200}} \right]^{1/2}$$

$$= [0.000422 \times 0.0112]^{1/2}$$

$$= 0.0022 \text{ mm}$$

□   □   □

The result clearly indicates the very small magnitude of the chip thickness which occurs when grinding. The depth of cut specified is about 10 times that used for finishing cuts on precision work, hence values much lower than the one determined can be expected for such operations. The very low value of chip thickness then occurring is obviously related to the high quality of surface finish produced by precision grinding operations.

The following data relating to the magnitude of forces and power which occur when precision grinding, has been extracted from a research paper by Grisbrook (1960):

Cutting speed 525 m/min
Work speed 10.98 m/min
Width of cut 12.7 mm
Depth of cut 0.0076 mm
Material cut, hardened steel of 800 HV
Wheel, WA 46 JV

Grisbrook developed a dynamometer for measuring the horizontal and vertical forces acting on the workpiece during a surface grinding operation. For the conditions quoted, the tangential force at the wheel was approximately 62.3 N, varying slightly with the amount of grinding done since the previous dressing of the wheel.

From estimates of the area of contact between wheel and work, and from values of C for 46 wheel, it is probable that some 18 or so grits would be cutting at one time and that the *average* tangential force/grit would be about 3.47 N.

It is most unlikely that the force is so evenly distributed. A dull grit will have a higher force than average acting upon it if the normal considerations for single point tools apply to grinding.

The metal removal rate achieved under the conditions quoted is given by:

$$w = 10.98 \times \frac{10^3}{60} \times 12.7 \times 0.0076 \text{ mm}^3/\text{s}$$

$$= 17.66 \text{ mm}^3/\text{s}$$

$$\text{Power at the wheel} = \frac{1525 \times 62.3}{60} = 1583 \text{ W}$$

$$\text{Power criterion} = \frac{1583}{17.66} = 89.7 \text{ W/mm}^3/\text{s}$$

This is a strikingly high value compared with the values for cutting with single point tools, or for milling operations, on steels. It is shown in Grisbrook's paper that the value is almost as high for unhardened steel of 200 HV. For precision grinding of steel the power criterion watts/w does not appear to depend much on the hardness value of the material cut.

Some research workers tend to express the above relationship in terms of specific energy; i.e. the energy required to remove one mm^3 of metal, denoted by

$$U_s = \frac{\text{Work done}}{\text{Volume}}$$

The units of U_s are

$$\frac{Nm}{mm^3} = J/mm^3$$

converted to specific energy units as follows:

$$U_s = 89.7 \ J/mm^3$$

Grisbrook was able to show that the value of U_s falls as the metal removal rate (w) is increased by an increase in work speed or in depth of cut, i.e. chip thickness (t) increasing. This fact should be compared with the observed fall of specific cutting pressure (P) with increase of chip thickness known to occur for single point cutting.

A fall in the value of U_s as the metal removal rate is increased means that the power requirements for grinding at heavy metal removal rates will be lower than the 89.7 $W/mm^3/s$ determined above, and an average figure of 30 $W/mm^3/s$ is about right for steels. The size of motor provided on a modern grinding machine reflects the high power requirement of the grinding process. One theory put forward to explain the rise in U_s which accompanies a fall in chip thickness is that the tangential force on the wheel contains a fairly large element of friction due to the rubbing of grits on the workpiece. The frictional element does not fall as the chip thickness is reduced, hence the energy consumed in removing a unit volume of metal rises.

Heating effects during grinding

The high value of U_s is an indication of the large amount of heat generated relative to the amount of material removed.

For the example considered about 90 J of heat are produced per mm^3 of metal removed.

If 20 per cent of this heat were to go into the metal removed, which, being steel, weighs 7740 kg/m^3 and has an average specific heat of around 0.2 over a temperature range up to melting point, the resulting temperature rise would be 2760 °C.

Specific heat capacity, steel at high temperature, is about $4200 \times 0.2 = 840$ J/kg/°C.

$$\text{Temperature rise} = \frac{90 \times 0.2 \times 10^9}{840 \times 7740} = 2760 \text{ °C}$$

Since steel melts at about 1530 °C it is obvious that only a fraction of the heat produced enters the chip. It is also known that melting is a time–temperature effect; microscopic examination of grinding swarf shows that melting does not occur to the extent which the above figures appear to suggest, and the short duration of the heating is probably one reason for this. The presence of grinding sparks and of some heat-fused globules among the grinding swarf is evidence of the very high temperature reached by the chips. The harder the metal cut, the greater the number of globules to be found in the swarf.

Independent attempts to measure the temperature of the work at the point of contact with the grinding wheel confirm that instantaneous values of the order of 1050–1650 °C are reached.

Dressing and trueing of grinding wheels

Dressing

This is a process used to clear the cutting surface of the wheel of any dull grits and embedded swarf in order to improve the cutting action. It may be done by means of a star-wheel dresser, by wheel crushing, or by means of a carborundum dressing stick.

Trueing

In this process a diamond is employed to bring the wheel to the required geometric shape, e.g. a true cylinder or a cylinder having a screw-thread form in an axial plane. It also restores the cutting action of a worn wheel as in dressing. The distinction between trueing and dressing is sometimes hard to draw; e.g. a wheel can be brought to the required 'form' by crushing.

The processes are effected by:

1. Dislodging whole grits from the bond.
2. Chipping the edges of the grits.

There is some difference in the manner of the achievement of (2) as between diamond trueing and crush dressing. Abrasive grit, being crystalline, tends to fracture along the most highly stressed crystallographic plane.

Diamond trueing tends to chip the grits along planes which make a small angle with respect to the direction of motion of the grit. Crush dressing may cause shear fractures along planes which make a large angle with respect to the direction of motion of the grit. As may be seen in Figures 12.8 and 12.9, crushed grit is likely to have more favourable cutting angles (greater clearance angles, lower negative rake angles) than diamond-trued grit.

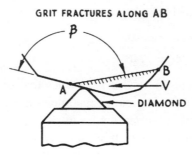

Figure 12.8 Diamond dressing of grinding wheel.

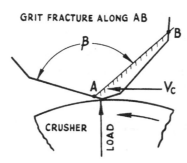

Figure 12.9 Crush dressing of grinding wheel.

Crush dressing tends to remove whole grits by fracture of the bond posts, especially in open-structure wheels where the 'voids' provide space for the collapse of grits, and it also tends to provide sharp cutting edges on any fractured grits. A crush-dressed vitrified wheel will have free-cutting properties but will not produce a surface finish on the work equal to that of the same wheel when diamond trued.

Diamond trueing at large depths of cut tends to remove whole grits; if the wheel is finished by taking several passes, each with an infeed of about 0.005 mm, followed by several passes with no additional feed, the trued surface will have its grit edges chipped away to give a very 'smooth' wheel which can then be used to obtain a fine surface finish on the work. The use of a somewhat rounded diamond for the final dressing of the wheel tends to result in a good finish on the work. Any heavy cuts taken with such a wheel will break down the fine surface of the grits (self-sharpening action), improve the cutting property and reduce the quality of the surface finish obtained.

Practical examples of dressing and trueing

Figure 12.10 illustrates a type of radius dresser used for precision toolroom work. Concave and convex curves may be dressed to various radii by adjustment of the lower slide (A). This slide can be controlled to swing through a predetermined angle by means of stops; slide (B) may then be brought into operation in order to dress a flank tangential to the radius as illustrated.

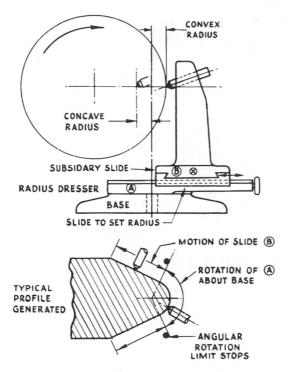

Figure 12.10 Radius dresser for precision wheel trueing.

The accuracy of the work is influenced by the sharpness of the point of the diamond employed because the action is a *generating* one.

Figure 12.11 shows a pantograph device as used to dress formed wheels for gear grinding. The pantograph copies the template profile on to the wheel at a reduced scale. Similar form dressing devices are available on toolroom surface grinders and are particularly useful in small press-tool manufacture for grinding the profiles of punches and segmental dies

Figure 12.12 illustrates a form crushing operation. Consideration of the various diameters involved will show that some relative sliding between the wheel and the crusher must occur and so cause crusher wear. It is difficult to maintain sharp corners of profiles by the crushing technique. Crushing of screw-thread forms is successfully employed on thread-grinding machines. A suitable crushing speed (V_c) is 25–30 m/min.

Self-sharpening action

The cutting life of a grit (see Figure 12.13) operating on steel, or on a similarly tough material, is comparatively short. Dressing the wheel each time the exposed grits had lost their cutting efficiency would be very time consuming and would greatly reduce the output from grinding operations. If a suitable balance between the cutting forces and the bond strength of the wheel is obtained, the need for continual redressing can be avoided.

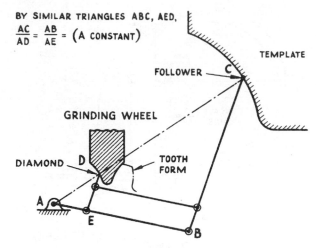

Figure 12.11 Pantograph method of wheel trueing used on gear grinding machine.

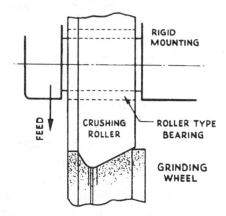

Figure 12.12 Form crushing of a grinding wheel.

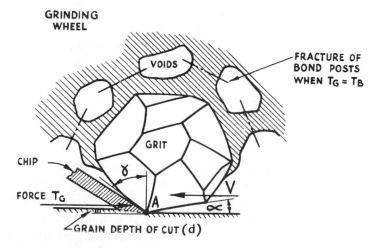

Figure 12.13 The cutting action of an abrasive grit.

It is reasonable to suppose that the forces acting on a grit will increase as the cutting edge deteriorates, other factors remaining constant. This may be seen to occur during cutting tests on single point tools. Gisbrook's work shows that increased cutting forces occur in surface grinding operations as the run proceeds. A particular set of the conditions investigated gave the following results:

(Newly dressed wheel)	T	F_N
Commencement of run	22.25 N	44.5 N
After 12.2 m of traverse	44.5 N	155.75 N

T and F_N, shown in Figure 12.14, are the components of forces acting on all the grits in contact with the workpiece. As these forces are seen to increase during cutting, increases in the force/grit must occur and the force/grit will rise for each particular grit as it gets duller. If the stress set up in the bond material by the cutting force is sufficient to fracture the bond posts (Figure 12.13, T_G reaching the limiting value T_B) the dull grit will be broken from the wheel and adjacent new sharp grits exposed for cutting. There will be a gradual loss of wheel diameter, and a continual replacement at the wheel cutting surface of dull grits by sharp grits. This is commonly called *self-sharpening action*.

An idealised concept of this action is represented in Figure 12.15.

Suppose the point T_G (1) represents the force on a grit newly exposed to cutting. As cutting proceeds there will be a progressive fall in the cutting efficiency of the grit and a consequent rise in force T_G. If a J grade wheel is employed, T_G will reach the limit imposed by the bond strength of the wheel somewhere near the end of the useful cutting life of the grit. At this point the grit will break from the wheel.

If, however, an L grade wheel is used, T_G will rise considerably higher before the grit breaks away. Should an N grade wheel be used, even higher values of T_G must occur before the grit is fractured from the wheel, but now there is a further possibility. Due to the elastic nature of the wheel, wheel spindle, workpiece, etc., the presence of dull grits will give rise to a very high value of F_N (Figure 12.14) and the wheel may deflect away from the work so that the value of T_G required to release the grit cannot be reached. Such conditions give rise to excessive heating of the surface ground, cause burn marks and

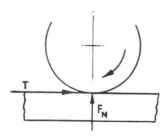

Figure 12.14 Components of grinding forces.

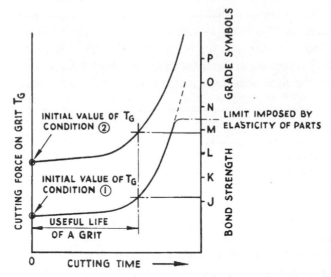

Figure 12.15 Graphical representation of self-sharpening action.

may cause small cracks in the surface of the metal which can be discovered afterwards by the use of crack-detection techniques. The condition is sometimes called 'glazing' due to the glossy polished surface of the wheel which results.

When it is not a simple matter to change the wheel grade, e.g. on machines employing large and heavy wheels, the conditions of grinding can be adjusted so that the initial force is raised, as (2) in Figure 12.15, or if necessary lowered. In this way a correct self-sharpening action can be achieved provided the grade of wheel is not grossly out of relationship to the grinding conditions.

Influence of the variables of the grinding process upon the cutting force per grit

Reference to Figure 12.6 and equation 12.1 shows that the expression

$$t = \frac{2v}{NV} \sqrt{\frac{d}{D_G}}$$

relates to the chip thickness cut when surface grinding. Symbol D_G has been substituted for D in order to distinguish between the wheel diameter and the work diameter (D_W) of cylindrical grinding. Since for similar figures, the ratio of the areas equals the square of the ratio of one of the linear dimensions, the area cut by a grit is proportional to t^2:

$$T_G = P \times \text{cut area}$$

and assuming that the specific cutting pressure (P) remains constant,

$$T_G \propto t^2 \propto \frac{v^2}{N^2 V^2} \times \frac{d}{D_G}.$$

D_G and N can be varied only by changing the wheel. V is not normally a variable except due to wheel wear. (Some grinding machines provide for increase of the wheel spindle speed to compensate for reduction of diameter as the wheel wears to keep V reasonably uniform.) The only variables which can be used to change the force/grit are work speed (v) and the depth of cut (d); of these v is seen from the above to be the more effective. J. J. Guest deduced this by a study of the geometry of chip formation during grinding.

If the more refined expression for chip thickness of equation 12.4 is employed we have

$$T_G \propto \frac{v}{V} \sqrt{\frac{d}{D_G}}$$

which again shows v to be a more effective variable than d in changing the cutting force on a grit.

Most modern grinding machines have a large range of work speeds; some have infinite variation of work speed throughout the range, generally by hydraulic means. The part played by v as a variable in obtaining self-sharpening conditions can then be fully exploited in selecting the grinding conditions.

Geometry of contact between wheel and work – external cylindrical grinding

Apart from the significance of v and d discussed above, the force/grit will depend upon the work diameter (D_W) and the traverse rate.

Guest showed that for external grinding, force/grit varies as

$$v^2 d \left[\frac{1}{D_W} + \frac{1}{D_G} \right],$$

hence the smaller the work diameter or the wheel diameter, the higher the force/grit.

The same result as is obtained by Guest's theory can be derived from the geometry of the chip shape shown in Figure 12.16. To an exaggerated scale KLMN is the chip cut by a grit, and the maximum area will occur across LN, where the thickness is maximum.

By ignoring the very small amounts of curvature which occur, KLN may be treated as a right-angled triangle, the right angle at L. KN is the distance moved by the workpiece during the period between the passage of successive grits, i.e the feed per grit represented by f_g. It can be seen that

$$t_{max} = f_g \sin(A + B) \text{ to a high order of accuracy.} \tag{12.5}$$

Figure 12.17 shows the geometrical relationships between D_W, D_G and the depth of cut d.

Consider the triangle in Figure 12.17. By the cosine rule,

$$\left(\frac{D_W}{2} + \frac{D_G}{2} - d \right)^2 = \left(\frac{D_W}{2} \right)^2 + \left(\frac{D_G}{2} \right)^2 - 2 \frac{D_W}{2} \frac{D_G}{2} \cos[180° - (A + B)]$$

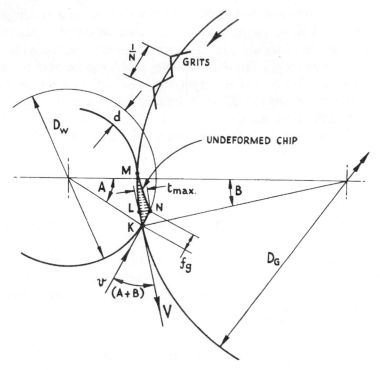

Figure 12.16 Geometry of chip formation four cylindrical grinding.

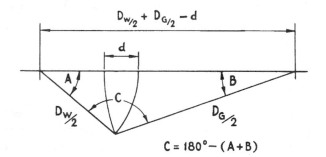

Figure 12.17

Since d is very small, terms involving d^2 are negligible and omitted from the expansion below. Hence

$$\frac{D_W D_G}{2} \cos(A + B) = \frac{D_W D_G}{2} - dD_W - dD_G$$

$$\text{or } \cos(A + B) = 1 - 2d\left(\frac{D_W + D_G}{D_W D_G}\right)$$

$$1 - \sin^2(A + B) = \cos^2(A + B) = 1 - 4d\left(\frac{D_W + D_G}{D_W D_G}\right) + \text{terms containing } d^2$$

Hence

$$\sin^2(A + B) \simeq 4d\left(\frac{D_W + D_G}{D_W D_G}\right)$$

$$\text{or } \sin(A + B) \simeq 2\left[d\left(\frac{1}{D_W} + \frac{1}{D_G}\right)\right]^{1/2}$$

Feed per grit, $f_g = v/NV$, and by substitution of these results in equation 12.5 we have

$$t = \frac{2v}{NV} \sqrt{\left(d\,\frac{D_W + D_G}{D_W D_G}\right)}$$

This is the form generally stated for the result of Guest's theory, but it may also be written as;

$$t = \frac{2v}{NV} \sqrt{\left[d\left(\frac{1}{D_W} + \frac{1}{D_G}\right)\right]}$$

Considering the main variables in external grinding,

$$\text{Force per grit} \propto t^2 \quad \text{and} \quad t^2 \propto \frac{v^2 d}{V^2}\left(\frac{1}{D_W} + \frac{1}{D_G}\right)$$

the pitch of the grits (N) being a constant for the grade of wheel selected. Clearly, raising of the work speed (v) is the most effective way of increasing self-sharpening action should dull grits not be broken from the bond.

Figure 12.18 shows the general conditions for external cylindrical grinding.

The traverse rate will influence the distribution of the cutting across the face of the wheel. If the traverse/rev of work (p) just exceeds $b/2$, where b is the width of the wheel, and cutting takes place in both directions of table reciprocation, wear is evenly distributed across the wheel. If $p < b/2$ a 'taper lead' will be produced by wheel wear, Figure 12.19. The effective cutting depth d_e is now smaller than the infeed. The traverse rate now influences the chip thickness; the effective cutting depth is given by $d_e = p \sin \phi$ and there is a lower force/grit.

Internal cylindrical grinding

The conditions are shown in Figure 12.20. The force/grit varies as

$$\frac{v^2 d}{V^2}\left(\frac{1}{D_W} - \frac{1}{D_G}\right)$$

and is much lower for internal than for external grinding for the same cutting speed (V). For internal grinding V is often much smaller than for external grinding due to the small diameters of wheel involved, and as V falls the force/grit rises. Note that

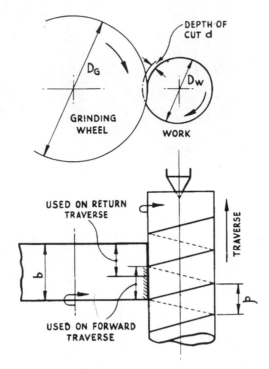

Figure 12.18 Conditions for external grinding.

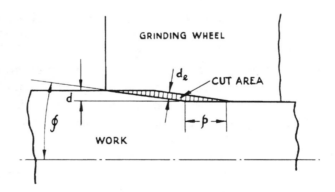

Figure 12.19 Effect of taper head on cutting conditions.

the arc of contact increases rapidly as D_G approaches D_W in magnitude. Generally somewhat softer grades of wheel are required for internal grinding than for external grinding.

Surface grinding

The general conditions for grinding on the periphery of the wheel have been discussed (p. 389). Grinding on the face of the wheel, Figure 12.21, gives rise to

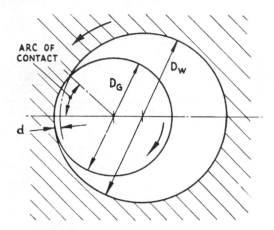

Figure 12.20 Arc wheel contact, internal grinding.

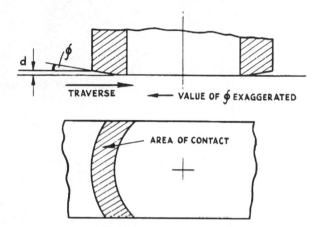

Figure 12.21 Contact between wheel and work, surface grinding on wheel face.

much smaller values of force/grit due to:

1. The geometry of the chip produced.
2. Taper wear at the edges of the wheel which reduces the effective cutting depth.

The cylinder wheels employed on a vertical spindle grinder need to be of a soft grade (G), large grit (36) and open structure (8 or 10); the two latter requirements help to provide chip clearance and also reduce the number of grits in contact with the work, thus increasing the force/grit. When a hard wheel has to be employed, e.g. to maintain a corner in slideway grinding, a saucer wheel having a thin edge is necessary in order to distribute the cutting load on fewer grits, Figure 12.22.

Cylindrical work – grinding of shoulders

The wheel head of some cylindrical grinding machines may be swivelled so that the spindle axis is inclined to the work axis as shown in Figure 12.23. A section in plane

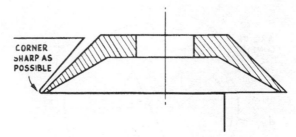

Figure 12.22 Use of 'saucer' wheel to reduce contact area.

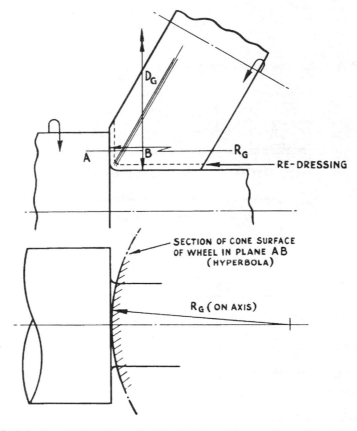

Figure 12.23 Grinding of shoulders of cylindrical work by angular setting of wheel head.

AB shows that the contact between the shoulder of the workpiece and the wheel is similar to that of a surface grinding operation where the wheel periphery is used. Ideally the same force/grit is desired when grinding the shoulder as when grinding the diameter, so that self-sharpening conditions occur at each grinding position. It is obvious that for most practical conditions of operating

$$\frac{1}{D_G} + \frac{1}{D_W} > \frac{1}{2R_G}$$

hence the force/grit will be higher for grinding on the work diameter than for grinding on the shoulder.

Excessive 'glazing' at the shoulder is less likely to occur for these conditions than when 'full contact' at the shoulder is made by using the wheel with its axis parallel to the work axis. A very important saving is made in wheel dressing, because both surfaces can be redressed continually without loss of the face width of the wheel as indicated in Figure 12.23.

Grinding and surface finish

When a surface is finished by grinding on the periphery of the wheel the *lay* will follow the work velocity relative to the wheel; for surface grinding it will be in the line of reciprocation of the table, for cylindrical grinding it will follow a helix defined by the work speed and traverse rate. Figure 12.24 illustrates a typical ground surface. The roughness is generally greater across the lay, where at some points the maximum depth of the grooves produced lie in the section plane, than along the lay. Waviness may occur along the lay; it will occur when the wheel is out of balance sufficiently to give rise to visible chatter patterns.

Example 12.1 shows that when surface grinding with a 46 grit wheel, depth of cut 0.025 mm, the chip thickness is 0.0022 mm. The maximum depths of the grooves cut in the work must be of this order, so that a peak-to-valley average reading of the surface finish across the lay will not exceed this value. Experience shows that for grinding, the surface finish average number (R_a) is between $\frac{1}{3}$ and $\frac{1}{5}$ of the peak-to-valley average number, and for the conditions of Example 12.1 a surface finish of between 0.8 μm and 0.4 μm is to be expected. Since the depth of the grooves cut will depend upon the value of t, where

$$t = \left[\frac{4v}{VCr} \sqrt{\frac{d}{D}} \right]^{1/2}$$

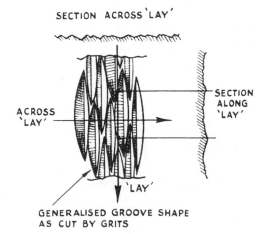

SECTION ACROSS 'LAY'

ACROSS 'LAY'

SECTION ALONG 'LAY'

'LAY'

GENERALISED GROOVE SHAPE AS CUT BY GRITS

Figure 12.24 Surface characteristics produced by grinding.

it can be seen that a low work speed and small depth of cut will tend to improve the finish, as also will a smaller grit size (for which C is relatively higher). This supposes a perfectly true wheel spinning about a perfectly constant axis, conditions which are never exactly satisfied in practice. The necessity of high-quality bearings to achieve high surface finish is obvious, and there is a limit set by the condition of the bearings to the improvements in finish which can be obtained by a reduction of the grit size, of the depth of cut or of the work speed.

It is of interest to examine the relationship between the out-of-balance forces acting on the wheel bearings and the surface-finish defects which may arise. Suppose a grinding wheel, rotating at 1200 rpm, is out of balance by 35 g acting at 200 mm radius.

$$\text{Centrifugal force} = M\omega^2 r$$

$$\omega = \frac{1200}{60} \times 2\pi = 40\pi \text{ rad/s}$$

$$\text{Centrifugal force} = \frac{35}{10^3} \times \frac{200}{10^3} \times (40\pi)^2 = 110 \text{ N}.$$

The centrifugal force gives rise to a force acting in a plane through the wheel and work axes. This force has a maximum value of 110 N and is reversed in direction at each $\frac{1}{2}$ revolution of the wheel. It would require a very stiff machine structure to prevent some very small decrease and increase in the mean distance between the axes under such conditions. Figure 12.25 illustrates the resulting chatter pattern. This may show up more distinctly when the ground diameter has been caused to slide in a close fitting hole. The pitch of the markings is given by

$$q = \frac{\pi v D_G}{V} \text{ mm}$$

and will be much smaller when grinding with small-diameter wheels than with large-diameter wheels due to the higher spindle speed.

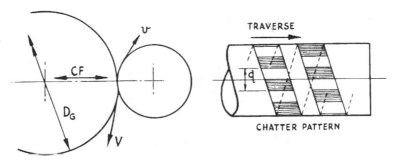

Figure 12.25 'Chatter' pattern resulting from unbalanced grinding wheel.

In general terms, the quality of surface finish which may be obtained from grinding operations is approximately as follows (R_a values).

1. General work of moderate size: 0.8–0.4 µm
2. Precision work of normal standard: 0.4–0.1 µm
3. Fine grinding: below 0.1 µm and under very good conditions below 0.02 µm. It is necessary to have an efficient coolant filter system in order to achieve these very fine finishes.

Economics of grinding

1. Grinding has two principal economic advantages over other methods of metal machining for the finishing of surfaces:
   (a) due to the large number of very small chips cut it is much easier to work to close tolerances and fine surface finishes;
   (b) due to the high work speed (compared with traverse rates for, say, milling) and to the very large number of cutting edges presented to the work, the time taken to produce a unit area of finished surface is lower than that for other machining processes.
2. Grinding tends to be restricted to finishing processes because the volume of metal to be removed is then relatively small. The main reasons for this are:
   (a) the large amount of power consumed per unit volume of metal removed by grinding as compared with alternative methods;
   (b) the relatively high cost of grinding wheels per unit volume of metal removed.

To compete in metal removal with a milling machine having a 7.5 kW motor, a grinding machine would require a 75 kW motor! It is generally uneconomic to remove large amounts of metal by grinding because the machines available lack the power and strength to remove the metal at time rates which compare with milling or turning.

Despite the drawbacks of high power consumption and wheel costs, certain developments in large quantity production techniques have been made in which grinding has replaced face milling. Datum faces may be prepared by such methods on castings and drop stampings, and the rotary table principle can be used to give a continuous output of such pieces. One advantage of such an arrangement is that the cutting tool can be serviced (dressed) on the machine spindle, so that loss of output from this cause is smaller than would arise from milling cutter changes.

12.4 Creep feed grinding

While grinding is normally thought of as a finishing process, techniques have been developed for bulk metal removal. Developed for the machining of nickel alloys in the aircraft industry, creep feed grinding calls for a deep cut at a very slow feed rate – the opposite of what is called for in normal grinding – and wheels of extremely hard material such as cubic boron nitrate or diamond, held in a metal matrix. These materials are prone to damage by vibration and chatter, but the large contact area of creep feed grinding reduces the propensity for this, while the harder material retains its shape better than with the conventional approach to grinding.

12.5 Honing

In this technique the abrasive is bonded into slab form, and is mechanically oscillated while pressed lightly against a surface. Used primarily to impart a high finish to a surface, the abrasive removes any 'highs' leaving a very smooth surface. To avoid grooving, the hone is also moved laterally to the direction of oscillation. The most frequent application of honing is in the production of finely finished bores, when this lateral movement is achieved by rotating the hones within the hole.

12.6 Lapping

This is a process similar to honing, but in this case the abrasive is not formed, but is introduced in the form of a slurry between the workpiece and a counterformed surface termed a lap, this latter being formed of a relatively soft and slightly porous material which serves to rub the abrasive medium against the workpiece in a controlled manner. In its most common industrial form, the lap is a rotating table and the parts being lapped have a planetary movement, resulting in a uniform surface of excellent flatness. If the lap is made three-dimensional, accurately curved surfaces can be lapped, a technique that is used in the manufacture of accurate lenses.

12.7 Ultrasonic machining

Another abrasive process using unbonded abrasive is ultrasonic machining. A piezoelectric transducer imparts low amplitude high frequency movement to a soft form tool while an abrasive slurry is fed to the interface between this and the workpiece, resulting in gradual erosion of the workpiece.

12.8 Barrel finishing (tumbling)

This is a completely random abrasive process of great value for the removal of burrs and flash from components, and for improving the surface finish. The workpieces are loaded into a barrel of multi-sided section, together with an abrasive slurry and a tumbling medium. As the barrel rotates the components drop and are rumbled against the medium and each other, removing surface protrusions and smoothing the surfaces. A variant of this is vibratory finishing, where a vibratory rather than rotational movement is used.

12.9 Grit (sand) blasting

In this the particles of abrasive are hurled at the surface, either in a stream of compressed air or as a slurry in a stream of liquid, and erode the material where they impact. The process has a variety of uses, from deburring and improvment of surface finish through the removal of scale and so on, from the surface, to the abrading of slots or holes in hard materials, and will be found in a variety of manufacturing industries.

Exercises – 12

1. (a) A grinding wheel is specified as A 461 8 V. Give the meaning of each symbol and explain the significance of each in relation to the cutting action of a grinding wheel.
    (b) Explain what is meant by self-sharpening action as applied to the grinding process and describe the various factors which give rise to this.

2. (a) Write short notes on the cutting action of a grinding wheel and state the factors affecting wheel selection.
    (b) Give typical values of peripheral wheel speed for cylindrical, surface and internal grinding, and compare the area of contact of wheel and work for these processes.

3. (a) How is the choice of grinding-wheel grade influenced by: (i) the area or arc of contact, and (ii) the nature of the workpiece material?
    (b) With the aid of a line diagram, explain how, in cylindrical grinding, the workface of the wheel may be preserved flat without recourse to dressing or crushing. Hence, find a suitable traverse rate for a 40 mm wide wheel grinding a 50 mm diameter shaft, if the work speed is 18 surface m per min.

4. (a) Using diagrams. explain why 'crushed' grinding wheels tend to be freer cutting than diamond dressed wheels.
    (b) Given Guest's criterion $v^2 t(D+d)/Dd$, explain the effect of a decrease of wheel diameter on cutting efficiency when surface grinding; surface speed of wheel to remain approximately constant.
    (c) What are the advantages of hydraulically operated table drives for reciprocating tables of surface grinding machines, as compared with rack and pinion drives?

5. (a) Explain the theory of the cutting action of a grinding wheel and define the characteristics by which the nature and action of a grinding wheel is assessed.
    (b) Define diagrammatically the geometry of a tool-and-cutter grinding machine and show how it would set to grind a face mill with inserted blades and 250 mm in diameter.

6. Sketch the conditions of chip formation for cylindrical grinding and deduce from these that the maximum chip thickness is given by,

$$\frac{v}{NV} \sin(A + B).$$

Given that

$$(A + B) \simeq 2\sqrt{\left(t\,\frac{D + d}{Dd}\right)}$$

where t = depth of cut, D = wheel diameter, d = work diameter, show that the force acting on a grit is proportional to

$$\frac{v^2}{V^2}\,t\left[\frac{D + d}{Dd}\right]$$

What bearing has the expression for the force acting on a grit, upon the conditions required for efficient grinding?

7. (a) Derive an approximate expression for the maximum thickness of the chip when a grinding wheel of diameter D is grinding a cylindrical workpiece of diameter d with a depth of cut t, the peripheral speeds of wheel and workpiece being V and v respectively.
   (b) Evaluate this maximum chip thickness if $D = 300$ mm, $d = 40$ mm, $t = 0.025$ mm, $V = 1500$ m/min, $v = 18$ m/min and the abrasive grains are 0.8 mm apart.

8. With reference to the grinding process, show how the wheel is prepared for (a) the grinding of an accurate radius; (b) thread grinding using a multi-ribbed wheel; (c) thread grinding using a diamond-trued single-form wheel. The geometry of the trueing and forming attachments should be given.

9. (a) Explain briefly why the mechanical efficiency of metal cutting by grinding is only about $\frac{1}{10}$ of the mechanical efficiency of metal cutting by turning. How does this fact influence the design of grinding machines?
   (b) Why does grinding give surface finishes superior to most other metal-cutting processes?
   (c) Compare the economics of the grinding process as a means of finishing cylindrical surfaces to close tolerances, with those of the turning process.

10. (a) Explain briefly why a cylindrical grinding machine should provide for a wide range of work speeds, preferably on an infinitely variable basis.
    (b) Explain briefly the significance of Guest's criterion

$$\frac{v^2 t}{V^2}\left[\frac{D \pm d}{Dd}\right]$$

    in relation to the grinding process.
    (c) When rough grinding steel bars, work speed 12 m/min, depth of cut 0.05 mm, the wheel was found to wear excessively. Use Guest's criterion to show that the operating conditions can be changed to correct this trouble without lowering the production rate. Suggest suitable values for the new cutting conditions.

13 Principles of Machining – Non-Traditional Methods

13.1 Introduction to non-traditional machining

In traditional methods of machining, material is removed from the workpiece by means of a tool of superior hardness. However, this concept of a hard tool cutting a softer workpiece cannot be applied to some machining processes developed in the last few decades.

Known scientific principles have been intensively developed and applied to material removal processes of unconventional nature in manufacturing industry. Chemical, electrochemical and thermoelectric sources of energy are tapped and used in machining processes which, in certain circumstances, provide either technological or economic advantages over traditional methods. In particular, it is feasible:

1. To machine the 'exotic' materials developed for high performance applications in the aerospace industry, e.g. very high tensile alloy steels and heat resisting alloys.
2. To produce design features which are costly, difficult or even impossible to make by traditional methods.

One very important feature of many of these newer processes is that the hardness of the workpiece material is of no significance from the machining point of view for it does not increase wear of the 'tool', where one is needed, nor does it reduce the rate of metal removal. Indeed, hardened components are sometimes found to 'machine' more readily than materials in a relatively soft condition. A number of industries have taken advantage of this characteristic. It is especially valuable in toolmaking, for the manufacture of dies and moulds, where greatly improved accuracy is achieved by finish machining after hardening, and expensive hand-working is reduced or eliminated.

Machining processes are categorised in Table 13.1 according to the fundamental energy source, with the exception of machining based on chemical sources of energy which has been omitted for reasons of space.

In addition to the processes shown in the table, a number of variants exist which are too numerous to deal with in this text.

Table 13.1 Categories of machining.

Fundamental energy source	Mechanical		Thermoelectric		Electro-chemical
Principles of material removal techniques	Shear	Erosion	Vaporisation (a)	(b)	Ion displacement
Medium employed for removal of material	Cutting tool	Abrasive particles at high velocity	Electrons – high voltage discharge	Radiation amplified light	High density current – electrolyte
Process	Conventional machining	Ultrasonic machining	Electro-discharge machining	Laser	Electro-chemical machining

13.2 Electro-discharge machining (EDM)

Introduction

Electro-discharge machining is one of several processes employing thermoelectric energy to remove material from a workpiece through melting or vaporisation of selected areas. Other processes in this group are: electron beam machining (EBM) which operates by transforming the kinetic energy of high speed electrons into thermal energy as they bombard and vaporise a local area of the workpiece: laser beam machining (LBM), which is dealt with in Section 13.3; and the lesser used plasma arc and ion beam processes. Of all these methods, EDM has the most practical advantages and is the one most widely applied in manufacturing engineering.

Basic principles of EDM

Briefly EDM is a process for producing holes, external shapes, profiles or cavities in an electrically conductive workpiece by means of the controlled application of high-frequency electrical discharges to vaporise or melt the workpiece material in a particular area. The electrical discharges are the result of controlled pulses of direct current and occur between the tool electrode (usually negative) and the electrically conductive material of the workpiece (usually positive).

The tool and workpiece are separated by a small gap, say 0.01 to 0.5 mm and are submerged or flooded with a dielectric fluid, e.g. paraffin, white spirit or light oil. A voltage exceeding the breakdown voltage of the gap (determined by the size of gap and the insulating resistance of the dielectric) is applied to initiate a discharge. A channel is ionised at the two closest points between the electrode and the workpiece, Figure 13.1, and a massive current flow results causing erosion of a particle of metal. The ionised channel is formed of a plasma (i.e. gas ionised at very high temperature, in this case 8000–12 000 °C) consisting of metallic atoms (M) vaporised from both the workpiece and the tool electrode, positive ions (M^+) and electrons (e^-) as shown in Figure 13.1 (inset). The instantaneous vaporisation produces a high-pressure bubble which expands radially into the dielectric. The discharge ceases with the interruption of the current. The metal is ejected leaving a small part-spherical crater in the workpiece and re-solidifies as globules in suspension in the dielectric. A sludge of black particles, mainly carbon, formed from the hydrocarbons of the dielectric is produced in the gap and is expelled by the energy of the discharge, remaining in suspension until removed by filtering. Immediately following the discharge, the dielectric surrounding the channel deionises and once again becomes effective as an insulator.

The rate of material removal from the tool and workpiece is asymmetric, it depends on the polarity, thermal conductivity and melting point of the materials as well as the duration and intensity of the discharges. By suitable selection of parameters, the rate of erosion of the workpiece can be increased whilst that of the tool is decreased. Since metal is removed by vaporisation, physical properties of the workpiece which are important in the cutting of metals, e.g. hardness and toughness, are not significant and hard metals, e.g. tungsten carbides, can be eroded without difficulty.

The ED machine

The general features of a basic ED machine are shown in Figure 13.2. Essentially they comprise a bed, table, slides, dielectric system and a column on which is

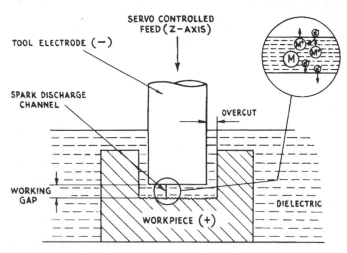

Figure 13.1 Basic principle of EDM.

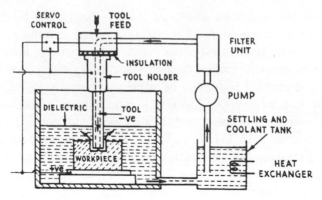

Figure 13.2 General features of ED machine.

mounted a head provided with a servomechanism to control the rate of feed of the quill in which the toolholder is mounted. Because the width of the working gap is of critical importance, the rate of feed is vital and control by a sensitive hydraulic servosystem or a directly coupled d.c. motor (which gives a faster response to signals) is essential.

The work table is mounted on orthogonal slideways, often provided with optical scales for positioning it to an accuracy approaching that of jig-boring machines. The whole working area is surrounded by a tank to contain the dielectric which circulates from the machine to a storage tank in which suspended particles of the erosion process settle (Figure 13.3). It is then pumped back into the working zone through a filter. To remove heat generated during erosion the tank is provided with a heat exchanger to maintain a dielectric temperature of 20 °C.

The performance of a basic ED machine can be improved by the provision of a small orbiting or translatory displacement of the electrode in a plane perpendicular

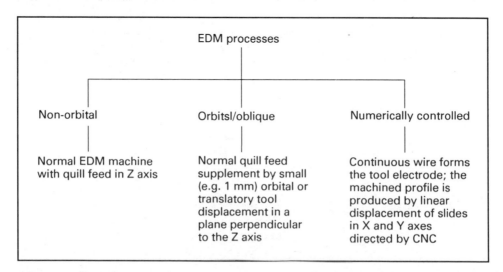

Figure 13.3 Categories of ED machine tools.

to the feed vector. This lateral movement improves the dielectric flushing action in the working gap, significantly reducing machining times and electrode wear. Alternatively, or additionally, for producing holes of circular section, the electrode may be rotated around its own axis. For producing these motions, an attachment may be fitted to standard machines, or the device may be an integral part of the machine.

The CNC continuous wire electrode ED machine, Figure 13.4, is an important development for producing complex shapes in through-hole applications e.g. extrusion or press-tool die cavities. The wire electrode, typically 0.25 mm diameter, is taken from a supply spool containing sufficient wire for 24-hour continuous operation, at a velocity of 0.1 to 8 m/minute. For highly accurate work, the wire is drawn through a sizing die and reduced in diameter by 0.015 mm; it is then annealed

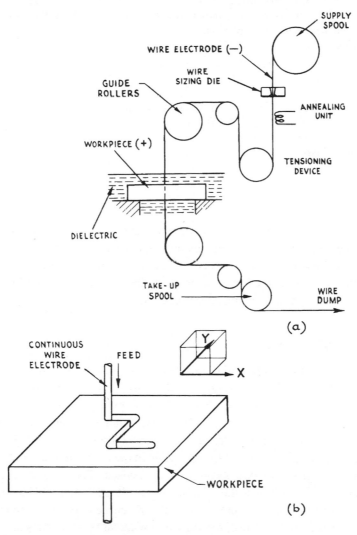

Figure 13.4 Principle of continuous wire electrode CNC electro-discharge machine.

and passed through sapphire guides to a device which stabilises the tension at 0.5–1.0 kg. In operation, de-ionised water is used as the dielectric and apart from its primary function, it may be used in a concentrated jet to carry the end of the wire, when setting up, through a pre-drilled hole in the workpiece and over the guide rollers to provide automatic operation. The electrode may be inclined at an angle of $1\frac{1}{2}°$ to the perpendicular to provide means for machining clearance in the profile of dies. Displacement of slides in the X and Y axes is under CNC control and is programmed as for conventional CNC machines.

Example 13.1

In a typical EDM operation on a CNC machine, a profile 100 mm long is machined in a steel extrusion die 20 mm thick by means of a 0.235 mm wire electrode. Given that the metal removal rate is 0.36 mm^3/s and the radial overcut 5 μm, determine the erosion time in minutes and state the rate of linear erosion.

Solution

$$\text{Erosion time} = \text{Vol. to be removed/metal removal rate}$$
$$= 100 \times 20 \times (0.235 + 0.01)/0.36$$
$$= 22.68 \text{ minute}$$

Linear rate of erosion is 4.4 mm/minute.

□   □   □

The capital cost of ED machines is comparable with traditional machines of the equivalent size and type. This factor renders comparison of economic performance with other machines fairly straightforward. A combination of the two methods is often practised, i.e. removal of bulk metal from the workpiece by drilling or milling, followed by roughing and finishing EDM processes.

Spark generating circuits

For metal to be eroded from the workpiece it is necessary for the ED machine to generate a spark whose characteristics can be controlled to provide the optimum conditions for a particular application, e.g. high metal removal rate or a fine surface texture. These requirements demand a voltage supply adequate to initiate and maintain the discharge process, also a system which provides the necessary control over the intensity, duration and cycle times of the discharges. These take place in the range from 2–1600 μs.

Many types of generator circuits have been devised. The resistance–capacitance (RC), i.e. relaxation circuit, was the original system. It is simple, reliable and provides fine surface textures, e.g. 0.25 μm R_a, but the discharges occur at relatively high voltages and are difficult to control, resulting in low metal removal rates and substantial tool wear. The most widely used system is the transistorised pulse generator which gives high metal removal rates and reduced electrode wear, through highly developed control of the machining parameters. To obtain both high machining rates and fine surface finishes, many machines are equipped with dual circuits.

The relaxation (*RC*) circuit

A basic form of *RC* circuit is shown in Figure 13.5(a). On commencing operation the capacitor is in the uncharged condition. Then it is charged from a d.c. voltage source (V_s) via the resistor which determines the rate of charging. The relationship between time (µs) and voltage is given in Figure 13.5(b), from which it will be seen

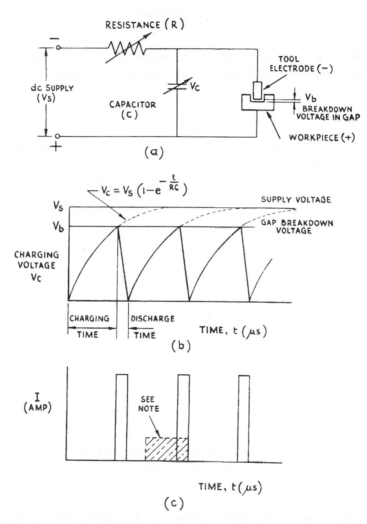

Figure 13.5 (a) Basic relaxation (*RC*) circuit; (b) relationship between charging voltage and time; (c) characteristics of current in working gap. Note the shaded portion represents a discharge of equal energy but of more desirable characteristics resulting in a higher metal removal rate and reduced tool wear.

that the voltage (V_c) across the capacitor increases exponentially as charging proceeds. When V_c has risen to the level of the breakdown voltage (V_b) existing in the working gap, the capacitor discharges across the gap and a particle is eroded from the workpiece. The spark is not sustained because the capacitor is discharged far more quickly than it can recharge via the resistor. The cycle is continuously repeated until the operation is complete.

This sequence can the examined in electrical terms since it is well known that the increase in voltage during the charging of a capacitor is given by the expression

$$V_c = V_s(1 - e^{-t/RC}),$$

where t = time (second)

RC = resistance (ohm) × capacitance (farad)

= time constant.

i.e. the voltage of the capacitor approaches the supply voltage with a time constant equal to RC, and the spark will be discharged when V_c reaches a value corresponding to V_b.

Example 13.2

1. Show that if $t = RC$, the voltage of the capacitor is 63.2% of the supply voltage after the elapse of time period, t.
2. What percentage of the supply voltage is reached by the capacitor when $t = 4RC$?

Solution

1. $V_c = V_s(1 - e^{-t/5RC})$

   hence, when $t = RC$, $e^{1t/RC} = e^{-1} = 0.368$

   $\therefore V_c = V_s(1 - 0.368) = 0.632V_s$
   i.e. $V_c = 63.2\%$ of V_s

2. When $t = RC$, $e^{-t/RC} = e^{-4} = 0.0183$

   $\therefore V_c = V_s(1 - 0.0183) = 0.982V_s$
   i.e. $V_c = 98.2\%$ of V_s

□   □   □

The theoretical energy (W joules) in an individual spark discharge is given by the expression

$$W = \tfrac{1}{2}V_b^2 C,$$

from which it follows that the greater the value of V_b with respect to V_s the greater the energy discharged. However, the exponential rate of charge exhibited by the curve, Figure 13.5(b), shows a marked decline as the value of V_c approaches that of V_s. Hence, if V_b were closer to V_s, fewer discharges would occur per unit time and in consequence the metal removal rate would suffer. The optimum value of V_b is usually taken at 73% of V_s for a high rate of machining.

Bearing in mind that the time cycle is important in allowing deionisation of the dielectric and for debris to be flushed from the working gap, it follows from the foregoing that there are optimum machine settings for voltage, R and C, for a given set of operating conditions.

The supply voltage usually lies between 200 and 400 V; an increase in V_s increases V_b and the machining rate, but produces a poor surface texture. A reduction in V_s enables a smaller working gap to be used, improving finish and accuracy, but reducing the machining rate. High rates of machining are obtained by reducing the time constant RC to give rapid charging; however, as R is reduced, the frequency increases and may reach a point at which de-ionisation is prevented from taking place and arcing occurs. Arcing causes effective machining to cease and creates thermal damage. Discharge energy (hence machining rate) and capacitance are related in the expression $W = \frac{1}{2}V_b^2C$, indicating that an increase in capacitance increases the discharge energy. The relationship established by Rudorff (D. W. Rudorff, Principles and Applications of Spark Machining, *Proc.I.Mech.E.*, **171**, 495 (1957)) is shown in Figure 13.6. It follows that the machine settings for optimum performance in a given set of machining conditions involve a compromise in selecting the parameters of the process.

The transistorised pulse generator circuit

The interdependence of parameters, a restricted choice of electrode materials and their high wear rate are among the disadvantages of the RC circuits. The introduction of semi-conductors has improved the design of pulse generators and allows the frequency and energy of discharges to be varied with a greater degree of control. Furthermore, the voltage of these machines is reduced to the 60–80 V range, permitting discharge characteristics with a lower profile, as shown in Figure 13.7, with the result that craters formed at each discharge tend to be shallower and wider.

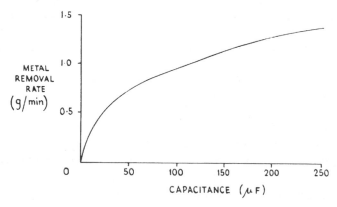

Figure 13.6 Metal removal rate as a function of capacitance (after Rudorff); brass electrode, hardened steel workpiece.

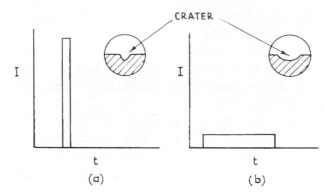

Figure 13.7 Comparison of characteristics for discharges of equal energy produced by (a) *RC* generator; (b) transistorised generator.

In a simple form of the transistorised circuit, Figure 13.8(a), the discharges are controlled by a switching unit on a fixed frequency selected by reference to machining parameters. However, conditions in the working gap vary, causing lags in ignition to occur. Since the ON-time and the OFF-time are fixed, the useful discharge period X varies and efficiency is diminished, Figure 13.8(b). An improved arrangement incorporating feedback is shown in Figure 13.9(a). In this circuit the conditions at the spark gap are monitored by a detector unit which determines the exact moment of current flow after the ignition lag. The time base for the ON-time then becomes effective, providing a constant discharge period. The time base for the OFF period ensures a constant interval for de-ionisation and flushing away of debris by the dielectric. Consistency in the energy of discharge as shown in Figure 13.9(b) results in a much more efficient operation.

Generators of the type described above are typically 25 A or 50 A and details of the performance of a 25 A generator are given below (courtesy: Charmilles).

Power	2 kW	Interval OFF-time: 2–1600 µs
Open gap voltage	80 V	Minimum discharge energy: 180 µJ
Maximum average current	25 A	Maximum discharge energy: 1.0 J
Discharge ON-time from	2–1600 µs	Finest surface finish: 0.4 µm R_a

An extension of the control over the process through the introduction of full adaptive control of an ED machine:

1. Continuously monitors online machining performance.
2. Provides online optimisation for machining parameters.
3. Provides progressively increased electrical power as the effective machining area increases.
4. Modifies characteristics of the discharge near the end of an operation to provide conditions for the finishing cut, which eliminates sub-surface damage and ensures the required surface finish is attained.

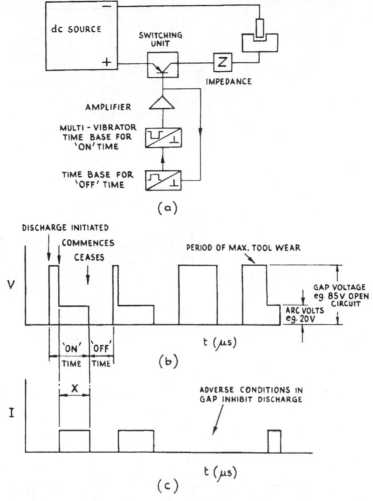

Figure 13.8 (a) Basic transistorised pulse generator circuit; (b) and (c) characteristics of discharges.

EDM process parameters

The tool electrode

Any material that is a good electrical conductor theoretically may be used as a tool electrode. Essentially the material should have:

1. Rigidity.
2. Low electrical resistivity.
3. A high melting point.

Low electrode wear is associated with a high melting point of the material: graphite has a very high value, 3500 °C, followed by tungsten, 3400 °C. However, in practice,

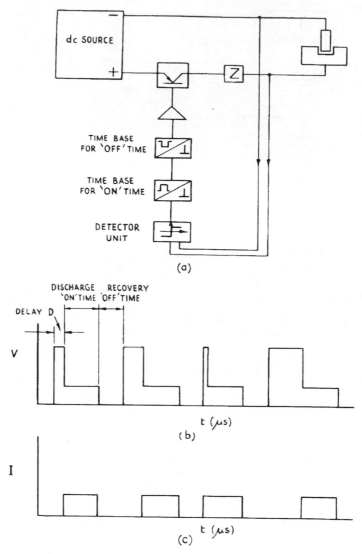

Figure 13.9 (a) Principle of the Isopulse generator (courtesy: Charmilles); (b) and (c) discharge characteristics.

considerations of cost (e.g. graphite is cheap, tungsten very expensive), rate of wear, ease of fabrication, availability, performance in terms of metal removal rate, surface texture produced, etc., combine to reduce the range of materials.

In view of the wear rate of electrodes, their replacement cost is an important factor. Tools are fabricated by a variety of methods – conventional machining, EDM, casting, etching, stamping, plating of moulded araldite, etc. In some applications, e.g. machining a forging die, inexpensive tools can be made from a previous component, i.e. an existing forging.

It is not possible to give firm data for tool wear rates. Manufacturers publish empirically derived tables for their machines, and the extract shown in Table 13.2 for three different settings of parameters when finish machining a steel component with a copper electrode illustrates the degree of variability, even within strictly limited conditions.

The choice of polarity on transistorised pulse generators has a very marked effect on wear of electrodes. The significance is shown graphically in Figure 13.10 for a machining operation using a 25 mm diameter graphite electrode on 1 per cent carbon steel. The band spread in each case is due to selecting different settings for the pause time.

Wear rates can be measured in linear or volumetric terms. In roughing operations, frontal wear of the tool is likely to be heavy and its expression in terms of a volumetric wear ratio would be appropriate. In finishing, a small loss of detail, significant in linear but not volumetric terms, may render the tool unsuitable for further use.

$$\text{Volume wear ratio, } \% = \frac{\text{volume of workpiece removed}}{\text{volume of electrode consumed}} \times 100$$

$$\text{end, side or corner wear ratio, } \% = \frac{\text{depth of machining}}{\text{end, side or corner wear}} \times 100$$

Table 13.2

Setting	R_a (μm)	Metal removal rate (mm^3/min)	Volumetric wear rate
1	0.40	0.2	20%
2	0.60	0.8	25%
3	1.60	2.1	26%

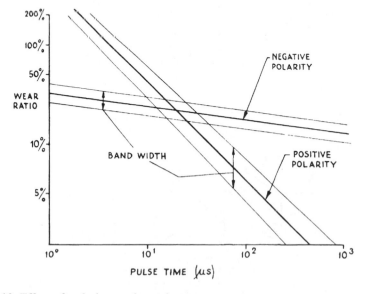

Figure 13.10 Effect of polarity on electrode wear.

An example of electrode wear resulting from the machining of a component predrilled to reduce EDM time is illustrated in Figure 13.11.

The dielectric fluid

In addition to providing suitable conditions necessary for discharges to take place, the dielectric fluid cools the electrodes and flushes away unwanted products of the process. The fluid must be of low unit cost and possess the following properties:

1. Low viscosity to ensure efficient flushing.
2. High flash point.
3. Non-toxic.
4. Non-corrosive.
5. High latent heat
6. A suitable dielectric strength, e.g. 180 V per 0.025 mm.
7. Rapid ionisation at potentials in the range 40–400 V followed by rapid de-ionisation.

From the wide range of dielectric fluids available, three classes are principally used:

1. Hydrocarbons.
2. Aqueous solutions of ethylene glycol.
3. De-ionised water.

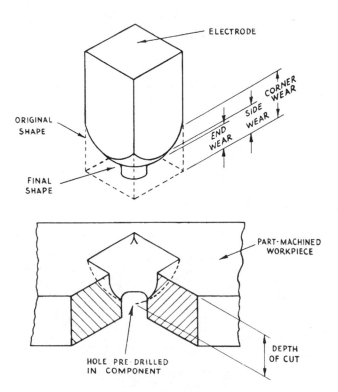

Figure 13.11 Typical electrode wear in a predrilled component.

Hydrocarbons are the major group and may be considered according to their viscosity, see Table 13.3. In finishing operations the working gap is small and the lighter oils are used to ensure flushing – a very important function – is performed satisfactorily. The gap is less restricted in roughing operations enabling higher machining rates to be achieved with heavier oils. Clearly, in EDM the flashpoint of the dielectric is important, also the machining rate is reduced if a low boiling-point fluid produces excessive gas on vaporising. The gases formed include hydrogen, methane, propane, acetylene etc. De-ionised water is widely used for EDM on continuous wire CNC machines equipped with a de-ionising unit.

Flushing techniques

The correct circulation and filtering of the dielectric fluid is most important in EDM. At the beginning of an operation with a fresh supply of dielectric, it is clean and has a higher insulation strength than one containing particles. However, debris is created immediately spark discharges commence, and dielectric strength is diminished by particles acting as 'stepping stones' in the tool–workpiece gap. If too many particles are allowed to remain, a 'bridge' is formed resulting in arcing across the gap causing damage to tool and workpiece. Therefore, the degree of contamination in the gap must be controlled to provide optimum conditions in which machining can take place (Figure 13.12).

The main techniques are:

1. Injection flushing (Figure 13.13).
2. Suction flushing (Figure 13.14).
3. Side flushing (Figure 13.15).

Table 13.3 Comparison of hydrocarbon dielectrics.

Dielectric fluid	Flash point	Typical application	Comments
White spirit	40 °C	Small work, close tolerances, sharp definition, e.g. blanking tools with fine detail	Low viscosity, useful in small working gaps or fine work not requiring high metal removal rate
Paraffin	50 °C	Medium work, e.g. plastic moulds, press tools, extrusion dies	Good general purpose dielectric; inexpensive
Light oil	130 °C	Large work, heavy rough machining, e.g. drop forging dies	Used in high power ED machines; does not filter so readily

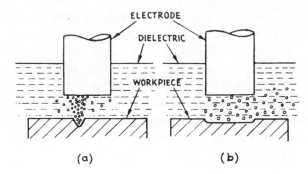

Figure 13.12 Contamination by debris in the working gap.

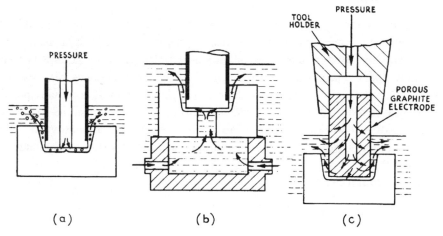

Figure 13.13 Injection flushing techniques.

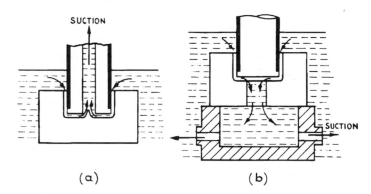

Figure 13.14 Suction flushing techniques.

Additionally, pulsating flushing coupled with a reciprocating movement of the tool in the Z axis, is used in combination with these techniques.

In the case of injection flushing, a slight taper is produced on the sides of the cavity due to the occurrence of lateral discharges as particles pass up the side of the tool. In Figure 13.13(a) the small 'pip' must be removed if the machined area is not

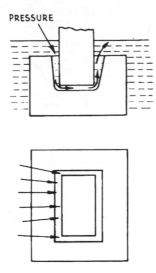

Figure 13.15 Side flushing.

a through cavity. The use of porous graphite electrodes permits a most effective flushing action, particularly useful when machining deep cavities.

Suction flushing (Figure 13.14) avoids the tapered side effect of the previous method because the debris is not drawn past the sides of the tool and avoids the occurrence of lateral discharges. Generally, the suction effect is evenly distributed over the face of the tool, tending to give a more efficient machining action. Care must be taken to avoid trapping gases in recesses or in the pot. An example of side flushing – which must not be directed from opposing directions – is shown in Figure 13.15.

Surface structure

Because the discharge temperature is of the order 8000–12 000 °C, the nature of the workpiece surface layers may be affected in some materials, e.g. steels, especially as carbon will be present due to its release from the molecular structure of the hydrocarbon dielectrics.

The depth of the heat-affected zone for a given material will depend on the characteristics and energy of the discharges, e.g. in a finishing operation it may be no more than 0.002 mm, whereas for a roughing cut it may be 0.2 mm. Typically the machined surface will exhibit three layers, (Figure 13.16) after exposure to high-energy discharges.

1. A recast layer, formed by molten metal particles being redeposited on the workpiece surface; this layer in steel workpieces may be harder than the parent metal.
2. A layer which has reached melting point, but not dispersed, and which remains as a recast layer.
3. An annealed layer, of hardness less than that of the parent material.

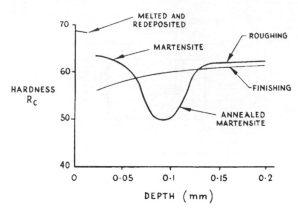

Figure 13.16 Typical surface structure subsequent to roughing and finishing operations by EDM.

In many instances these effects will be minimised by the finishing cuts and will not be detrimental to the function of the component. However, for some components, e.g. in aerospace, the modification to the surface layers may be a critical factor, especially as microcracks may be present. Experimental findings indicate that fatigue endurance limits are less than would be expected from similar components not subjected to EDM (see Figure 13.17). In applications where the presence of a highly stressed surface layer is undesirable it may be relieved by shot-peening, grinding, electropolishing or chemical machining.

Summary of approximate operating data

As a guide to the operational results in the application of ED machines the following data are given. It must be recognised that these values may not apply in specific instances, but they are included here in order to convey some idea of the magnitude of the values likely to be experienced in practice.

Metal removal rate	Typically 1 mm³/second, but since roughing and finishing operations with different electrode/workpiece materials are involved and machine sizes and powers cover an extremely wide range, the rate may occur in a range from 0.01 to 150 mm³/s.
Overcut	Typically 0.005–0.2 mm per side (minimum corner radius = size of overcut).
Taper	Typically 0.05–0.5 mm per 100 mm per side.
Accuracy of workpiece	Tolerances of ±0.05–±0.15 mm easily achieved; tolerances of ±0.005–0.01 mm obtainable with close control.
Surface texture	1–2 μm R_a easily achieved; 0.25–1.1 μm R_a obtained with care.

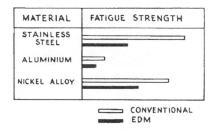

Figure 13.17 Comparison of fatigue strength of identical components produced by conventional and electro-discharge methods.

13.3 Laser beam machining (LBM)

Introduction

LASER is an acronym for Light Amplification by Stimulated Emission of Radiation. Since the invention of lasers in the early 1960s they have found wide application in engineering. The output radiation from lasers is in the form of a highly collimated beam of electromagnetic radiation at wavelengths in the infrared, visible or ultraviolet regions of the spectrum. Many lasers can produce extremely high power densities when focused using convex lenses and these systems are used for micromachining, heat treatment, welding and cutting. Lasers can offer many advantages over conventional methods of heating. Other lasers have very low power output but have very stable and precisely defined output wavelengths. These lasers can be used for metrology using interferometry and for holography.

Principle of operation of the laser

There are a very large number of different kinds of laser. They differ in the active medium used and consequently have different output characteristics and require different services to run them. The main part of the laser is the medium where excited atoms or molecules emit radiation. These media can be a gas, a solid or liquid and energy must be supplied to the atoms to raise their internal energy. Gases are excited by passing a current through them and solids can be excited by focusing light from a flashtube (similar to that used in photography) or from another laser. When the electron in an atom is excited its potential energy is precisely defined or quantised and when this electron drops to a lower energy state a precise packet of energy is emitted as a photon. The frequency, ν, and so the wavelength, λ, of the emitted photon is given by the energy difference between the energy states by

$$E_2 - E_1 = h\nu \frac{hc}{\lambda}$$

There are of course many millions of atoms contributing photons to the emitted radiation and they, once excited, will emit the photons at random. The light emitted is:

1. Isotropic: of equal intensity in all directions.

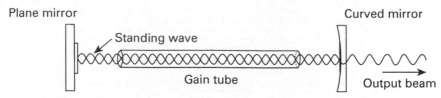

Figure 13.18 Principle of a laser.

2. Unpolarised.
3. Incoherent: the emitted waves are jumbled up in phase.

This is called spontaneous emission and it is the way we observe light from normal light sources.

In 1917 Einstein proposed another way in which atoms could emit light. He called this stimulated emission. In this excited atoms are forced or induced to drop to a lower energy state, and so give off a photon, by another external wave of exactly the same wavelength as the atom will emit. These emitted photons have some very special properties which make the laser so different from normal light sources. The emitted photons are:

1. In the same direction as the stimulating wave.
2. Of the same polarisation as the stimulating wave.
3. In phase with the stimulating wave, i.e. coherent.

Thus as a wave of the correct wavelength passes through an excited medium it will interact with many millions of excited atoms and will grow in amplitude by this process of stimulated emission. An important factor about stimulated emission is that the rate of amplification is proportional to intensity of the stimulating wave while spontaneous emission does not depend on any stimulating wave. Thus for stimulated emission to be important the stimulating wave must be of high enough intensity to make the stimulated emission rate larger than the spontaneous emission rate. This is achieved by confining the light between two mirrors.

Some of the atoms in the medium will be in the lower state and when the wave interacts with these atoms a photon is absorbed from the wave by these atoms which are excited to the higher state. This mechanism competes with stimulated emission and so, to have gain, the stimulated emission must be larger than the absorption. To achieve this the excitation must produce more atoms in the higher state than in the lower state. This is called population inversion.

Figure 13.18 is a schematic diagram of a laser. It shows the gain medium between two mirrors aligned perpendicular to the axis to the gain medium. When the atoms are excited spontaneous emission will produce photons which in turn are amplified by stimulated emission. The feedback provided by the mirrors means that only one photon is initially required for the light intensity to grow by a very large amount. The atomic excitation is therefore being extracted by the wave as a coherent beam. In the laser one mirror is highly reflecting while the other is partly reflecting so allowing a fraction of the radiation into the outside world for application. The beam is of high intensity simply because atoms which would emit radiation by spontaneous emission

Table 13.4 Power densities and industrial applications.

Power density (W/mm^2)	Condition of material	Typical application
$<10^4$	Heat absorbed	Heat treatment
$10^4–10^6$	Molten	Welding
$10^6–10^8$	Vaporised	Cutting, drilling

in all directions are being forced to emit light into the narrow beam a few millimetres in diameter. We often use the term wall-plug efficiency which is simply the ratio of electrical power in to the optical power out.

Another feature of the laser is that the light is travelling in both directions between the mirrors and consequently a standing wave is set up. This standing wave will, since it is resonating, be of a very precise frequency and it is this which makes the laser so useful as a source for interferometry.

The output from lasers which is continuous in time is called *continuous wave* (CW). In other lasers the output can be pulsed in time either as a single pulse or as a train of pulses. A technique, known as Q-switching, is often used in which the cavity is closed by a shutter which is subsequently opened when the gain is at maximum. The energy in the population inversion can then be released into the beam in a giant pulse with duration of about 50 ns. This increases the peak power of the pulse by perhaps 1000 times. It is often useful to supply energy to a target as a very high power pulse rather than continuously.

The following lasers find application in engineering.

Gas lasers

1. Helium neon lasers. These output CW power in the range 1–50 mW and are not used for heating. Their output wavelengths can be very precise and stable and

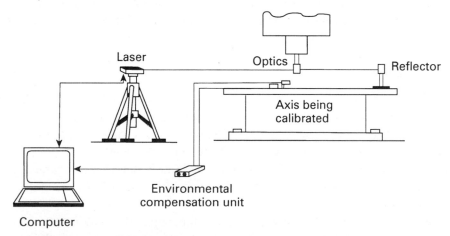

Figure 13.19 Laser calibration of machine tool axis. (Note environmental compensation unit to measure temperature, air pressure and humidity.)

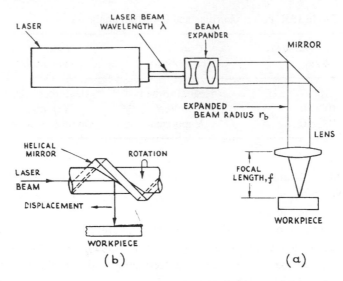

Figure 13.20 (a) Principle of laser machining processes; (b) a method of displacing beam with minimum inertia for high speed linear welding or cutting.

are used for interferometry or holography. Wall plug efficiency is about 0.02 per cent. The helium neon laser at 633 nm when stabilised using iodine absorption, produces a frequency which is internationally agreed, and is directly traceable to the definition of the metre. (See Figure 13.19.)

2. Argon ion lasers. These produce CW power in the blue/green of many watts and are normally used for holography. These lasers require three phase supply and clean cooling water and have an efficiency of about 0.01 per cent.

3. Carbon dioxide lasers. These are widely used for heat treatment and produce CW output power in the range 10–6000 watts in the far infrared at 10.6 microns. These lasers have an efficiency of up to 20 per cent.

4. Excimer lasers. These lasers can give average power of about 1000 watts in 50 ns pulses at pulse repetion rates of 1000 Hz. The output wavelengths are in the ultraviolet about 300 nm with efficiency of about 1 per cent. Excimer lasers use a hazardous gas such as fluorine.

Solid state lasers

The most important of these are the neodymium (Nd) lasers. The Nd atoms can be seeded into certain crystals such as YAG (yttrium aluminium garnet) or even glass. These lasers have output wavelengths of 1.06 microns and can be pulsed or CW. The CW lasers output power in excess of 100 watts and the pulsed lasers can produce Q-switched pulses in excess of 1 Mwatt for about 50 ns at pulse repetition rates of 50 Hz. Higher pulse energies up to a few joules are achieved from Nd:glass lasers with efficiencies of a few per cent.

By a choice of suitable power densities, the required condition of a metal workpiece in the area of the spot can be selected. It should be noted that the surface condition of the workpiece as regards reflectiveness will affect efficient absorption.

Table 13.5 Characteristics of lasers used in production processes.

Type	Electrical efficiency (%)	Wave length (nm)	Continuous power (W)	Focused peak power (W)	Applications in manufacturing
CO_2	>10	10 600	up to 1000	10^8	Cutting, drilling, heat treatment, scribing, welding
He–Ne	<0.1	633	0.002	10^{-3}	Alignment, displacements, interferometry
Nd in glass	>1	1 060	up to 1000	10^9	Drilling, welding
Ruby	<0.15	694	0.05	10^8	Drilling

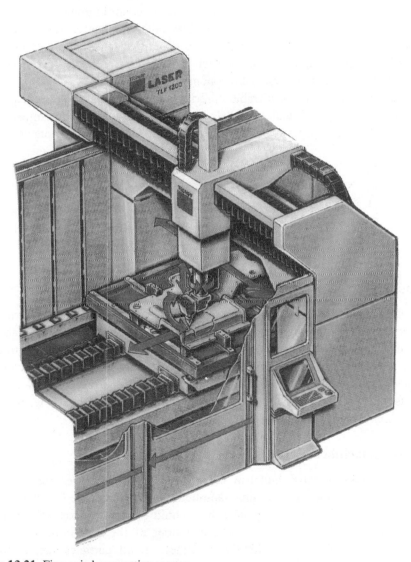

Figure 13.21 Five axis laser cutting centre.

Although metal removal rates can be quite high, the extremely low efficiency of the laser, see Table 13.5, ensures that the input power requirement is also very high, making the process uncompetitive with traditional methods for the bulk removal of metal (note that conventional chip removal techniques require energy for mechanical shear whereas the cutting action of the laser requires energy for vaporisation).

Industrial applications

Despite limitations, especially of low overall efficiency, the laser has advantages which ensures its increasing value and use. It will melt or vaporise any known material including diamond, produces narrow kerfs, and small heat-affected zones with negligible effect on adjacent areas. The beam is easy to control through optical systems or by CNC displacement of slides, and enables otherwise inaccessible areas to be machined or welded, or for these processes to be carried out in specially controlled environments. It can be applied to small batch production, e.g. the production of sheet metal parts in quantities too small to be economic for a press tool.

The laser has made a considerable contribution in two particular areas, in the cutting or scribing of difficult materials, e.g. ceramics, and in micromachining, especially in the microelectronics industry, producing circuits, transistors and diodes on silicon wafers.

The laser is a robust piece of equipment and provided suitable safety precautions are taken it can be used under normal operating conditions by skilled or semi-skilled workers (see Figure 13.21).

13.4 Ultrasonic machining (USM)

Introduction

Unlike the processes previously described, USM removes material from the workpiece in the form of chips, and has the advantage that it is suitable for machining materials whether or not they are electrically conductive. In particular it provides a method for machining hard, brittle materials, e.g. diamond, ceramics, glass, and those with special physical or mechanical properties, e.g. titanium and heat resisting alloys. USM can be used for machining cemented carbide, but at the cost of a high rate of tool wear.

The basic principle of USM

A tool with a cross-section similar in shape to that required to be produced in the workpiece is axially vibrated at a small amplitude in the ultrasonic frequency range, Figure 13.22. Abrasive grits, suspended in a liquid carrier, are passed through the gap between the tool and workpiece. The vibrations of the tool tip transmit energy into the grits and on impact with the workpiece small particles, i.e. chips, are removed from the workpiece. A gradual downward feed of the tool maintains a static load between the tool and the workpiece during the cutting operation.

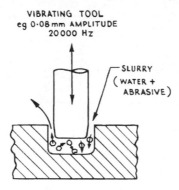

Figure 13.22 Principle of ultrasonic machining.

The material removal process is observed to be due to several factors:

1. Hammering by the tool of abrasive grits which are in contact with the workpiece.
2. Impact of free abrasive grits on the surface of the workpiece.
3. Cavitation.

The mechanism of (1) is thought to account for the greater part of the material removal.

The ultrasonic machine

The machine tool is compact and may be floor or bench mounted. It has a table capable of orthogonal displacements in the X and Y axes on which a rotating table may be fitted. The tool vibrator spindle is mounted in ball or roller slides, usually in the vertical plane, with a feed mechanism controlled to provide a steady working force during operation and a woodpecker action to facilitate removal of retained grits or debris from the working cavity. The basic elements of an ultrasonic machine tool are as follows:

1 Main frame, table and slides.
2. Acoustic elements.
3. Slurry system.
4. Feed mechanism.

The acoustic elements

The basic elements in the vibrating unit are illustrated in Figure 13.23 and comprise the generator, transducer unit with toolholder, velocity transformer and the tool. Magneto-strictive transducers are generally employed in USM, but piezoelectric transducers may be used. The generator unit converts power from the supply at 50 Hz to high frequency energy in the range 20000 to 25000 Hz. The audio threshold is approximately 15000 Hz, hence the practical minimum working frequency for the machine is about 16000 Hz with the upper limit of frequency being imposed by considerations of heating in the transducer.

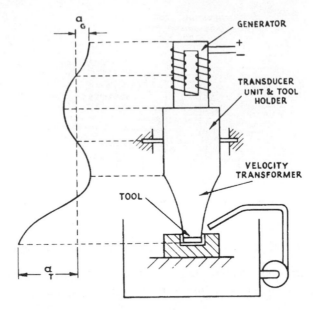

Figure 13.23 Basic elements in the acoustic unit related to the distribution of the vibration amplitude in the system.

In the magneto-striction transducer, the high-frequency vibration is produced by passing a current at a particular frequency through a coil wound around the transducer core. This core is composed of a stack of nickel plates and small changes in dimension (i.e. magneto-striction) are obtained when the laminae are magnetised. The dimensional change of the transducer is of the order of 0.025 mm although both frequency and amplitude may be adjusted within limits, and the toolholder is so designed that the transducer amplitude is increased by gains of up to six or seven times. Maximum fatigue strength is obtained by interposing a velocity transformer, manufactured from monel metal, between the transducer and the tool. The transducer, transformer, toolholder and tool must be in resonance to achieve optimum tool amplitude and power output. The shape of the tool unit is important in this connection (see Figure 13.24).

The power input to the transducer is termed the machine power rating, and ranges from 0.05 kW for a small USM drill to 4 kW; a typical machine has a rating of about 0.5 kW.

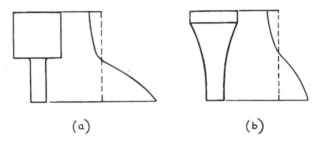

Figure 13.24 Examples of types of velocity transformers and their respective vibration amplitudes.

Process parameters

There are a large number of variables in USM and the results of research investigations into their effects are inconclusive; however, the process parameters may be grouped as follows:

1. Materials – tool and workpiece.
2. Acoustic system – amplitude, frequency, working force.
3. Abrasive slurry – abrasive material, size of grit, concentration of grits in slurry.
4. Geometry of machined area.

The tool

Carbon steels and stainless steels are frequently used for tool materials. When cutting hard materials, e.g. ceramics, a hardened tool steel would be used for roughing cuts and a medium carbon steel for finishing operations. Examples of some tool/workpiece wear ratios are given in Table 13.6.

A typical example of the characteristics of tool wear is illustrated in Figure 13.25. The working force exerted by the tool is closely controlled by hydraulic means and varies from about 1 N when machining fine holes to several hundred newtons for large work.

Table 13.6 Tool wear ratios.

	Workpiece material	
Tool material	Glass	Cemented carbide
Brass	1:50	1:0.75
Copper	1:200	—
Mild steel	1:100	1:1
High carbon steel	1:250	1:3
Cemented carbide	1:1000	1:0.85
Stainless steel	1:150	1:2.5

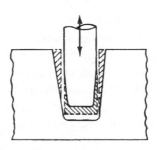

Figure 13.25 Typical characteristics of tool wear and taper in a workpiece.

The abrasive

The abrasive grits are suspended in water to form a slurry. The main materials used for the abrasive in ascending order of hardness, are aluminium oxide, silicon carbide and boron carbide, the latter being the most expensive. The machining rate, which ranges from, say, 0.01 to 2 mm/minute, is affected by the grain size. In general, the maximum rate is achieved as the grain size approaches the magnitude of the tool amplitude. The abrasive size also affects the surface finish; typical examples are:

280 grit = 0.6 μm R_a
800 grit = 0.25 μm R_a

Material removal rate decreases as the sharpness of the grits declines and the abrasive must be replaced in order to maintain the machining rate.

13.5 Water jet cutting

A high pressure jet of water will erode any surface on which it impinges. This has been applied to the cutting of a wide range of materials which are difficult to cut by normal methods. Most high energy cutting processes generate intense heat, resulting in a heat affected zone adjacent to the cut. Water jet cutting, on the other hand, takes place at low temperature, and introduces only low stresses in the workpiece. The addition of abrasive materials to the jet improves the cutting efficiency. Typical materials cut with water jet techniques include glass and ceramic, and sophisticated alloys where heating is undesirable.

Figure 13.26 shows diagramatically the construction of an abrasive water jet cutter. In the mixing chamber (5) a precise dosage of abrasive material is added to the water jet (3). The water jet accelerates this abrasive material to high speed through the abrasive nozzle (6). The high kinetic energy of the individual particles effects a microcutting process which allows even the hardest materials to be worked.

Figure 13.27 shows the principle of water pressure creation. In a regulated hydraulic system (2) an oil pressure of 25 to 200 bar is created. In the pressure converter (3) the oil pressure generates a water pressure of 500 to 4 000 bar. The required cutting pressure is kept constant by the pulsation damper (6) and taken to the cutting head (8) by a high pressure line. Regulation of the hydraulic system provides for smooth control of the cutting pressure from 500 to 4 000 bar.

Figure 13.28 (a) shows water jet cutting samples in metal and metal alloys. Water jet cutting is used for materials and thicknesses which are not possible to machine with thermal processes, for example steel, stainless steel or non-ferrous metals, such as aluminium, copper or brass. Filigree structures and fine webs can be manufactured without a problem. Cutting edges are burr-free and without heat-affected zones. Figure 13.28 (b) shows samples of composite and honeycomb materials. Many

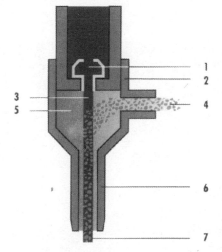

1 Water jet nozzle
2 Abrasive head
3 Water beam
4 Abrasive substance
5 Mixing chamber
6 Abrasive nozzle
7 Water abrasive beam

Figure 13.26 Production of the abrasive water jets.

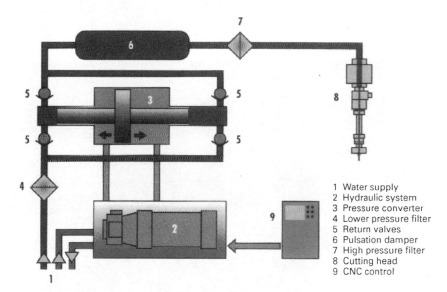

1 Water supply
2 Hydraulic system
3 Pressure converter
4 Lower pressure filter
5 Return valves
6 Pulsation damper
7 High pressure filter
8 Cutting head
9 CNC control

Figure 13.27 High pressure water system.

material combinations or shapes are often difficult or impossible to cut with conventional processes, for example aluminium alloys like Alucobond. Layered materials or moulded laminated material such as compressed chipboard or printed circuit boards may also be easily processed with the abrasive water jet.

Figure 13.28 Water jet cutting examples: (a) metals and metal alloys; (b) composite and honeycomb materials.

Exercises – 13

1. (a) Describe the principle of operation of a relaxation generator used for electrical discharge machining and explain how its disadvantages are overcome in other types of generator.
   (b) Explain with reference to the relaxation generator why a servo-mechanism must be used to feed the tool and show how an error signal may be obtained when using either this generator or a pulse generator.

2. (a) Draw a graph to show the relationship between the breakdown voltage of the dielectric fluid used in a spark erosion machine and the distances between electrode and workpiece. Discuss the influence of this relationship upon the design and operation of such machines.
   (b) Show that for a spark machine operating on a relaxation circuit the breakdown voltage, U, is given by

   $$U = E(1 - e^{-t/RC})$$

   where E = supply voltage;
   $\quad t$ = charging time;
   $\quad R$ = resistance of circuit;
   $\quad C$ = capacitance of circuit.

   Hence, determine for the above machine the average power output, given that the resistance $R = 3.2$ ohms; capacitance $C = 150$ μF; supply voltage $E = 200$ V and the breakdown voltage $U = 160$ V.

3. (a) Compare and contrast EDM and ECM with respect to:
   (i) principle of operation;
   (ii) applications.
   (b) When considering the use of electromachining and the more traditional metal removal processes, what factors would influence the decision in deciding which techniques to apply?

4. (a) Explain the principles of metal removal by the EDM method and show by means of a well proportioned diagram the general features of an ED machine.
   (b) Compare the following processes in terms of metal removal rate, surface finish and dimensional accuracy:
   (i) electro-discharge machining;
   (ii) ultrasonic machining;
   and for each process indicate a typical application which would justify its use, giving reasons for your choice.

14 Screw Threads Specification, Tolerancing, Gauging and Measurement

14.1 Introduction

Manufacturing methods for screw threads exemplify the principles of cutting and forming previously described; the main methods are summarised below.

Taps and dies

The tap is a fluted screw of hardened steel with its leading threads tapered off and ground back to provide cutting clearance. It cuts threads in a plain hole generally drilled somewhat larger than the minor thread diameter. The threads of the tap which follow the cutting portion have no cutting clearance and act as a guide to control the pitch. Dies, based upon a circular nut, cut external threads using similar principles.

To reduce the idle time needed to unscrew a tap or die, the cutting portions (generally four) may be fitted into a suitable holder and automatically withdrawn from the component thread before rapid retraction of the tool. The self-opening diehead has no lower limit of component diameter; the collapsible tap is only feasible for the larger internal threads.

Screw cutting and thread chasing

The principles are of historical significance; the screw-cutting lathe was an important development of the early machine tool. The method controls the two basic elements of a thread independently: the pitch by the kinematics of the lathe, and the form by the plan shape of the cutting tool.

Disadvantages which make the method relatively slow are:

1. Waste material is removed in long thin strands at relatively low cutting speeds and a lot of idle return motions are necessary.
2. A vee-form tool cannot be advanced radially to cut a chip round its entire form because material from both flanks crowds onto the tool rake face causing torn component material and possible tool breakages.

To overcome delays due to (1) many production lathes incorporate threading cycle mechanisms which give automatic retraction of the tool at the end of the cut and rapid return and advance (including the required element of in-feed) for the subsequent cut. Control elements of the more sophisticated NC lathes incorporate 'canned-cycles' for screw cutting.

Problems associated with (2) can be overcome by in-feeding the tool in a direction inclined by the semi-angle of the vee, instead of perpendicularly to the component axis. This results in one flank being 'formed' and the other 'generated' so that a chip is cut from one flank only (see Figures 14.1 and 14.2).

The screw thread

Imagine that a right-angled triangle is wound around a straight cylinder in such a manner that the base of the triangle coincides with the circumference of the cylinder. The hypotenuse (Hx) of the triangle will then describe a helix around the cylinder. A thread is obtained in the same manner, the profile of the thread following a helix. The height of the right angled triangle is equal to the lead of the thread, or pitch (p) as it is also called, in the case of the single start thread. The pitch is also defined as the distance between corresponding points on adjacent threads and is therefore expressed in terms of a unit of length (mm) or in terms of the number of threads per unit of length (threads per inch).

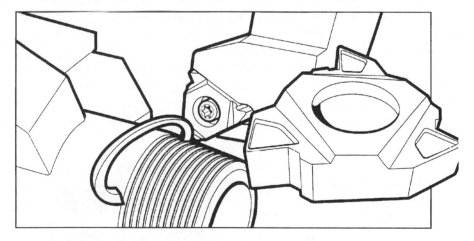

Figure 14.1 Threading today is largely performed by indexible inserts as part of a rapid turning process.

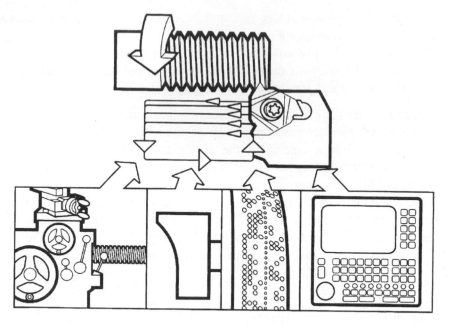

Figure 14.2 Tool motion in threading by lead screw, cam or CNC.

The lead of the thread is defined as the axial distance through which a point on the thread advances during one turn of the thread. The angle (φ) is called the lead angle, but is also known as the helix angle, and thus describes the relation between pitch and diameter (see Figures 14.3 and 14.4).

Multiple start threads

Threads can have two or more parallel thread grooves, which means two or more starts. The lead of a thread with two starts will be twice that of a single start screw. A large lead permits the thread to be screwed in or out faster. The axial load on the thread increases, however, with increasing lead angle. This load is distributed among more thread grooves on a multiple start thread

The lead increases in relation to the pitch by a multiple equal to the number of starts. On a single start thread, the pitch and the lead are equal; on a two start the lead is twice the pitch, with three starts, the lead is three times the pitch, etc. (see Figure 14.5).

Multiple start threads can be manufactured in two ways. The first and the most

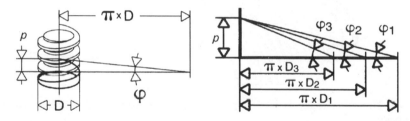

Figure 14.3 Screw thread geometry.

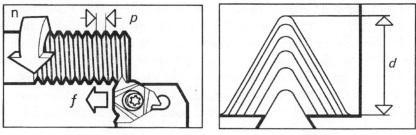

Figure 14.4 The thread pitch is a product of workpiece rotation and tool feed. Thread height/cutting depth is achieved in several passes.

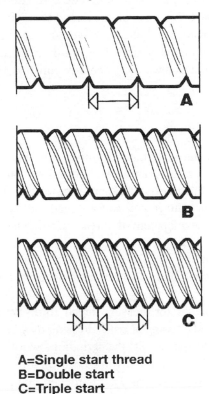

A=Single start thread
B=Double start
C=Triple start

Figure 14.5 (A) Single start thread; (B) double start thread; (C) triple start thread.

common way is to complete a single thread groove in a number of passes and then start with the next groove. Another alternative is to adjust the first infeed and machine one pass on a groove, and then go back and machine the first pass on the next groove, after which the second infeed is set and the next pass is machined on each groove, and so on until all the grooves are finished.

Thread milling

By replacing the single point tool method of screw cutting and thread chasing by a form milling cutter (hob) set to cut the full depth of thread, short chips will be cut as described in Chapter 11 and the thread completed in one pass. For short lengths of thread, especially on large diameters, a form relieved cutter with a number of threads (annular grooves) can be advanced to the depth of cut while the work is held stationary. The component can then be rotated about $1\frac{1}{3}$ times while the cutter is traversed at the appropriate rate of thread pitch. This, known as the 'plunge-cut' method, is much faster than thread chasing for work of large diameter or of coarse pitch; however, the faceted milled surface is unlikely to be acceptable for work of high quality or precision.

Single form cutters, e.g. Acme, can be used to rough out long transmission screws such as machine tool leadscrews. The axis of cutter rotation needs to be inclined to the screw axis by the mean helix angle of the thread to minimise interference effects. Interference effects arise whenever helical grooves are cut with tools having the form of a rotating disc because mutual contact does not lie in an axial plane of the screw. Thread milled transmission screws are generally finished by thread chasing on a screw-cutting lathe.

Thread grinding

The process originated in early attempts to finish grind pre-cut hardened blanks for use as screw gauges. Modern taps are ground from pre-fluted hardened HSS blanks in a single pass, which illustrates subsequent progress. Kinematically the methods are either 'pass-through' or 'plunge-cut' as for thread milling.

Important stages in the development of thread grinding have been:

1. Manufacturing of grinding wheels of composite structure capable of removing large amounts of material while also retaining accurate thread form and giving a good surface finish.
2. Development of techniques for dressing thread forms on single and multi-ribbed wheels in which diamond dressing or crush dressing may be used (see page 383).
3. Provision of adequate power at the grinding wheel spindle coupled with arrangements for varying wheel and work-head rotation speeds. Thread grinding can achieve very high pitch accuracy on hardened components (e.g. micrometer screws) and is now the basic method for completing the manufacture of the essential equipment for screw thread production.

Thread rolling

By deforming a plain cylinder within the plastic range, threads can be produced without cutting away waste material. The flat plate method, rolling the blank under pressure between two suitably profiled flat plates, is a rapid method of producing bolts of relatively low quality such as ironmongery bolts. If the flat plates are replaced by cylindrical rollers (two or three according to the system employed) the rolling time can be extended and tougher metal worked to closer tolerances and better surface finish.

Only a brief summary of screw-thread manufacture is possible here as an introduction to the more specialised topics of this chapter. Full descriptive treatments and explanatory diagrams will be found in books listed in the general section of the Bibliography.

Screwed assemblies have more complex fitting requirements than plain assemblies but basic size, tolerance and fundamental deviation remain the controlling factors.

14.2 Nomenclature and specification

BS 2517: *Definitions for Use in Mechanical Engineering* gives the standard definitions applicable to threads. Details of the principal thread forms are contained in:

> BS 3643: *ISO Metric Threads*.
> BS 4827: *ISO Miniature Threads*.
> BS 4846: *ISO Trapezoidal Threads*.
> BS 21: *Pipe Threads for Tubes and Fittings*.

Because they are likely to remain in use for servicing purposes the following are also of interest:

> BS 84: *Whitworth Form Threads*.
> BS 93: *British Association (BA) Threads*.

The well-known *Machinery's Screw Thread Book* is another handy reference.

Certain ideas about specification, tolerancing, gauging and measurement of screw threads of vee form (with a few minor exceptions) are common to all the above thread systems; these will now be explained with special reference to ISO metric threads.

Figure 14.6 shows general features of screw threads of vee form. The *nominal* diameter is generally the *major* diameter of the basic form (exception, pipe threads). The 'form' of a thread lies in a section plane along the axis.

Figure 14.7 gives the basic form for ISO metric threads and proportional dimensions in terms of the pitch (p). For bolts (external threads) the actual form lies on or below the basic form; for nuts (internal threads) it lies on or above basic form. The design form of the external thread is shown in Figure 14.8. The root of the vees is rounded, an important factor in relation to impact loading and fatigue strength; it also has relevance to the cutting tools. The crests may be either flat or rounded as shown. Die-threading and thread-rolling produce rounded crests, but for single point thread cutting, the flat crest involves less trouble. In terms of pitch these radii are:

> Root radius 0.1443P.
> Crest radius 0.1082P (also the minimum radius allowed by tolerance at the root).

A reasonable tolerance at the root is necessary in order to allow for normal tool wear.

The design form of the internal thread as given in BS 3643 has a radius at the major diameter (root of the vees for a nut). This radius is large enough to clear the flat crest of the basic form and provides for wear of screwing taps at their major

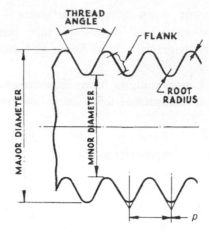

Figure 14.6 General features of a screw thread of vee from.

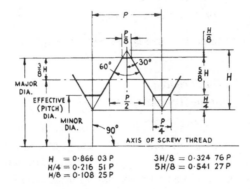

$$H = 0\cdot 866\ 03\ P \qquad 3H/8 = 0\cdot 324\ 76\ P$$
$$H/4 = 0\cdot 216\ 51\ P \qquad 5H/8 = 0\cdot 541\ 27\ P$$
$$H/8 = 0\cdot 108\ 25\ P$$

Figure 14.7 Basic form of ISO metric threads.

diameter. Essentially vee threads are dimensioned to fit along the flanks of the vees, the most important dimension being the *effective* (or pitch) diameter.

Figure 14.9 shows the three essential dimensions which determine the fit of mating threads. The *effective diameter* (or pitch diameter) is defined as the diameter of the pitch cylinder. This is an imaginary cylinder, co-axial with the thread, which intersects the thread flanks in such a manner that the intercept on a generator of the cylinder, between the points where it cuts the opposite flanks of the thread groove, is equal to half the basic pitch.

The combined significance of the semi-angle, pitch and effective diameter in determining the relative position of adjacent flanks is shown in Figure 14.10. The pitch may be measured along any line parallel with the axis and intersecting the flanks, such as AB; dimensions p_{AB}, p_{CD}, etc., determine the relative position of points on the flanks along the thread. A dimension of similar significance to the effective diameter may be measured along any line perpendicular to the axis and intersecting the flanks, such as *EF*; dimensions E_{EF}, E_{GH}, etc., determine the relative

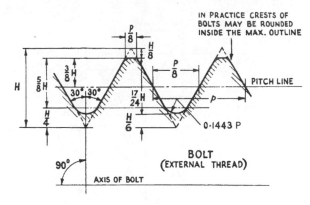

Figure 14.8 Design of ISO metric external thread.

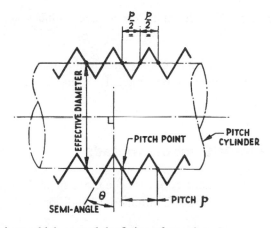

Figure 14.9 Dimensions which control the fitting of vee threads.

position of diametrically opposed points on the flanks of the thread. It is important to realise that the common elementary concept of screw-thread dimensions given in Figure 14.1 is inadequate as a specification for manufacturing purposes.

The helix is a three-dimensional curve, and the geometry of a screw thread cannot be completely specified by reference to a section plane containing the screw axis, as shown in Figure 14.9. It is an implied condition that the developed helix follows a straight line. Departures from this condition are often referred to as *drunkenness*. Equipment capable of measuring small drunkenness errors is somewhat elaborate and treatment of this topic may be found in specialised texts on metrology.

BS 3643 provides for different *qualities* of screwed work as below:

Close fit	designation 5H (nut)	4h (bolt)
Medium fit	designation 6H	6g
Free fit	designation 7H	8g

Numbers relate to tolerance magnitude, letters to the deviation from basic size.

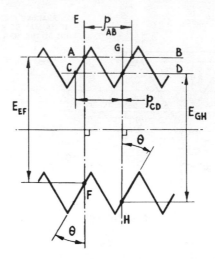

Figure 14.10 Dimensions which determine the relative positions of the flanks of a vee thread.

The full designation M8 × 1.25 – 6H/6g stated on a drawing would imply an ISO metric thread, nominal diameter 8 mm, pitch 1.25 mm, nut to 6H tolerance, bolt to 6g tolerance (i.e. medium fit).

14.3 Tolerance for ISO metric threads

Figure 14.11 illustrates tolerance zones for the close fit (5H/4h). The H/h symbols are for zero deviation so the maximum material condition is controlled by the basic form. Notice also:

1. The tolerance zones are greater at the root and crest than along the flanks.
2. A *unilateral* system is used, bolt tolerance below the basic size, nut tolerance above.
3. *A clearance exists at the minor diameters.*
4. The upper limit of major diameter of the nut is not specified; it is controlled by tap dimensions conforming to BS 949.
5. Because the basic form is not symmetrical about the pitch line (see Figure 14.8) relatively large diameter tapping holes are permissible, which reduces the torque when tapping and so minimises breakages. Figure 14.12 shows the large amount of extra material which needs to be removed to tap a symmetrical form. (This is the principal difference between the ISO metric form and the earlier metric SI thread form.)
6. From the tolerance zones shown in Figure 14.11 it follows that three different types of error are allowed:
   (a) error of the flank angle;
   (b) error of pitch over the length of fitting;
   (c) error of effective diameter.

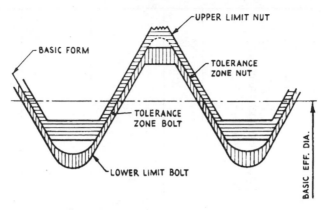

Figure 14.11 Tolerance zones for close fit (5H/4h).

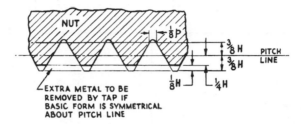

Figure 14.12 Materials to be removed by a tap.

It is equally necessary to realise that any combination of these errors should not cause the profile to lie outside the tolerance zone at any point along the fitting length of an assembly.

Magnitudes of tolerances and deviations

As for plain work, tolerances and deviations are in proportion to the basic size to maintain constant quality for different work sizes. The rational basis of the system used is complicated by the fact that the coarser the pitch the more serious flank angle error becomes, and the greater the fitting length of screwed parts the more serious any progressive pitch error becomes. It is not possible to relate tolerances only to the work tolerance as for plain work.

The tolerances and deviations given in BS 3643 are based on ISO specification 965/1–1973(E) and are derived as shown in Table 14.1.

The value of d in tolerance formula $90p^{0.4}d^{0.1}$ is the geometric mean of the diameter range. The calculated values have to be rounded according to the R40 series. However, the information given here is to show the fundamentally logical basis of a screw thread system of limits and fits and reference to BS 3643 should always be made to obtain actual dimensions.

Figure 14.13 shows the three different grades of work by tolerance zones drawn to scale The 'close' class should only be used where the highest quality is essential because it is relatively expensive to produce. The 'medium' class is suitable for most

Table 14.1 Screw thread tolerances.

	Tolerance grade	Major diameter (µm)	Effective diameter (µm)	Minor diameter (unit as shown)
BOLTS	4	$0.63 \times$ grade 6	$0.63 \times$ grade 6	(i) Upper limit nominal $- (1.2268p \times FD)$ mm
	6	$180p^{2/3} - 3.15/p^{1/2}$	$90p^{0.4}\,d^{0.1}$	(ii) Tolerance $0.0072p + \left[\dfrac{\text{effective diameter tolerance}}{\text{for the grade}}\right]$ mm
	8	$1.6 \times$ grade 6	$1.6 \times$ grade 6	$0.8 \times$ grade 6 µm
NUTS	5	(i) Lower limit = basic size	$1.06 \times$ grade 6 bolt	(i) pitches 0.2–0.8 mm $433p - 190p^{1.22}$ µm (ii) pitches 1 mm and above $230p^{0.7}$ µm
	6	(ii) Upper limit undefined (Controlled by tap)	$1.32 \times$ grade 6 bolt	
	7		$1.7 \times$ grade 6 bolt	$1.25 \times$ grade 6

FD (deviation) g (for bolts), $FD = 11p$ µm   (p = pitch)

The above apply where the nominal length of engagement L_N has limits of:
L_N minimum $2.24pd^{0.2}$ mm
L_N maximum $= 6.7pd^{0.2}$ mm
where d is the smallest diameter of the range for which the tolerances are common.

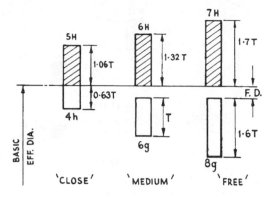

Figure 14.13 Three classes of fit represented to scale.

work; the 'free' class meets conditions where quick and easy assembly is required and where threads may become dirty or damaged.

Example 14.1

Working from basic principles, find limits for the three important diameters of an M36 × 4 'free' class screw thread assembly. The diameter falls within the range 30–39; the length of engagement is within L_N limits. Find the least possible depth of engagement and express this as a percentage of the basic depth.

Solution

The required assembly is 7H/8g.

Bolt dimensions
$$FD = 15 + (11 \times 4) = 59 \ \mu m$$
(rounded value 0.060 mm)

Mean diameter of range $(d) = \sqrt{(30 \times 39)} = 34.2$ mm

Effective diameter tolerance, grade 6 = $90 \times 40^{0.4} \times 34.2^{0.1} = 223 \ \mu m$

Tolerances (grade 8)

$$\text{Major diameter} = 1.6 \left(180 \times 4^{2/3} - \frac{3.15}{\sqrt{4}} \right) = 724 \ \mu m$$

(rounded value 0.750 mm)

Effective diameter tolerance = $1.6 \times 223 = 357 \ \mu m$
(rounded value 0.355 mm)

Minor diameter tolerance = $(0.072 \times 4) + 0.355 = 0.643$ mm

Upper limit, minor diameter of bolt = $36 - \{ (1.2268 \times 4) + 0.060 \}$
$= 31.033$ mm

Nut dimensions

Tolerances (grade 7)

Effective diameter tolerance = $1.7 \times 223 = 379 \ \mu m$
(rounded value 0.375 mm)

Minor diameter tolerance = $1.25 (230 \times 4^{0.7}) = 758 \ \mu m$
(rounded value 0.750 mm)

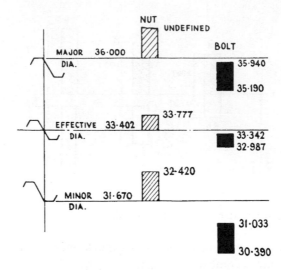

Figure 14.14 Screw-thread tolerances drawn to scale.

Limits for M36 × 4 – 7H8g thread assembly.

Diameter	Nut	Bolt
Major	36.000 +	35.940/35.190
Effective	33.777/33.402	33.342/32.987
Minor	32.420/31.670	31.033/30.390

These limits are shown to scale in Figure 14.14.

Least depth of engagement $= \frac{1}{2}(35.190 - 32.420)$

$$= 1.385 \text{ mm}$$

Depth of basic form $= 0.5413p = 2.165$ mm

Minimum depth ratio $= \dfrac{1.385}{2.165} \times 100 = 64\%$

□   □   □

14.4 Screw-thread gauging

Due to the complex geometry of a helical groove, a full assessment of the accuracy of a screw thread by direct measurement is a lengthy procedure. For this reason, and especially during manufacture, the accuracy of screw threads is generally controlled by limit gauging. Figure 14.11 illustrates the tolerance zones within which the parts must lie. A gauging system based upon Taylor's principles may be employed to restrict the work to the tolerance zones. Screw gauges, limits and tolerances are given in BS 919.

If a tapped hole is considered, the gauges required are seen to be:

1. A *full-form go* gauge, made to basic sizes.
2. An *effective diameter not-go* gauge.
3. A *minor diameter not-go* gauge.

The full form *go* gauge is a screw plug of length equal to the work length of engagement; it defines the maximum metal condition of the tolerance zone. The effective *not-go* gauge has restricted contact with the workpiece, as shown in Figure 14.15; it defines the minimum metal condition at the pitch points of the flanks.

The full truncated form is rather expensive to manufacture, and is used only for very coarse pitches where, due to the longer length of thread flank, angle error is of greater significance than for fine pitches. Ideally, the effective *not-go* gauge should not be influenced by errors of pitch or angle of the thread gauged; in its practical form these effects are minimised, but not entirely eliminated.

The minor diameter *not-go* gauge is a plain cylindrical plug sometimes called a *core not-go* gauge, or *core plug*.

Reference to Taylor's principles of gauging will show that the following gauges are required for a complete check of an external thread:

1. A *full-form go ring* gauge.
2. An *effective diameter not-go* gauge.
3. A *major diameter not-go* gauge.
4. A *minor diameter not-go* gauge.

Screw ring gauges are expensive and rather slow in use; they have a limited application for final inspection purposes, and for gauging screw threads on thin-walled components which distort easily under the contact forces exerted by a caliper-type gauge. Caliper-type gauges having roller or 'edge' type anvils are more commonly used.

The effective diameter *not-go* gauge must be of a caliper type, and have a thread form similar to that shown in Figure 14.15 in order to conform to Taylor's Principles. For the reasons given above there is an occasional need of ring-type *not-go* gauges where the walls of the workpiece are very thin.

The major diameter can be gauged by using a plain caliper gauge, made to the lower limit of size, as the *not-go* gauge; alternatively, blanks may be limit gauged for size prior to threading, a method which generally provides sufficient control.

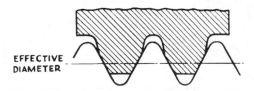

Figure 14.15 Truncated form of effective not-go gauge.

The minor diameter of the thread is rarely tested by means of a separate *not-go* gauge. Optical projection of the thread form will show whether the root radius of the thread is being satisfactorily produced relative to the flanks. If this is so, work lying within the effective diameter limits is bound to lie within the minor diameter limits.

Internal and external thread inspection is illustrated in Figure 14.16.

The common form of caliper gauge for external screw-thread work is shown in Figure 14.17. For very large work this type of gauge becomes unwieldy, and gauges

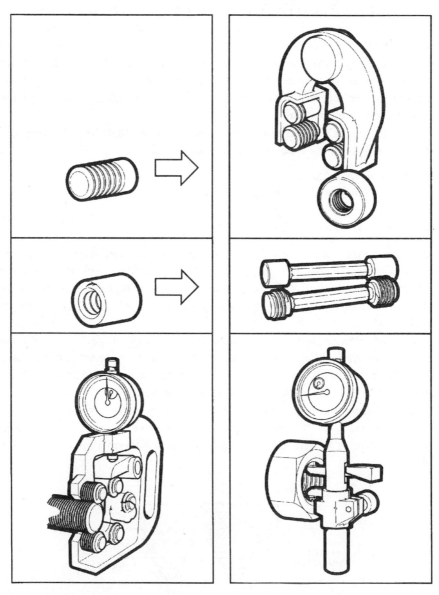

Figure 14.16 External and internal thread inspection.

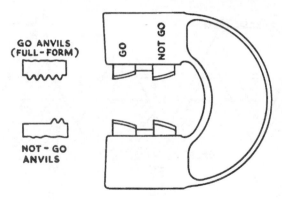

Figure 14.17 Screw thread caliper gauge.

which measure effective diameter in terms of the radius of an arc are then substituted.

The effectiveness of the gauging method

The gauging methods described restrict the work to the tolerance zone, because the magnitude of the zone is fixed by the major diameter, minor diameter and effective diameter limits. Gauging methods do not, however, enable the particular sources of error to be distinguished, and for this purpose direct measurements are necessary.

Figures 14.18 and 14.19 illustrate the effect upon the fitting conditions of errors of pitch, and errors of flank angle. Parts having such errors will assemble, provided there is a sufficient difference between the simple effective diameters to absorb these errors.

Such errors give rise to the concept of *virtual effective diameter*. As defined in BS 2517, virtual effective diameter is the effective diameter of an imaginary thread of perfect form and pitch, having full depth of thread but clear at the crests and roots, which will just assemble with the actual thread over the prescribed length of engagement. This diameter exceeds the simple (measured) effective diameter by an amount relating to the combined effects of errors of pitch and errors of flank angle.

The gauging value of a screw plug or ring will be influenced by errors of pitch or flank angle so both must be very accurately measured, pitch directly on a special type of measuring machine, angle by optical projection or measuring microscope.

It follows directly from the geometry of Figure 14.18 that for errors of pitch

virtual change in effective diameter = $\delta p \cot \theta$

where δp is the maximum pitch error over the length of engagement and θ the semi-angle of the vee. For ISO metric the virtual change in effective diameter is $1.732\ \delta p$.

Figure 14.19 shows the effects arising from angle errors $\delta\theta_1$ and $\delta\theta_2$. Because the depth of thread is not symmetrical with respect to the pitch line the effects will differ

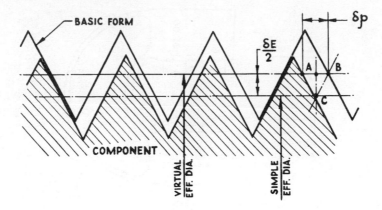

Figure 14.18 Influence of pitch error on the effective diameter required in a mating part.

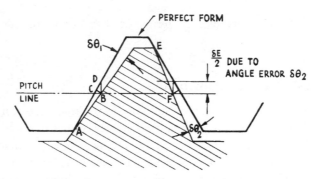

Figure 14.19 Influence of angle of error on the effective diameter required in a mating part.

slightly according to whether contact is made at A or E.

Considering point A,   AB = 0.25p from the basic form.

BC ≃ AB × $\delta\theta_1$ rad   BD = 2BC

hence BD ≃ 2 × 0.25p × $\delta\theta_1$ rad

Separation by amount BD would occur on both flanks of the thread form and so accommodate an error $\delta\theta$ on *both* flanks.

It follows from the geometrical conditions of error $\delta\theta_1$ on *both* flanks that the resulting change of effective diameter would need to be:

$$\delta E_1 = 2 \times 0.25p(\delta\theta_1 + \delta\theta_1)\ \frac{\pi}{180}\ \text{(angle error in degrees)}$$

$$= 0.0087p(2\delta\theta_1) \tag{14.1}$$

If angle error of opposite direction is considered, the contact point will be at *E* and the expression changes to

$$\delta E_2 = 2 \times 0.375p(\delta\theta_2 + \delta\theta_2)\ \frac{\pi}{180}$$

$$= 0.0131p(2\delta\theta_2) \tag{14.2}$$

This higher value can only arise if the crests of the thread are not rounded as allowed by the specification.

In general the non-symmetrical aspect is ignored and positive or negative angle errors are regarded to have equal effect; hence, from equations 14.1 and 14.2,

$$\delta E \text{ (average value)} = \tfrac{1}{2}(0.0087 + 0.0131)p(\theta_1 + \theta_2)$$
$$= 0.0109p(\delta\theta_1 + \delta\theta_2)$$

Flank angle errors are to be measured in degrees and added together regardless of their direction.

Note: The formula $E = 0.0115p(\delta\theta_1 + \delta\theta_2)$ quoted for metric SI threads in the NPL booklet *Notes on Screw Gauges* is based on the early thread form which was symmetrical relative to the pitch line.

Clearly the above concepts apply when measuring or gauging a screwed part. A full form screw gauge represents the imaginary form of perfect pitch and angle etc. introduced in BS 2517 to define virtual size.

If a component having pitch or angle errors is produced it is possible to change the effective diameter sufficiently for the full form *go* gauge to accept the work, by cutting beyond the maximum metal condition an amount equal to the effective diameter equivalent of the errors, as given above. If, however, the pitch and angle errors are too large, the effective diameter equivalent to be removed is sufficient to enable the effective *not-go* gauge to pass over the work, and thus reject it.

14.5 Measurement of the effective diameter

Screw-thread measurement, as distinct from gauging, is a large topic, and one aspect only can be treated here, the measurement of the effective diameter. The handbook *Gauging and Measuring Screw Threads*, prepared by the National Physical Laboratory (NPL), Metrology Division, and published as *Notes on Applied Science* No. 1 by HMSO, remains one of the most useful and authoritative sources of information, and should be available in every standards room and metrology laboratory. Some of the more significant points relating to the measurement of effective diameter set out here, exemplify principles of measurement already discussed in Chapter 15.

Figure 14.20 shows an NPL type screw-thread measuring machine of floating carriage construction. Certain features incorporated in this design are of interest.

1. The machine has two kinematic slides to give displacements along AA and BB, as shown in Figure 14.21. When work is supported on-centres, its axis lying along AA, the micrometer lying along BB must be correctly aligned so that there are no sine or cosine errors introduced (see Figure 15.6). In order to achieve this, the lower slide has one of the location pins which slide in the vee groove eccentrically mounted; a small rotation of this pin causes the slides to rotate relative to AA, and in this manner BB can be aligned at an exact right angle to AA. After adjustment, this pin is sealed and stamped to prevent tampering.

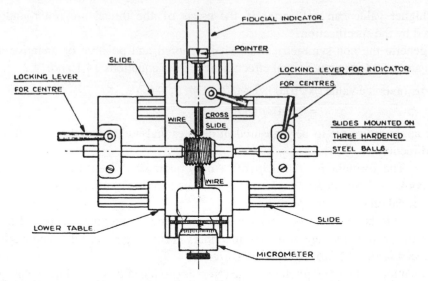

Figure 14.20 Thread diameter measuring machine.

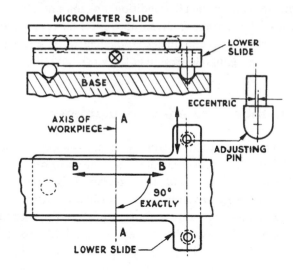

Figure 14.21 Adjustment of kinematic slides of screw thread diameter measuring machine to obtain true alignment.

2. The micrometer has a drum of large diameter which enables readings to be made direct to 0.002 mm, and by vernier scale to 0.0002 mm. Parallax error of reading is very small indeed. As a result of this high magnification factor, the system has a very large mechanical advantage: a very small torque applied at the drum produces a considerable force at the measuring contacts. A fiducial indicator is fitted to the 'fixed' anvil so that the force at the measuring contacts is a small and constant one. The 'moving' anvil of the micrometer does not rotate, the position of the measuring cylinder in the thread vee is then not affected by any tendency to turn, as the drum is finally adjusted.

3. The NPL procedure for effective diameter measurement requires that the machine is set from a plain cylindrical standard held between the machine centres. The thread measuring cylinders are included in the setting dimension, as shown in Figure 14.22. The geometry of the contact points for setting the machine is thus similar to the geometry of the contact points when measuring the effective diameter (see Section 14.5). Any error due to dissimilarity of contact geometry, arising from elastic compression at the contact points, is thus very small. The contact points for setting and measuring are shown in Figure 14.22. A correction for the difference in contact geometry can be made, by using the information given in the NPL screw thread book (correction for elastic compression).

Figure 14.23 shows the basic geometry of the NPL P value method of effective diameter measurement. The dimension from the underside of the measuring wires, taken on each side of the thread, is denoted by the letter P. If this value is added to the value representing the setting standard when the micrometer is first set, i.e. the micrometer set to read $(S + P)$, the machine will automatically read the measured effective diameter of the thread.

Let d_c = the mean diameter of the pair of measuring cylinders employed, p = pitch of the thread, θ = semi-angle of the vee, E_s = the simple (measured) effective diameter.

From Figure 14.24

$$\frac{P}{2} = \text{AB} - \text{AD} \qquad\qquad \text{AB} = \tfrac{1}{4}p.\cot\theta$$

$$= \tfrac{1}{4}p.\cot\theta - \frac{d_c}{2}(\operatorname{cosec}\theta - 1) \qquad \text{AD} = \frac{d_c}{2}\operatorname{cosec}\theta - \frac{d_c}{2}$$

$$P = \tfrac{1}{2}p.\cot\theta - d_c(\operatorname{cosec}\theta - 1)$$

$E_s = (T + P)$, where T is the dimension under the cylinders.

A further advantage of the NPL procedure is now revealed. The value of d_c cannot be known exactly, but when the machine is set, the actual cylinders are

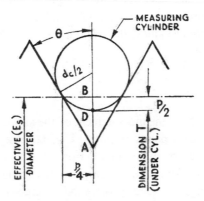

Figure 14.22 Setting and measuring on an NPL type screw diameter measuring machine.

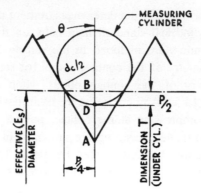

Figure 14.23 Geometry of 'P' value.

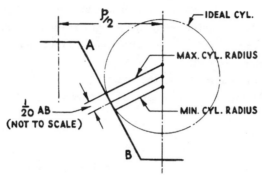

Figure 14.24 Geometry of 'best-size' cylinder.

incorporated into the dimension (see Figure 14.22). The error in E_s arising from uncertainty in the value for d_c has a value represented by

$$\delta E_s = (\text{cosec } \theta - 1) \, \delta d_c$$

which is smaller by $2\delta d_c$ than the error would be if the machine were set without including the measuring cylinders.

Accurate measurement depends upon the design of the equipment and upon the methods employed; accurate trigonometrical calculations cannot of themselves do more than provide numerical values, the accuracy of which may not be reflected in the practical measurement.

Best-size cylinders

Ideally, the cylinders chosen for the measurement of the simple effective diameter should contact the flanks at the pitch points (Figure 14.9), because this will make the value obtained for E_s independent of flank angle error. The required diameter is given by $\frac{1}{2} p \sec \theta$. In practice, *best-size cylinders* are permitted to have a small manufacturing tolerance, as will be seen by reference to the NPL screw thread book, and for such cylinders the effect of flank error on the value of E_s can be safely ignored.

Example 14.2

Best-size cylinders for measuring the effective diameter of an ISO metric thread are required to touch the flank within $\frac{1}{20}$ of the flank length on either side of the pitch point. Determine the upper and lower limits of size for such cylinders in terms of the pitch of the thread.

Solution

From Figure 14.24, by properties of the fundamental triangle,

$$AB = 5/8p$$
$$= 0.625p$$
$$\tfrac{1}{20}(AB) = 0.0313p$$

Ideal size cylinder, diameter $= \tfrac{1}{20}p \sec 30°$
$$= 1.1547p/2 = 0.577p$$

Limits of best-size cylinder $= 0.577p \pm (2 \times 0.0313p \times \tan 30°)$
$$= 0.577p \pm 0.036p$$
$$= 0.613p/0.541p$$

Note: A tolerance of $\pm0.043p$, often quoted, relates to the metric SI thread. The ISO metric thread has a shorter flank length.

□   □   □

The above account of screw thread effective diameter measurement by means of cylinders is not exact, because the conditions represented in Figure 14.23 could occur only if the vee groove of the thread were annular instead of helical. For an M24 × 3 screw, a discrepancy of about 0.0023 mm on E_s is introduced on account of the helix angle effect. For combinations of pitches and diameters which give rise to large helix angles a correction will be necessary; the correction formulae are given in the NPL screw thread book. For most threads the correction due to the difference in contact geometry mentioned in item (3), p. 449 is approximately equal and opposite to the correction required on account of the helix angle effect, and for work where an accuracy of ±0.0025 mm is sufficient, both effects may generally be neglected.

The measurement of effective diameter by means of cylinders, should not be used on *acme* or *buttress* thread forms without reference to the NPL screw thread book, because of the large errors which may then arise from the helix angle effect.

Exercises – 14

1. Figure 14.8 gives the design form of the ISO metric external threads. Show that:

    nominal depth of thread $= 0.61344p$
      allowable crest radius $= 0.0361p$
        effective diameter $=$ nominal diameter $-0.64952p$

2. (a) Write, in symbolic form, the three tolerance grades and fundamental deviations applicable to ISO metric threads as specified in BS 3643. Show how they are associated to provide three different classes of fit.

(b)   Using the information given in Table 14.1 show, for an M80 × 3–8g thread, that the major and minor diameter tolerances are 0.600 mm and 0.551 mm respectively, the effective diameter tolerance being 0.335 mm.

(c)   Why does the formula for effective diameter tolerance have terms in both p and d?

3. (a)   Show diagrammatically the tolerance zones associated with a nut and screw of Whitworth form and explain why the tolerance on the effective diameter is less than that on the other elements of the thread form.

(b)   How does the gauging of external screw threads present special problems in the design of gauges? Illustrate a design which meets the requirements, commenting on the special features necessary.

4. (a)   What is understood by the term 'Taylor's principle of inspection'?

(b)   How is this principle employed in the inspection of external and internal screw threads?

(c)   Briefly describe the methods used for inspecting screw ring and screw plug gauges.

5. What is meant by 'virtual' or 'compound' as applied to the effective diameter of screw threads?

The basic form of the British Association (BA) thread is shown in Figure 14.25. Show that an error of pitch of 0.01 mm requires an adjustment of 0.0227 mm at the effective diameter and that flank angle errors of +0.5°, −1.5° respectively on a 0BA screw (pitch = 1 mm) require an adjustment of +0.018 mm to accommodate them.

6. A special form of buttress thread has a sharp crest and root, and an included angle of 50°. The leading flank is inclined at 5° and the trailing flank at 45° to a line perpendicular to the thread axis. If this thread is measured by the NPL method show that the P value is given by

$$P = 0.9195 \times \text{pitch} - 1.2233 \times \text{cylinder diameter}$$

7. (a)   Outline the NPL method for measuring the effective diameter of a screw plug gauge. What are 'best-size' cylinders and why are they used?

(b)   Derive an expression for the P value for measuring ISO metric threads. Find the P value for measuring an M8 × 1.25 thread with cylinders of 0.7160 mm mean diameter.

(c)   Indicate why rake angle correction is necessary when measuring some threads.

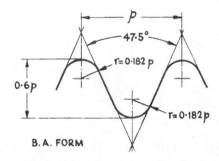

Figure 14.25 Basic form of BA thread.

8. (a) Describe, with sketches, the method used to check the simple effective diameter of a plug screw gauge on a floating carriage micrometer.

  (b) For the expression $E = T + P$, state what the letters represent and explain why it is not used in this form when measuring the effective diameter of a precision screw.

  (c) Two thread measuring needles having diameters of 0.8761 mm and 0.8733 mm respectively, have a value of the constant $P = 0.5051$ mm when used to check a screw of 16 tpi BS Whitworth. Why will the value of this constant be different if the needles are used to check a screw of any other form? Find the new value for a screw of 1.5 mm pitch with an included angle of 60°.

9. Working from first principles, show that the limits of diameter for 'best-size' cylinders for measuring threads of BA form are $(0.546 \pm 0.019)p$, and that for No. 1 BA the limits are 0.5085/0.4743 mm.

15 Precision Measurement

15.1 Introduction

Engineering dimensional measurement involves the Euclidean concepts of the straight line and plane. Linear measurements are ratios expressed in terms of some arbitrary length standard, e.g. the metre. Primary length standards are defined in terms of the wavelength of monochromatic light, for reference an inch is defined legally as 25.40 millimetres. The establishment of an absolute length standard belongs to the realm of physics rather than engineering.

The international metre is defined in terms of the wavelength of monochromatic light. The International Committee of Weights and Measures recommended Krypton 86 as the radiation source: under standard conditions the metre equals 1 650 763.73 wavelengths. The metre had its origin in the International Prototype Metre of 1889, a line standard of the form shown in Figure 15.1. The principal working standards of industry are the well-known slip (or block) gauges together with length bars and, less frequently used, laser interferometers.

15.2 Length standards

The International Prototype Metre (Figure 15.1) is a *line standard*. The length is defined by two fine terminal lines. A form of line standard now commonly used in machines employs an optical measuring system. Sources of measuring error can easily arise from:

1. Temperature variations, which may arise from the residual heat of the cutting process, or from environmental factors such as draughts, sunshine on the workpiece or the machine tool, or on the co-ordinate measuring machine.
2. Flexure variation, caused by changes in weight distribution, i.e. by changing the points of support of a part on a machine.
3. Secular change, dimensional change occurring over long periods of time due to internal changes of the stress distribution and of the grain structure of the material.

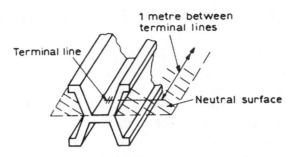

Figure 15.1 International standard metre of 1889.

Recent tests on precision machining of large aerospace components made from aluminium alloys showed that a part would often take eight hours to return to the temperature of the measuring room. Further tests showed that opening the door of the room for two minutes caused a temperature drop of 10 °C in the room temperature. The heating system brought the room temperature back up to 20 °C within ten minutes, but the component on the co-ordinate measuring machine (CMM) lost 2 °C and then took three hours to regain the normal room temperature. It must be mentioned that the component was 1.8 m across, and of thin-walled construction, so had little mass, and hence a faster response to temperature drop than a smaller component. The main point to note however, is that the change of 2 °C in component temperature took up 66 per cent of the available tolerance on dimensions. It was thus not safe to take measurements for this two-hour period.

In general the CMM suppliers recommend three grades of temperature control, depending upon the degree of accuracy required. These are as follows.

Class A

A thermostatically controlled (and air conditioned) room held to 20 °C ± 1 °C, with thermal gradients of 1 °C over eight hours. It is important to remember that the CMM itself will expand and contract, and it should be held at a constant temperature to avoid distortion. Lamps (and people) generate heat, and should not shine directly onto the machine. A component to be measured should be left to cool down for at least twelve hours in the temperature controlled room before measurements are taken. Its temperature can be checked by using contact thermocouples at several places to ensure that it is safe to measure it.

Class B

- The room must be maintained at a constant temperature by being positioned at the centre of the building to prevent heat transfer to or from the outside environment.
- Heating systems must *not* be installed within the CMM room. If hot air blowers are used the air flow must *not* impinge on the machine.

- If the room does have openings to the outside they should be double glazed and face north.
- All room openings should be opened for the *minimum* possible time. If a door has to stay open for five minutes then the machine should not be used for *several hours* to allow the machine to stabilise.
- Precision temperature monitoring systems should be installed to verify that the machine is used at standard temperatures – the room may take some time to reach acceptable temperatures on a Monday morning.
- It is important to note that a large machine may experience considerable temperature differences at different heights – these may well be unacceptable and should be carefully monitored.

Class C

Medium accuracy may be obtained with a CMM installed in the factory provided some sensible safety precautions are taken:

- Do *not* put CMM near doors (even internal doors).
- Avoid any factory heating system sources.
- Avoid thermal radiation from processes, heating systems, or sunlight.
- Shielding panels all around the machine will often minimise the impact of the local factory environment on the machine.

Obviously the component sizes and materials are significant functions to consider, temperature errors on small steel parts will cause much smaller errors than when large aluminium parts are being measured (Table 15.1).

Interferometry

Light is a form of energy radiation having wave properties. Suitable sources can emit monochromatic rays (rays confined to a very narrow spread of wavelength λ), which provide a basis for interferometric measurement.

As they leave a common source, rays are 'in phase', but by causing them to traverse paths of differing length before they re-combine, two such rays can interfere. Newton's rings provide a well-known example of this principle. Figure 15.2(a) shows, in simplified form, optical conditions for interference. Rays PQR and PR emerge from a common source but travel by different paths to the eye. Ignoring phase change at the reflecting surface, the intensity of the re-combined rays depends upon the difference of their path lengths (PQR – PR) and their wavelength λ. Figure 15.2(b) illustrates graphically the results of re-combination.

1. When the path difference is $N\lambda$ (i.e. in phase).
2. When the path difference is $(N + \frac{1}{2})\lambda$ (i.e. out of phase).

The distinct dark bands caused by interference provide a physical basis for precision measurement.

Table 15.1 Coefficients of linear expansion.

Material	Approx. expansion per deg C, units $\times 10^{-6}$
Aluminium	22–24
Brass	18–20
Bronze	16–18
Cast iron (grey)	9–10
Copper	16–17
Magnesium	28–30
Nickel	12–13
Steel	11–12
Titanium	9–10

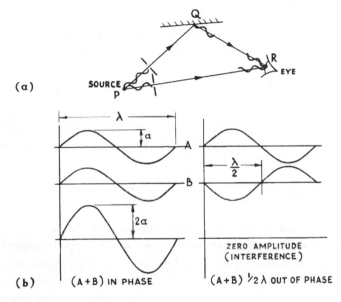

Figure 15.2 See text for description.

The simplest practical method of using the interference effect is by means of an optical flat (or proof plane). This is a thick disc of either glass or quartz with parallel faces ground and polished flat to a very high order of accuracy. Usually only one face is of specified flatness; it may be coated to increase the light reflected from the surface (it then becomes a more efficient beam splitter).

Figure 15.3 shows an optical flat resting at a very small angle α (exaggerated in the diagram) to the upper lapped surface of a gauge block. Incident rays, R_1, R_2, ... from a common monochromatic source are partially reflected at the lower (coated) face of the flat while the remaining portion of the beam is reflected at the surface of the gauge. The two reflected portions of each ray (af and bcd for R_2) re-combine on the retina of the eye, giving rise to apparent dark bands wherever the air gap

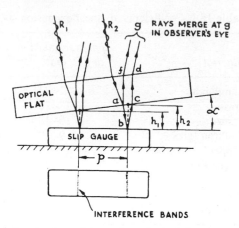

Figure 15.3 Interference bands viewed through an optical flat.

between the slip and the plane gives rise to a path difference of $\frac{1}{2}\lambda$. As the air gap is crossed twice, interference bands will be seen where the width of the air gap is $(N\lambda) + \frac{1}{4}\lambda$, $+\frac{3}{4}\lambda$, $+\frac{5}{4}\lambda$, etc.

Two deductions can be made from the appearance of the bands:

1. As they are similar to contour lines on a map, any curvature or any irregularity in their successive pitches (p) indicates that the slip gauge surface is not flat.
2. From the pitch of the interference bands and known wavelength of the light, angle α can be determined:

$$\alpha \text{ radians} = \frac{\lambda}{2p}$$

This principle is sometimes used for fine measurement of angles.

Figure 15.4 shows the optical arrangement of an NPL type of interferometer for testing the flatness and parallelism of slip gauges. The slip is wrung to a flat-lapped rotatable platen. Comparison of the interference bands on the face of the slip and on the platen monitors the parallelism of the slip gauge. Insets (a) (b) and (c) of Figure 15.4 show varying types of error deducible from a comparison of the two interference patterns.

Laser interferometry

(Laser = light amplification by stimulated emission of radiation) (see p. 417).

A monochromatic laser source greatly increases the distance over which interference effects can be used; up to 50 m is possible. A helium–neon source, wavelength 0.6328 µm, is common. Coupled to a digital readout display the system can provide direct readings.

The equipment, shown in Figure 15.5, has three main items, a sensor unit, a reflector unit and a control display unit. The sensor is fixed, the reflector can be displaced horizontally as shown.

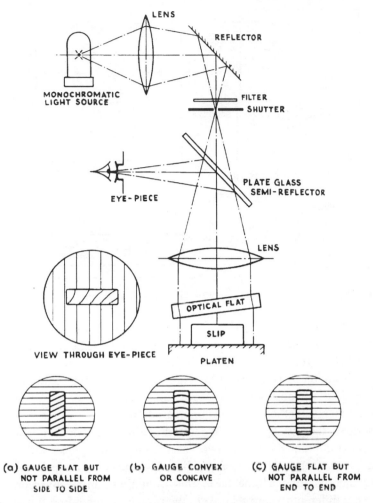

Figure 15.4 NPL type interferometer for flatness and parallelism of slip gauges.

A beam emitted from the laser source divides at the beam splitter at point A. One portion follows AB and, by reflection at B finally arrives at the photoelectric sensor at C. The path length of this beam has to be kept constant.

A second portion of the beam passes right through the beam splitter and enters the prismatic reflector. By internal reflection via D and E this beam reaches the sensor unit at F, whence it is reflected back to the beam splitter, where a portion of it is finally directed to C on the photoelectric cell. Any displacement of the reflector unit will double the amount of the change in path length A D E F and back to C.

Both portions of the original beam return some radiation to point C. As explained on p. 457 the intensity of the combined rays at C will vary from a maximum to a minimum (interference) according to the phase difference of the combining rays. Displacement of the reflector unit will cause a series of electrical pulses to be generated by the photoelectric cell; these can be counted electronically and processed

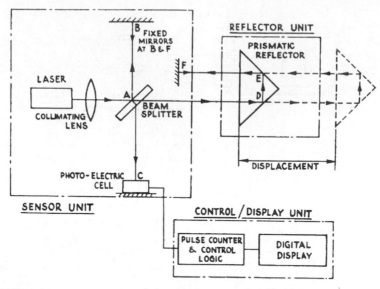

Figure 15.5 Equipment for measurement by laser interferometer.

to give a digital display in any desired unit of length. A helium–neon source will give rise to a pulse for each 0.3164 μm displacement. This is not, of course, as fine a degree of resolution as is achieved by using sources of three different wavelengths and the method of coincidences. However, for many purposes, it provides a sufficiently accurate method of precision measurement.

As described, the system could not detect any difference between +ve or −ve shift of the reflector unit. By fitting two laser sources of phase difference $\frac{1}{4}\lambda$ and collecting pulses from each at separate photo-electric cells, the pulses being fed to a logic unit incorporating a time base, this limitation can be overcome. Incoming pulses have a different pattern relative to the time base according to the direction of the displacement so enabling pulses to be added or subtracted as required. Such systems are sometimes incorporated as a means of measuring displacement for CNC machines. The time base signals enable displacement rates to be controlled.

Slip gauges: BS 888 and BS 4311

These are the working length standards of industry and require no further description here. BS 888 contains a useful appendix on the care and use of slip gauges, and the advice given should be followed. There are also details of several useful accessories; measuring jaws for internal and external work, scribing and centre points, and holding devices, all of which contribute to a wide application of slip gauges in industry.

An appreciation of the accuracy to which such gauge blocks are manufactured can be obtained by noting the permissible errors laid down in BS 888. For gauges up to and including 25 mm the permissible errors are as shown in Table 15.2.

Table 15.2 Maximum permissible errors of slip gauges up to and including 25 mm (Unit 0.000 01 mm).

	Workshop grade	Inspection grade	Calibration grade	Reference grade
Length	+20 − 10	+20 − 10	±12	±5
Flatness	25	10	8	8
Parallelism	25	10	8	8

In the writer's experience few slip gauges of any new set supplied lie at the extreme limits. A set of workshop grade slips, calibrated immediately after purchase, contained only 3 gauges near the limits, and only one where the calibration was 0.000 02 mm above the high limit of length. If four gauges from a new workshop grade set, each under 25 mm, are wrung together, then by the principles outlined on p. 565, the overall size is unlikely to be outside limits of +0.0005−0.0002 mm.

15.3 Some sources of error in linear measurement

Accuracy in measurement depends as much upon method and cleanliness as upon the equipment available. Where slip gauges are used as a basis of reference, it is likely that errors arising from these sources exceed the error of the reference standard for a majority of the precision measurements made in average inspection departments.

Apart from temperature effects, errors can arise from the following causes:

1. Flexure at contacting surfaces.
2. Errors of alignment.
3. Errors of reading, i.e. parallax effects and vernier acuity.

Figure 15.6(a) illustrates the difference in the contact geometry between the flat anvils of a measuring machine or comparator, a slip gauge and a precision ball. Since there must be some contact force, there must be some deflection due to stress in each instance. For the same measuring force the reading obtained for the ball will contain a larger error, due to elastic deflections at the contact points, than will the reading obtained for the slip gauge. It is an important general principle in the use of sensitive comparators, that the measuring force must be maintained at a small constant value, and that the geometrical conditions of contact which occur for each of two comparative readings should be as alike as is possible. The first object is achieved for the bench micrometer, by the use of a fiducial indicator, and for comparators, by the spring load on the moving anvil; the second is achieved by providing different shaped measuring tips as part of the equipment of high-class measuring machines or comparators.

Figure 15.6(b) illustrates errors of alignment with respect to work measured between the flat anvils of a bench micrometer. These errors are frequently called the sine and cosine errors.

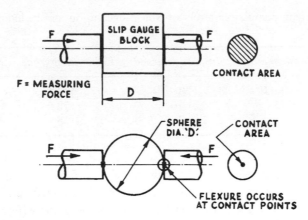

Figure 15.6 (a) Geometry of connecting surfaces.

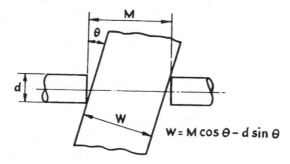

$$W = M \cos \theta - d \sin \theta$$

Figure 15.6 (b) Sine and cosine errors of measurement.

Figure 15.7 illustrates parallax error for the reading of scales. It should be obvious that errors from this source tend to fall as the magnification factor of the instrument rises. Reference to BS 887 *Vernier Calipers* and BS 870 *External Micrometers*, will show that the dimension here indicated as *t* is controlled by the specification so as to limit parallax effects. One way of defeating the parallax effect is to project the scale and the index on to the same plane, as is done in most optical measuring equipment.

Figure 15.8 illustrates the well-known principle of double graticule lines and what is meant by vernier acuity. Most people can judge spaces *a* and *b* to be equal to a higher order of accuracy than that with which they could position one line directly

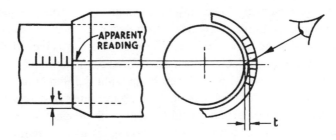

Figure 15.7 Parallax error of micrometer reading.

Figure 15.8 Double graticule lines.

over another. For a similar reason many people prefer to judge contour accuracy, tested by optical projection (shadowgraph), by leaving a very thin band of light between the shadow and the master outline.

15.4 Angular measurement

The basis of most angular measurement is the divided circle as exemplified by the scale of a vernier protractor. In its most refined form this circle is a silver-coated or glass disc upon which the division lines are ruled or etched, and the scale is read through an optical system of considerable magnifying power. Optical dividing-heads and circular dividing-tables operating on this principle are now obtainable reading to 1 second of the arc, and of maximum cumulative error not exceeding 5 seconds of arc.

A sense of proportion is brought into precision angular measurement if it is realised that a penny at 4 km distance subtends an angle of approximately 1 second of arc. The writer once saw a working drawing in which a taper shoulder of 3.2 mm depth, on a ground spindle, had been given an angular tolerance of ±5 seconds of arc!

As an alternative to the divided circle, angle slip gauges are available (see Table 15.3), generally made to a tolerance of 2 seconds of arc. Since they may be wrung together additively or subtractively, a small number of gauges can give a large range of combinations.

Figure 15.9 shows how an angle of 14°24'9" can be built up from such gauges. The precision polygon shown in Figure 15.10 is a further piece of basic equipment in angular measurement; it is the solid angle equivalent of the divided circle, and has the advantage that the accuracy is independent of the axis about which it rotates, provided that this is parallel to the lapped faces.

There are two main sources of error in angular measurement:

1. Error of centering, e.g. of a divided scale.
2. Error between the plane in which an angle is defined and the plane in which it is measured.

Figure 15.11 shows a circular scale correctly divided about axis A, and rotating relative to the index line about axis B at eccentricity (e) from A. It can be seen that the error, Δ radians, between the reading and the actual, angle of rotation is given by

$$\Delta = \frac{e}{R} \sin \theta \text{ (very nearly)}$$

The sine curve of errors has been plotted.

Table 15.3 Angle slip gauges.

Degree series	1° 3° 9° 27° 41° 90° (square)
Minute series	1' 3' 9' 27'
Second series	3" 9" 27"

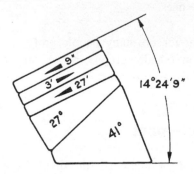

Figure 15.9 Angle slip gauges.

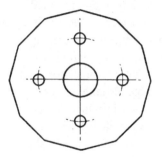

Figure 15.10 Twelve-sided precision polygon.

There is a phase shift of the sine curve as the position of B moves with respect to the numbering of the divided scale.

Example 15.1

An accurately divided circular scale 100 mm diameter, rotates about a centre displaced 0.0025 mm from the centre of graduation on a line joining the 150° and 330° positions, and is nearest to the 150° position. Draw a graph from which the errors due to the eccentric mounting of the sale can be read to the nearest second of arc.

Solution

$$\frac{e}{R} = 0.000\ 05 \text{ radians}$$

$$= 10 \text{ seconds of arc (very nearly)}$$

Figure 15.12 shows an easy graphical construction for the sine curve, and the method of positioning the phase shift involved.

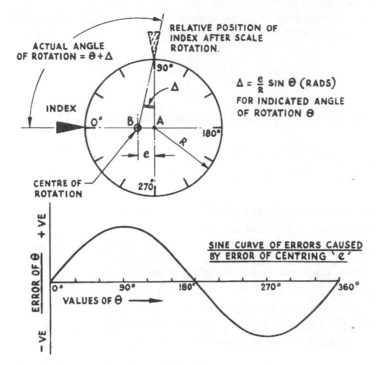

Figure 15.11 Centering error in circular division.

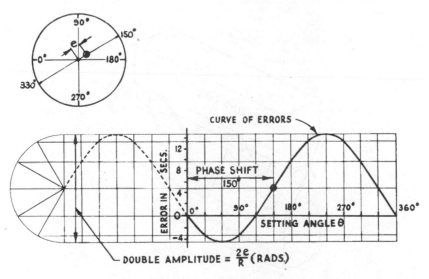

Figure 15.12 Graphical construction of the curve of errors caused by an error of centering.

The influence of such error of centering upon circular division is not confined to circular scales; it is equally important in relation to gearing, and to spline and serration fitting.

☐　　☐　　☐

15.5 Measurement of small linear displacements

A comparator is an instrument which magnifies small linear displacements in order to make them visible. Apart from high magnification, a comparator must have the following qualities:

1. It must be robust and give repeat readings consistently.
2. The magnification factor must be constant.
3. It must operate from a small uniform force exerted at the moving anvil.

There are many types of linear comparator now available, the main types are: *mechanical, optical, pneumatic* and *electrical*. Examples of each should be studied in a Metrology Laboratory and their relative advantages and disadvantages assessed. The following discussion of some of the main operating principles is not intended to be exhaustive, but is to draw attention to a few of the most important points.

Mechanical

Levers in some form or other are among the chief means of magnification. Crossed strip hinges, Figure 15.13, are frequently employed in place of pivots in order to avoid 'play'. Figure 15.14 illustrates the operating principles of one of the most successful types of mechanical comparator. The unit A is displaced against a light spring by the movement of the measuring anvil. The knife edge of this unit causes B

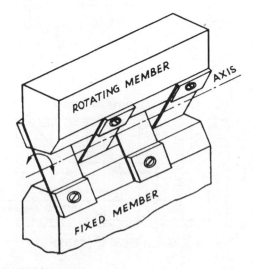

Figure 15.13 Crossed strip hinge.

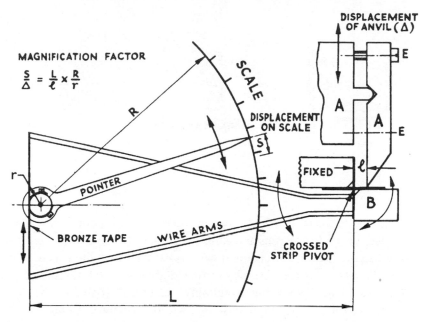

Figure 15.14 Operating principle of Sigma comparator.

to rotate about the centre of the 'crossed-strip' hinge and so to rotate the long arms attached to it. These arms tension a bronze tape which is part wound round, and secured by screws, to the spindle which carries the pointer. The pointer moves against a suitably divided fixed scale. An interesting feature of the arrangement is the method of mounting the knife edge of A such that dimension *l* can be adjusted by means of the clamping screws E. This enables the desired magnification factor to be set. An electromagnetic eddy-current damping device (similar in principle to that fitted to domestic electric current meters) is attached to the pointer spindle and makes the instrument 'dead-beat'.

Optical

The main advantage of optical type comparators is that a beam of light, which can be used as a magnifying lever, has no inertia and may be contained within a compact space by reflecting it between mirror surfaces. Most of the comparators which use a beam of light are refinements of the optical lever illustrated in Figure 15.15. Since the change of angle on reflection is twice the change of the angle at which the incident ray enters, there is a multiplying factor of two each time the beam is reflected.

Pneumatic

The underlying principle is illustrated in Figure 15.16. Air supplied at a constant pressure P_1 passes through a control orifice and into a chamber having an escape

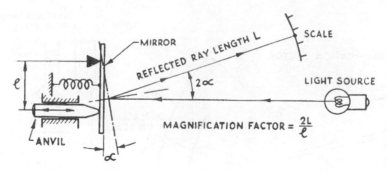

Figure 15.15 Principle of optical lever.

orifice called the measuring jet. If the surface of the workpiece closes the escape completely, $t = 0$, the pressure in the chamber will rise to P_1. As distance t is increased, pressure in the chamber will fall to P_2 and there a relationship between t and P_2. If P_2 is measured by some suitable pressure measuring device, t can be measured directly using a suitably calibrated scale on the pressure measuring meter. The system measures without metallic contact at the workpiece, and is particularly suited to measurement of the work during a grinding process as a basis for automatic size control. The scale is approximately linear over a small range of t (a few hundredths of a mm only). The special equipment needed for each bore-size or other application makes the method more suitable for use on long runs of repetition work than for general purpose measurement.

Practical tests show the system has an approximate linear relationship between the dimensionless ratios. P_2/P_1 and A_m/A_c (see Figure 15.16(b)), where A_m is the area through which air escapes and A_c is the area of the control jet. Linearity extends approximately over a range of P_2/P_1 of 0.6 to 0.8. Over this range

$$P_2/P_1 = kA_m/A_c + 1.1 \qquad\qquad (15.1)$$

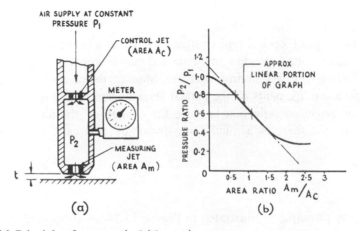

Figure 15.16 Principle of pneumatic (air) gauging.

The sensitivity of the system is given by

$$\frac{\text{scale displacement } (\delta R)}{\text{change in } t \ (\delta t)} = \frac{\delta A_\text{m}}{\delta t} \frac{\delta P_2}{A_\text{m}} \frac{\delta R}{\delta P_2}$$

where $\delta A_\text{m}/\delta t$ is the sensitivity of the measuring head,
  $\delta P_2/\delta A_\text{m}$ is the pneumatic sensitivity,
  $\delta R/\delta P_2$ is the pressure gauge sensitivity.

Since $A_\text{m} = \pi d_\text{m} t$, $\delta A_\text{m}/\delta t = \pi d_\text{m}$,

and from equation 15.1

$$\delta P_2/\delta A_\text{m} = k P_1/A_\text{c}$$

the pneumatic sensitivity is shown to be directly proportional to the supply pressure and inversely proportional to the area of the control jet.

For the middle of the linear range equation 15.1 becomes

$$0.7 = k \frac{A_\text{m}(\text{mean})}{A_\text{c}} + 1.1$$

or

$$A_\text{c} = \frac{-k A_\text{m}(\text{mean})}{0.4}$$

which shows that for high pneumatic sensitivity A_m must be small. This implies that linearity can only be obtained over a very small range of t. The linear range can be increased by fitting a measuring head with an orifice as shown in Figure 15.17. The parabolic end of the plunger maintains a linear relationship between A_m and t over an extended range t.

Example 15.2

A back pressure air gauging system has a linear range between values of the pressure ratio from 0.6 to 0.8, the linear relationship being represented by

$$P_2/P_1 = -0.5 A_\text{m}/A_\text{c} + 1.1$$

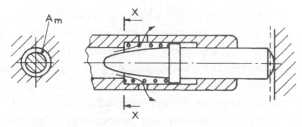

Figure 15.17 Example 15.2.

The control jet has 0.4 mm diameter, the measuring jet 0.65 mm diameter. Air is supplied at a pressure of 3 bars and the measuring indicator displaces 2 mm per 10^{-3} bar change of pressure.

1. Show that the linear range extends over 0.025 mm.
2. Find the overall sensitivity of the equipment within this range.

Solution

Linear range

$$A_m = 0.65\pi t \qquad A_c = 0.2^2\pi$$

$$0.5A_m/A_c = 8.125t$$

$$0.6 = -8.125t_1 + 1.1 \qquad t_1 = 0.062$$

$$0.8 = -8.125t_2 + 1.1 \qquad t_2 = 0.037 \qquad \therefore \text{Linear range} = 0.025 \text{ mm}$$

Sensitivity,

$$\frac{dR}{dt} = \frac{dA_m}{dt}\frac{dP_2}{dA_m}\frac{dR}{dP_2}$$

$$\frac{dA_m}{dt} = 0.65\pi \qquad t_{mean} = 0.0495$$

$$\frac{dP_2}{dA_m} = \frac{-0.4P_1}{A_m(\text{mean})} \qquad A_m(\text{mean}) = 0.65\pi \times 0.0495$$

$$\qquad = \frac{-0.4 \times 3}{0.0322\pi} \qquad\qquad = 0.0322\pi$$

$$\frac{dR}{dP_2} = 2 \times 10^3 = 2000$$

$$\frac{dR}{dt} = 0.65\pi \times \frac{1.2}{0.0322\pi} \times 2000$$

$$\qquad = 48\,450$$

Magnification is $\times 48\,450$

□   □   □

Electrical

There are numerous electrical principles which can be applied to the measurement of small displacement, e.g. strain gauges may be used. Electronic amplifying devices can give magnifications of extremely high order. However, for stability and reliability the most successful comparator of this type operates on variable inductance measured via a bridge network. Figure 15.18 shows the measuring head. Displacement of the iron armature between the inductance coils L_1 and L_2 puts the bridge circuit out of balance, causing the ammeter to move and to indicate the magnitude of the displacement, Figure 15.19.

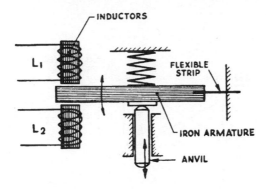

Figure 15.18 Measuring head of electrical comparator.

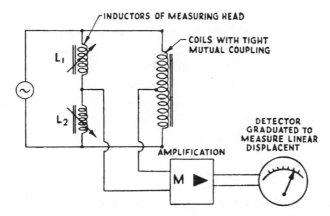

Figure 15.19 Bridge network of electrical comparator.

Optical magnification of the workpiece

The comparators described above magnify small linear displacements in order to make them visible; an alternative to this is to magnify the workpiece so that direct measurement may be made to a high order of accuracy. There are alternative ways of achieving this end, both of which have particular advantages.

The measuring microscope

Figure 15.20 shows the optical principle of a microscope. The objective lens is a magnifier which produces an image CD of workpiece AB; the eyepiece is a further magnifier which makes the image CD appear as the virtual image EF. If the objective lens magnifies 6 times, and the eyepiece 10 times, the virtual image will be 60 times full size.

In a toolmakers' or measuring microscope the work is mounted on a rectangular co-ordinate table having micrometer control of the displacements made by either slide. Cross lines in the focal plane of the eyepiece provide a datum against which

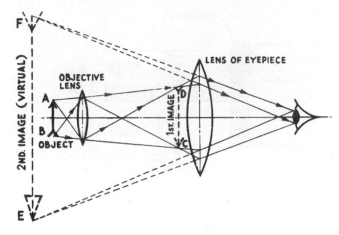

Figure 15.20 Optical principle of a microscope.

displacements can be measured. Angular measurements can be made either by having a graduated circular table as part of the work stage, or by having a rotatable graticule line in the eyepiece (goniometric eyepiece). It is generally more convenient to measure very small work by means of a measuring microscope than by contact methods, as instanced by the measurement of the smaller BA screw threads.

Optical projection

The somewhat simpler optical system of optical projection is illustrated in Figure 15.21. The degree of magnification depends upon the distance l between the focal plane of lens P, and the screen. Instruments which can handle work of a moderate size at 50 or 100 magnifications tend to be rather bulky, but optical projection, resulting in a magnified shadow outline of the workpiece, has a number of attractions. Direct measurements by rule can be made; at 50 magnifications 0.02 mm is represented by 1 mm on the rule. Profiles can be drawn at 25 or 50 times full sizes for

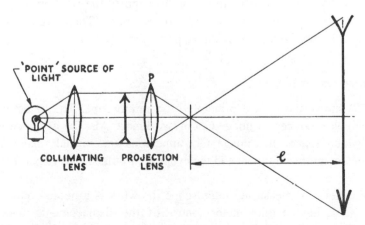

Figure 15.21 Optical projection.

such things as press tool dies, and this is the cheapest method of profile inspection. Most screw gauge profiles are checked by the method. At large magnifications small errors in the lens systems give proportionally large distortion, also the sharpness of the shadow outline is affected by the thickness of the workpiece projected, but such disadvantages are relatively small in relation to the general utility of the method for work where the accuracy required is not closer than about 0.02 mm. Angles can be measured by means of a suitably large vernier protractor.

15.6 Measurement of small angular displacements

Fundamentally, the amplification of small angular displacements is not greatly different from the amplification of small linear displacements; mechanical, optical and electrical devices may be employed.

A mechanical dividing head is comparable in operating principle to a micrometer, a large angular rotation is used to cause a small angular displacement.

The optical device for amplifying an angular displacement most frequently employed is the autocollimator. There are versions of this instrument which read directly to minutes of arc without the aid of a measuring microscope, but in its most refined form the microscope is essential, and enables an accuracy of the order of one second of arc to be obtained. Figure 15.22 illustrates the optical principle upon which the instrument works. Suppose a ray of light is emitted from source S at the middle of the focal plane of the collimating lens. The lens will convert this into a parallel beam which is then reflected from some working surface such as CD. If CD is inclined at angle α, the reflected ray makes an angle of 2α with the incident ray. The reflected beam re-enters the collimating lens to be refocused at some new point in the focal plane such as T. Distance h is proportional to the angle, and measurements of h in plane AB enable values of small angles to be determined. Note that the value of h is independent of the distance m of the instrument from the reflecting surface, although of course it is directly dependent upon l (a constant for the lens) and upon α. The full optical system of a typical instrument is rather more complex and is represented in principle only by Figure 15.23. The autocollimator and

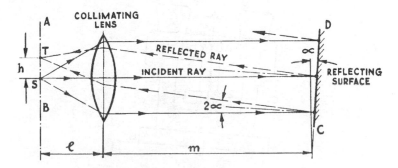

Figure 15.22 Optical principle of the autocollimator.

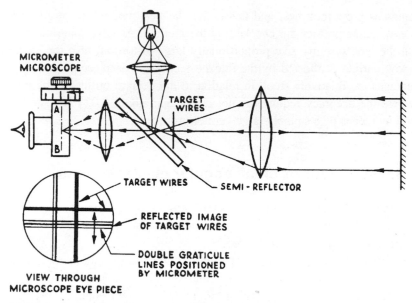

Figure 15.23 Optical system of the autocollimator.

angle slip gauges fulfil for angular measurement the same purpose as is fulfilled by the comparator and slip gauge blocks for linear measurement.

The gravitational pull of the earth has a fixed direction for any comparatively small area, and may be employed as a datum for the measurement of angles. The plumb-bob has its modern counterpart in instruments based upon a pendulum.

Figure 15.24 shows such an instrument. At the end of the pendulum there is a soft iron portion A which displaces, under gravitational force, between inductance coils C_1 and C_2 depending on angle θ. A bridge circuit of similar kind to Figure 15.19 feeds a signal to the meter which measures displacement from a datum in either minutes/seconds of angle or millimetres per metre. The pendulum and indicating meter have a damping system; the instrument has a range of about $\pm 2°$ and can be read to increments of one second of arc.

A spirit level is an alternative instrument for measuring small angular displacements relative to a horizontal datum, the level of a liquid at rest. Figure 15.25 shows the main

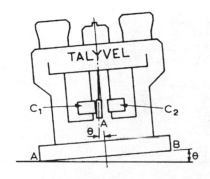

Figure 15.24 Talyvel (courtesy of Taylor–Hobson).

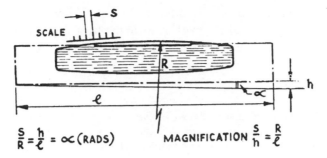

Figure 15.25 Precision spirit level.

features of this instrument. A 20 second level has a displacement of 2 mm for a tilt of 0.01 in 1000 and is representative of the precision class of this inexpensive and very useful piece of equipment. The principal use of both items of equipment is that of testing straightness and flatness as described on p. 479.

15.7 Indirect measurement

Many measurements made in precision engineering work are obtained indirectly, generally by calculation from other directly determined dimensions.

Figure 15.26 illustrates an indirectly made measurement. A disc of known radius is placed in the vee, and dimension h is measured as a means of determining dimension l. There are two very important points to observe:

1. Since l is calculated from values of r, α and h (and the 90° corner angle also forms part of the solution), l cannot be accurately found unless all the features used in its determination are known to be correct.
2. Error of l ($=\Delta l$) is determined from the apparent error of h ($=\Delta h$); if $\Delta h/\Delta l > 1$ the method is critical and a reliable result can be obtained, but if $\Delta h/\Delta l < 1$ the result is less certain, since the accuracy to which l is known is lower than the accuracy to which h has been measured.

For an indirectly determined dimension the geometrical conditions upon which the result depends should be carefully studied; calculations should be arranged to give

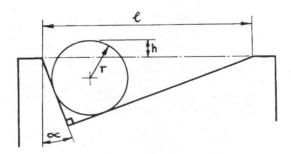

Figure 15.26 Use of roller for measuring a dimension indirectly.

values of high numerical accuracy, and the method should be analysed to show that the result is not greatly affected by small errors in the basic dimensions from which it is calculated. These points are best illustrated by a few worked examples.

Example 15.3

The dimensions of the arrangement shown in Figure 15.26 are: $\alpha = 20°$, $r = 11.996$ mm, $h = 5.208$ mm. Use these dimensions to determine dimension l correct to the nearest 0.002 mm and comment upon the geometric features which must be checked in order to prove the result reliable.

Solution

From triangles A, B, C and D of Figure 15.27.

$$l = (r - h)\tan \alpha + r \sec \alpha + r \csc \alpha + (r - h) \cot \alpha$$
$$= r(\sec \alpha + \csc \alpha) + (r - h)2 \csc 2\alpha$$
$$= (11.996 \times 3.987\,98) + (6.788 \times 2 \times 1.555\,72)$$
$$= 68.9602, \text{ i.e. } 68.960 \text{ to the nearest } 0.002 \text{ mm}$$

Geometric features which require testing in order to prove that the calculated solution is valid are:

1. The straightness of the inclined edges on which the disc rests, and of the edges from which dimension h is measured.
2. The accuracy of the 20° angle, which should be tested between edges referred to in (1).
3. The accuracy of the right angle between the edges on which the disc rests.

☐   ☐   ☐

Example 15.4

Figure 15.28 represents a bore of large diameter D measured by means of a pointed end bar, length L, which is swung about point A to touch the bore on the opposite side at two points distance W apart.

Obtain an approximate expression for D, in terms of L and W, which is sufficiently accurate when W is small relative to L. If $L = 500$ mm and the calculated value of D is required to be known to an accuracy of ± 0.01 mm, find the maximum value which W can have, assuming that this dimension is obtained by a rule measurement to an accuracy of ± 0.50 mm.

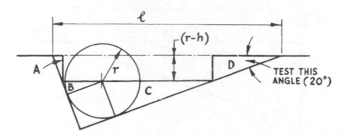

Figure 15.27 Example 15.3.

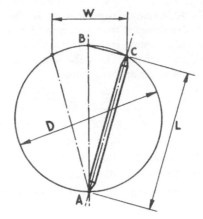

Figure 15.28 Measurement of a large bore with a point gauge.

Solution

From Figure 15.28, since ACB is a right angle (angle in a semicircle)

$$(AB)^2 = (AC)^2 + (BC)^2$$

If dimension W is small, $BC \simeq W/2$

hence $$D^2 = L^2 + \frac{W^2}{4} \text{ (very nearly)} \qquad (15.2)$$

Let $D - L = \Delta$, then from equation 15.2

$$(L + \Delta)^2 = L^2 + \frac{W^2}{4}$$

and $$\Delta = \frac{W^2}{8L} \text{ if second powers of } \Delta \text{ are neglected}$$

$$D = L + \frac{W^2}{8L} \text{ (very nearly)} \qquad (15.3)$$

Calculation of the degree of approximation involved in any specific instance is left as an exercise for the reader.

We must now consider the influence of errors in the value of W upon the value obtained for D, by application of the above formula. By differentiation,

$$\Delta D \simeq \frac{2W}{8L} \Delta W \text{ (}L\text{ regarded as a constant) or for small errors;}$$

$$\text{error in } D = \frac{W}{4L} \times \text{error in } W, \text{ very nearly.}$$

For the values given in the question, when $D = \pm 0.01$ and $W = \pm 0.50$.

$$W = \frac{4L \times \Delta D}{\Delta W}.$$

$$= \frac{4 \times 500 \times 0.01}{0.5} = 40 \text{ mm}$$

i.e. provided $L > 500$ and $W < 40$, a rule measurement for W, in error by ± 0.05 will not give rise to an error in the value of D calculated from equation 15.3 greater than ± 0.01 mm. By an exact method of calculation and assuming length W is known exactly, $D = L/\cos (\sin^{-1} W/2L)$ giving $D = 500.4003$. Using equation 15.3 the same values of L and W result in $D = 500.4$, showing the closeness of the approximation.

Methods such as the above are of importance in metrology, and the degree of accuracy obtained in the final result should never be taken for granted. Analysis of the method employed is an important step in deciding the validity of a particular result.

Example 15.5

Show, for a sine bar, that the accuracy of the angle set is a function of the accuracy of the centre distance between the rollers and of the setting height. How should an angle of 80° be set up in order to minimise errors?

Solution

Figure 15.29 shows the usual arrangement of a sine bar in which angle θ is obtained from dimensions h and l. It will be assumed that the rollers are of identical diameter and have their centre line exactly parallel with the edges of the sine bar.

From the diagram $\sin \theta = \dfrac{h}{l}$

Assuming h to vary, by differentiation

$$\cos \theta \, d\theta_1 = \frac{dh}{l} \qquad \text{or} \qquad \Delta\theta_1 \simeq \frac{\sec \theta}{l} \Delta h \tag{15.4}$$

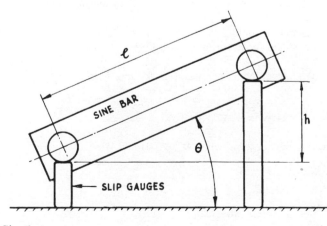

Figure 15.29 Sine bar.

assuming l to vary,

$$\cos\theta \, d\theta_2 = -\frac{h}{l^2}\,dl \quad \text{or} \quad \Delta\theta_2 \simeq -\frac{\tan\theta}{l}\,\Delta l \qquad (15.5)$$

from equations (15.4) and (15.5)

$$\text{Total error in }\theta: \quad \Delta\theta = \frac{\sec\theta}{l}\,\Delta h - \frac{\tan\theta}{l}\,\Delta l \text{ (radians)} \qquad (15.6)$$

a result which can be obtained more directly by the use of partial differentiation.

The following deductions can be made from equation 15.6:

1. The higher the value of l, the greater the accuracy of angular setting, other things remaining constant.
2. The higher the value of θ, the lower the setting accuracy, since as $\theta \to 90°$, $\sec\theta \to \infty$ and $\tan\theta \to \infty$, and it is unlikely that dimensions l and h are entirely free of error.

The most satisfactory method of setting up an angle of 80° is to set the sine bar to 10°, clamp it to the face of a high-grade cube and then to rotate the cube through 90° in a vertical plane, from its initial position on the surface plate.

□   □   □

The relationships between linear and angular dimensions discussed in this example illustrate another aspect of the relationships between errors with which one is concerned when making indirect measurements.

15.8 Straightness testing

Level method

This method is frequently used for cast-iron straight edges; it is also suitable for testing the straightness of the 'ways' of machine-tool beds. A level of appropriate sensitivity is mounted upon a 'bridge-piece' as shown in Figure 15.30. The vee grooves in the ends of the bridge-piece enable it to be used on the inverted vee guides of machine-tool slides when required.

Figure 15.31 shows the method of working. The bridge-piece and level are placed in consecutive positions along the work, and the 'slope' at each position determined

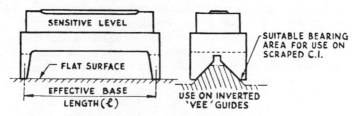

Figure 15.30 Mounted level for straightness and flatness testing.

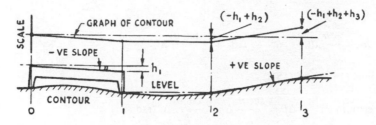

Figure 15.31 'Level' method of straightness testing.

from the reading of the level. It is necessary to convert these values into linear distances, the heights h_1, h_2, etc., at the ends of length of the base representing the 'slope' indicated by the level. It can be seen from Figure 15.31 that the readings must be summated (added with due regard to sign), because although the slope of the second reading is zero, the distances of points 1 and 2 of the surface below the horizontal datum through 0 are equal.

There are two methods of presenting results:

1. Graphically, which has the advantage of giving a pictorial impression of the errors but is subject to considerable scale distortion.
2. In a tabular form, in which the values are treated in a manner similar to (1) but without the actual drawing of a graph.

Example 15.6

A 20-seconds level, used on a bridge-piece for which $l = 130$ mm, is employed to test the straightness of a surface 650 mm long. The level readings obtained are: $+0.2$, $+1.4$, 0, -1.2, $+2$, where unit displacement of the bubble represents a slope of 20 seconds of arc. Find the errors of straightness of the surface.

Solution

Let h be the height at the end of a 130 mm length to produce a slope of 20 seconds of arc.

$$h = \frac{20}{3600} \times \frac{2\pi}{360} \times 130 = 0.0125 \text{ mm (very nearly)}$$

Figure 15.32 shows a graphical solution. It is clear that the surface tested is somewhat inclined to the horizontal and that a straight line through the end points is a more reasonable datum from which to express the errors of straightness.

The tabular method (see Table 15.4) of solution is easy to follow once it is realised that the process is equivalent to the above graphical one. The heights of the points are first obtained using a horizontal datum. The heights of points on a straight-line datum passing through the end points of the surface are then written down; the final column gives the differences between points on the surfaces tested and this new straight line datum.

□   □   □

Table 15.4 Values of h, and subsequent values, unit = 0.01 mm.

Position	0	1	2	3	4	5
Reading	–	+0.20	+1.40	0	−1.20	+2.00
Value of h	–	+0.25	+1.75	0	−1.50	+2.50
Summation of values	0	+0.25	+2.00	+2.00	+0.50	+3.00
New straight-line datum	0	+0.60	+1.20	+1.80	+2.40	+3.00
Errors from new datum	0	−0.35	+0.80	+0.20	−1.90	0

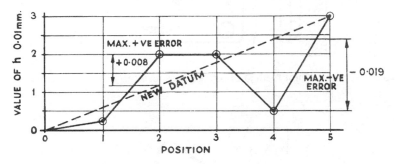

Figure 15.32

Straightness testing may be done in a similar manner by employing alternative types of instrument to measure the slopes. Figure 15.33 shows an arrangement for testing a rectanglar-type straight edge by means of an autocollimator. A 'level' operating on the pendulum principle of p. 474 may be substituted for the spirit level, and this instrument will give readings much more rapidly.

Other practical methods of testing straightness for which there is no space to give detailed explanation here are:

1. The beam comparator.
2. The alignment telescope.
3. The taut wire.
4. The water trough.
5. The electronic level (see p. 474).
6. The laser interferometer (see p. 460).

Some of these methods have special advantages where the surfaces involved are very large, e.g. for testing the beds of very large planing machines.

15.9 Roundness

The deviations of a hole with respect to roundness are normally caused by deflection, vibration, insufficient lubrication, wear etc. Out of roundness normally consists of waves. The number of indentations and bulges can vary from two to several hundred.

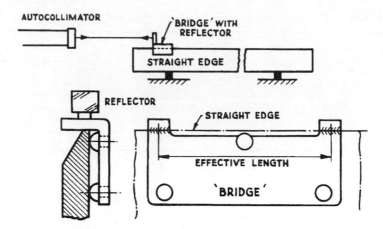

Figure 15.33 Autocollimator method of straightness testing.

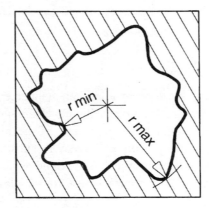

Figure 15.34 Deviations of a hole with respect to roundness.

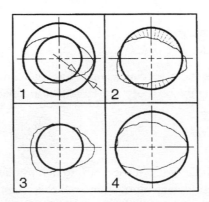

Figure 15.35 (1) Minimal radial separation (MRS); (2) least centre circle (LCC); (3) maximum inscribed circle (MIC); (4) minimum circumscribed circle (MCC).

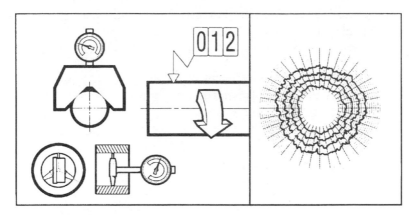

Figure 15.36 Measuring roundness.

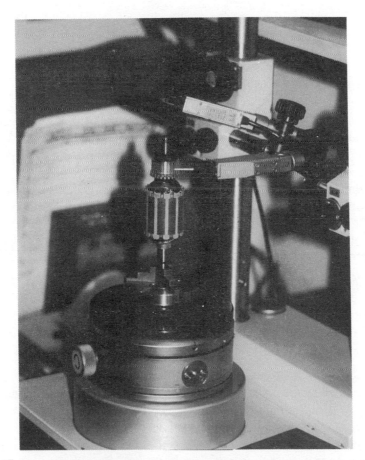

Figure 15.37 Roundness measurement of electric motor commutator at Black & Decker.

Out of roundness is specified as the difference between the largest and the smallest measured radii, measured from a defined centre point (Figure 15.34). However, there are various ways of defining the centre of a hole (Figure 15.35).

1. The most common method is to define the centre as the point at which the smallest radial deviation, is obtained. (This method is known as MRS – minimum radial separation, or TIR – total indicator reading).
2. The least centre circle (LSC) method where the centre point is the centre of the circle where the sum of the squares on the radial coordinates gives the smallest value.
3. The maximum inscribed circle (MIC).
4. The minimum circumscribed circle (MCC).

If nothing else is stated, the roundness value refers to the measurement in accordance with TIR, which gives the smallest value. Roundness can be measured in various ways but principally a pin is used which follows the internal diameter of a hole or the external diameter of a shaft when the workpiece rotates. The pin senses the variations in shape and a polar diagram can be drawn and studied (Figure 15.36).

Figure 15.37 shows the roundness checking of a Black & Decker motor commutator.

15.10 Measurement of surface texture

The measurement of surface texture is fairly complicated, and only a general outline can be given here. It is advisable to study BS 1134 Parts 1 and 2 in order to get some appreciation of the difficulties involved, and of the terminology applied to surface finish measurement.

Surface finish measurement is complex, because the character of a machined surface involves the three dimensions of space. Any attempt to express the quality of a surface by means of a number is necessarily, rather limited. Figure 15.43 shows, grossly exaggerated, some of the main features of a machined surface; note the primary texture of short wavelength, usually termed *roughness*, the secondary texture or *waviness* and the directional nature of the texture pattern which is covered by the term *lay*. Any numerical assessment of a surface finish will be influenced by the direction in which measurements are taken relative to the lay, and by an arbitrary distinction between roughness and waviness.

There are two main methods for the assessment of surface texture.

1. *Comparison with roughness standards*. Flat and cylindrical sample surfaces, finish machined by the common workshop processes and then calibrated, are used as a basis of comparison. Work produced is compared visually and by 'feel' with the sample surfaces acting as 'standards'. The method depends upon individual judgement and lacks precision on this account, but it is relatively cheap to apply and is a convenient one for general purposes.

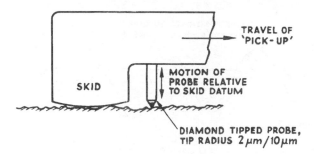

Figure 15.38 'Pick-up' of surface-finish recording instrument.

2. *Probe-type instruments*. Figure 15.38 shows the 'skid' and probe of an instrument of this type. The head is traversed over the finished surface; the probe rides up and down the roughness undulations relative to a datum set by the skid. The relative displacement is highly magnified electronically and the results presented either as a surface finish graph, Figure 15.39, or as a roughness value. Figure 15.40 gives details of one widely used instrument for surface finish testing. There are certain limitations of the method which are not always appreciated.

(a) Use of a skid for setting the datum is convenient, as it is then not necessary to line up the work to a high order of accuracy, but in riding over the surface undulations the skid modifies, in some small degree, the wave pattern recorded by the probe.

(b) The probe must have a point of finite radius, and so cannot penetrate to the full depth of the finest scratches, a fact which is significant in the measuring of very high-grade finishes.

(c) Graphs produced by such instruments are subject to considerable scale distortion, e.g. vertical axis × 50 000, horizontal axis × 100. They do, however, give a reasonable impression of the character of the surface tested if this fact is kept in mind.

(d) Instruments which give numerical readings are influenced by the sampling length traversed by the measuring head (pick-up) and by the particular starting and finishing positions used. The greater the wavelength of the main roughness markings, the longer must be the sampling length in order to give a satisfactory value.

Surface finish numbers cannot show the differences of texture as between surfaces of equal roughness value produced by different methods, e.g. between, say, grinding and honing. For such purposes a graph is essential.

R_a values

The numerical form of assessment given in BS 1134 is known as the arithmetical mean deviation R_a value. The physical meaning of an R_a value depends upon the graphical form of the surface tested. Figure 15.39 shows a typical surface finish graph with the sampling length L marked off. To obtain the R_a value, the line YY'

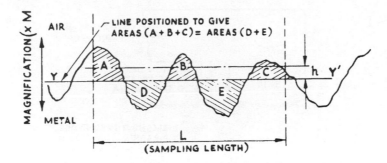

Figure 15.39 Surface-finish graph.

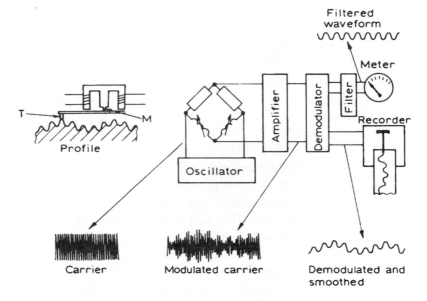

Figure 15.40 System diagram of Taylor–Hobson Talysurf.

must be drawn so that the shaded areas above and below it, enclosed by the graph, are equal in value, i.e. areas $A + B + C$ = areas $D + E$. The required value is then determined from

$$h = \frac{A + B + C + D + E}{L} \times \frac{1000}{V}\ \mu m$$

where V is the vertical magnification of the graph and L the actual measured base of the graph, i.e. 100 mm to represent 1 mm. Instruments of the type shown in Figure 15.40 make this calculation electronically.

R_a values should be specified according to the 'preferred' series shown in BS 1134, the finer values being 0.025, 0.05, 0.1, 0.2, 0.4 etc. Numerical assessment of surface finish is not of so great a precision that a difference between, say, 0.05 and 0.06, has much significance.

Surface texture and machining factors

Each type of cutting tool will leave more or less unique markings on the machined surface. The direction of the dominating surface pattern, lay, will be influenced by the machining method. As mentioned previously, however, conventional process to surface texture relationships have been changed through developments in cutting tools and machinery. Milling and turning can machine surfaces that were previously ground etc. This means also that the conventional cost picture between the level of surface texture and manufacturing costs has changed.

The theoretical surface texture can be calculated for milling and turning operations. The result will give an approximate value of what can be achieved under ideal conditions. The practical result will be affected by a number of different factors in the processes. Moreover, the resulting dynamic and static stability of the total process-system is of vital consequence to the quality of surface texture achieved.

The major factors of the cutting tool are:

- Stability.
- Overhang.
- Cutting geometry.
- Workpiece material.
- Tool wear.
- Cutting data.
- Chip formation.
- Machining temperature.

The major factors affecting the machinery are:

- Stability.
- Machining environment.
- Coolant application.
- Machine condition.
- Power and rigidity.

The major factors affecting the workpiece are:

- Stability.
- Material quality.
- Design.
- Clamping.
- Blank condition.
- Previous machining process.
- Tolerances on dimensions and form.

Terminology

The following is a basic overview of the typical terminology used within surface texture evaluation relating to the most common standard methods based on the centre or mean line system. Some methods evaluate the profile–height dimensions some the longitudinal dimensions and some the form of the irregularities.

In Figure 15.41, describing the centre or mean line system of evaluation there are the following definitions:

L   – evaluation length
P   – profile
T   – top-line within sampling length
B   – bottom-line,
       within sampling length
M   – mean line
LN – sampling length (cut-off)
LT – traverse length
C   – profile section level

The arithmetical mean line of the profile (also called centre line) is the reference line parallel to the general direction of the profile throughout the sampling length. The areas limited by the mean line and profile are equal on both sides. The profile section level is the distance between the profile top-line to the parallel line, sectioning the profile where the bearing capability of the surface is assessed

The evaluation length is the measured length along the mean line and is made up of several sample lengths (cut-offs), usually five in number. The traverse length includes also the start and finish distances.

Of the following more commonly occurring parameters described, the majority describe height limitations of surface irregularities while the others describe spacing longitudinally and bearing characteristics (Figure 15.42).

Surface texture defined

The surface texture in metal cutting is the resulting irregularities arising from the plastic flow of metal during a machining operation. It varies mainly with the method of machining, type and condition of tool, cutting data, workpiece material and overall stability.

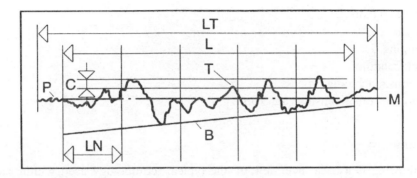

Figure 15.41 Surface texture evaluation.

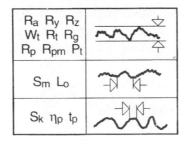

R_a R_y R_z W_t R_t R_g R_p R_{pm} P_t	
S_m L_o	
S_k η_p t_p	

Figure 15.42 Common parameters.

Surface texture is defined through the elements (Figure 15.43).

1. *Roughness (R)* is the smaller irregularities. These are finely-spaced micro-geometric deviations measured along the smallest sampling length on the workpiece.
2. *Waviness (W)* is the larger irregularities, within the next level up of the sampling length on the surface. The spacing of peaks and valleys is larger and sampling length is therefore longer than that of roughness. The roughness irregularities are superimposed on the macrogeometric wavy profile.

On the next level up are the deviations from actual component form, such as straightness, roundness, etc. Both the waviness and roughness irregularities may be superimposed on this considerably longer inspected surface. These deviations are not normally classified under surface texture whereas waviness and roughness are, along with the two following elements.

1. *Surface lay (L)* is the orientation of surface pattern. This describes the direction of the dominating pattern, generated by the machining method.
2. *Flaws (F)* are faults not included in the actual measurement of the surface but indicated separately as regards design and inspection. These include material inclusions, scratches, cracks, holes and other unintentional deformations of the surface.

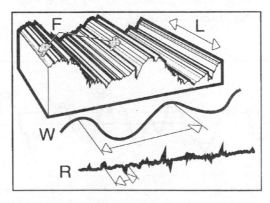

Figure 15.43 Enlarged view of a machined surface.

In order to measure roughness, which might typically be within 10 microns in height and very short spacings, a relatively short evaluation distance is sufficient. For waviness which might be 30 microns in height and much longer spacings, a longer distance is needed.

Parameters describing profile height

R_a–arithmetic average roughness

There are quite a number of surface texture parameters and some are used a lot more than others. R_a is the arithmetic average of filtered roughness deviations in relation to a centreline, along an established evaluation length (Figure 15.44). It is by far the most commonly used all over the world. Although popular and widely adopted it is vital to be aware of the limitations of R_a and to use a suitable combination of surface texture parameters.

R_a is one of several parameters relating to the height of profile irregularities. It gives an average of irregularities, distributing material from excessive peaks to be within a rectangular area where the cut-off is the base-length. It will as such not give a clear indication of the shape of the physical surface. The same value can be measured for various types of surfaces. Its use is also limited as very rough, fine and short surfaces are not suitably assessed through the R_a value.

The profile for R_a can be obtained through measuring with a stylus instrument for mathematical assessment of the values registered. The response characteristic has to be selected to limit irregularity spacing for roughness measuring, large enough to include detail but should exclude waviness. Standards specify suitable data and previous names for this evaluation included: *Centre Line Average (CLA)* and *Arithmetic Average (AA)*.

R_y (R_{max} and R_{ma}) – maximum individual peak-to-valley height

This is the maximum distance between the top-line at the peaks of the profile and the bottom-line of the valleys. These are obtained from five smaller sampling lengths within the evaluation length (Figure 15.45).

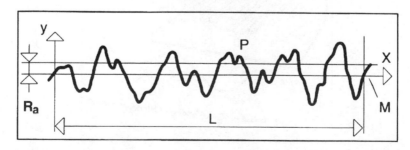

Figure 15.44 Profile height parameters (R_a value).

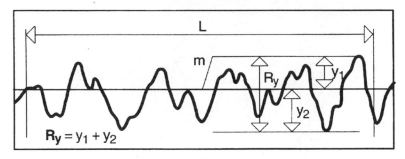

Figure 15.45 R_y value.

R_{ZISO} – mean peak-to-valley height

This is the arithmetic average of the five highest peaks and the five deepest valleys over the profile within the sample length. Depending upon the profile form, extra assessment may be needed if insufficient peaks and valleys prevail or in combination with waviness. It should also be noted that the ISO and DIN standards are not the same. (R_{ZDIN}) indicates maximum peak-to-valley height of roughness values of five consecutive sampling lengths over the cut-off profile.) (See Figure 15.46.)

R_q (RMS) – root mean square roughness value

This is the square average value of the profile deviations of the filtered roughness profile within the evaluation length.

R_p – single highest peak above the mean line

This is the value of the highest single peak above the mean line of the filtered profile (as taken from R_{pm}).

R_{pm} – mean peak height value above the mean line; the mean levelling depth

This is the arithmetic value of the five single highest peaks above the mean line, similar to R_{ZDIN}, of the filtered roughness profile from each sampling length (Figure 15.47).

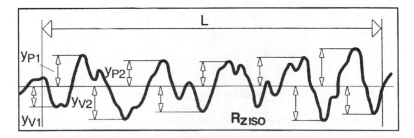

Figure 15.46 R_z value.

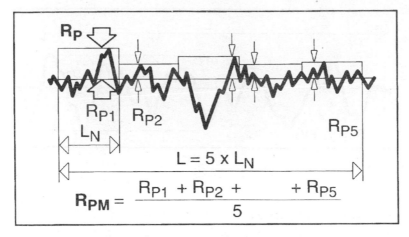

Figure 15.47 R_{pm} value.

P_t – maximum profile depth

This is the maximum distance between two parallel lines containing the filtered profile within the traverse length.

W_t – waviness depth

This is the maximum peak-to-valley distance of levelled waviness with roughness eliminated within the evaluation length (Figure 15.48a).

R_t (R_h R_d) – maximum peak-to-valley height; the greatest roughness depth

This is the maximum peak-to-valley height of the filtered profile over the evaluation length, without reference to sampling lengths (Figure 15.48b).

Parameters describing spacing and bearing

D – profile peak density

This is a peak count over the evaluation length of the filtered roughness profile. A peak is only included after the profile has passed through a lower and upper variable threshold, parallel to the mean line.

S_m – mean spacing of profile irregularities

This is the mean distance between points on the negative flanks of the profile as they cross the mean line, along the evaluation length (Figure 15.49).

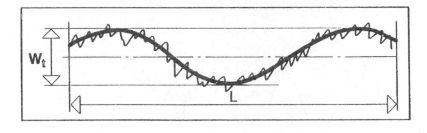

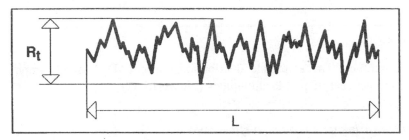

Figure 15.48 (a) W_t value; (b) R_t value.

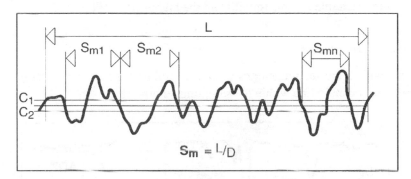

Figure 15.49 S_m value.

L_o (l_{mo}) – developed profile length

This is the actual length measured of the filtered roughness profile when all peaks and valleys over the traverse length have been levelled to represent the straight line.

I_r (l_o, l_r) – profile line ratio

This is the ratio of the developed profile length to the sampling length (Figure 15.50).

S_K – skewness of the profile; amplitude distribution curve

This is a measure of the asymmetry of the distribution density as obtained from the filtered roughness profile. For instance, a negative skewness value represents good

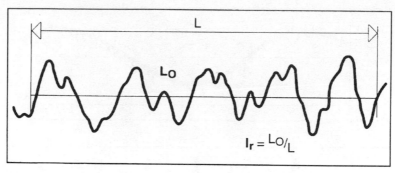

Figure 15.50 I_r value.

bearing properties. The amplitude distribution curve (ADZ) is the graph of the frequency in per cent of profile amplitudes (Figure 15.51).

t_p – profile bearing length ratio; surface ratio

This is a ratio in per cent of bearing lengths (b) to the sampling length. The level is parallel to the mean line and measured from the higher peak (Figure 15.52).

h_p – profile bearing length

This is the total of the section lengths when the profile peaks are cut by a section line (C) parallel to the main line within the sampling length.

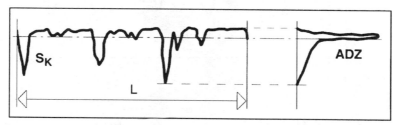

Figure 15.51 Skewness of surface profile.

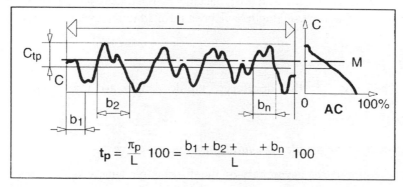

Figure 15.52 t_p value.

Bearing ratio curve, Abbot curve

This is a graphical (AC) representation of the ratio between the bearing surface and the profile depth.

Specification of surface texture

Figure 15.53 illustrates the specification of surface texture. The symbols in the right-hand column are used to define the direction (the 'lay') of machine marks on the

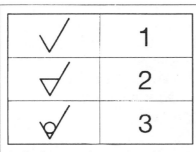

1 Surface may be produced by any method.

2 Machining is required. Allowance required.

3 Material removal prohibited.

a roughness value R_a in micro-meters or roughness grade number N1 to N12.

b production method or surface treatment

c sampling length in mm

d direction of lay

e machining allowance

f other parameter than R_a and value in micrometer (in brackets)

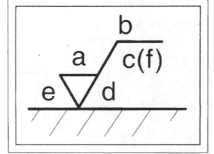

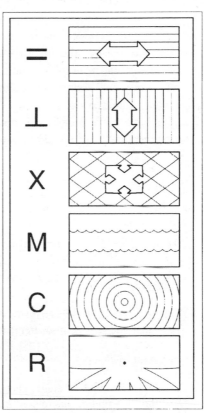

R_a μm	Roughness grade
50	N 12
25	N 11
12.5	N 10
6.3	N 9
3.2	N 8
1.6	N 7
0.8	N 6
0.4	N 5
0.2	N 4
0.1	N 3
0.05	N 2
0.025	N 1

Figure 15.53 The specification of surface texture.

surface. The appropriate symbols should be inserted at 'd' in the surface finish diagram.

15.11 Practical metrology

Precision measurement such as has been discussed in this chapter is best studied in conjunction with practical measurement work, preferably in a Metrology Laboratory. Reference to the appropriate British Standard is always advisable. Development of equipment is continuous, and equipment manufacturers generally provide full technical information concerning the operating principles and applications of their products. For this reason, the information in this chapter has been restricted very largely to fundamental principles which remain valid, even if future changes in the design of the equipment modify the methods employed.

The working of numerical examples serves a useful purpose by making sure that the principles have been understood, but in no sense can this type of exercise be a satisfactory substitute for practical measurement work.

Exercises – 15

1. A slip gauge, nominal size 8 mm, measured on an NPL-type interferometer, gave the following results:

Wavelength (λ)	Fraction displacement (f)
0.643 851 μm	0.85
0.508 586 μm	0
0.467 818 μm	0.50

    (a) Find the values of f for a gauge of exact size.
    (b) Show that the above measurements satisfy conditions for a gauge 0.000 2 mm undersize.
    (Use an 8-digit calculator).

2. (a) Explain the essential differences between linear measurement and circular dividing.
    (b) Describe the dividing instrument known as a circular division tester and show the effects of eccentricity of the divided circle and how these are eliminated. State the average accuracy of this type of instrument.

3. A 150 mm diameter circular divided scale is mounted with a slight eccentricity. When tested against a 12-sided precision polygon, the errors in minutes of arc read from an Angle Dekkor were:

Position	0	30°	60°	90°	120°	150°	180°
Error	0	+1.8	+2.6	+1.8	0	−2.5	−5
Position	210°	240°	270°	300°	330°	360°	
Error	−6.8	−7.5	−6.8	−5	−2.5	0	

    Determine the amount and the angular position of the eccentricity.

4. Describe with the aid of a sketch the optical principle of an autocollimator.

    When testing the straightness of a surface 1550 mm long, a reflecting surface supported vertically on a 'bridge' having feet at 150 mm apart is used in conjunction with an autocollimator, and the following results are obtained:

LH										RH
Position	1	2	3	4	5	6	7	8	9	10
Reading (sec)	+12	+16	0	−24	−20	−8	0	−12	+16	+12

The +ve sign indicates work rising towards RH end.

Find the maximum errors of straightness of the surface from a datum line through the end points.

5. The following results were obtained from a test on a straight edge using a 125 mm–10 seconds precision level. Slip gauges of equal size were wrung on the base of the level at a pitch of 25 mm.

Reading position	1	2	3	4	5	6	7	8	9	10
Divisions on level	+4	+1	0	−1	−3	+1	0	+2	+2	−1

Obtain the linear equivalent per division and hence plot a graph showing the variations of the edge from true straightness.

6. (a) Illustrate the use of calibrated balls, rollers and slip gauges for the measurement of:

    (i) plain internal diameters ranging from 6 mm to 25 mm;
    (ii) taper plug gauges having a large taper;
    (iii) radius gauges above 300 mm radius.

    (b) Develop the geometry for one of these examples and comment on the need for sufficient checks to give a good knowledge of the overall accuracy of the gauge.

7. (a) The 32 mm dimension of the profile shown in Figure 15.54 is checked, using a standard 25 mm roller, by measuring dimension h. Find this dimension correct to the nearest 0.002 mm.

    (b) For a certain gauge, made to this profile, h was found to be correct, but the 120° angles were both 4 min of arc oversize. Find, to the nearest 0.002 mm, the error introduced into l.

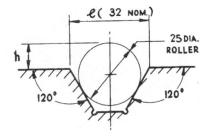

Figure 15.54

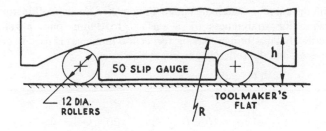

Figure 15.55

8. Figure 15.55 shows a method used to determine the radius R of a circular arc. If h measures 17.120 mm find R.

   If the reliability of the measurement of h is ±0.005 mm, to what degree of accuracy has the value of R been found?

9. (a) Show, for a sine bar, that the error of angular setting θ arising from errors of the dimensions l and h is given by:

$$\Delta\theta \text{ (rads)} = \frac{\sec\theta}{l}\Delta h - \frac{\tan\theta}{l}\Delta l$$

   (b) If, for a 100 mm sine bar, the setting error $\Delta\theta$ is not to exceed 15 seconds of arc when $\Delta l = +0.004$ mm and $\Delta h = -0.002$ mm, what is the maximum value of θ to which the sine bar may be set?

10. (a) What is the meaning of the following terms: surface texture; surface waviness; arithmetical mean deviation; root mean square; peak to valley; 0.8 mm and 2.4 mm wavelength?
    (b) What are the relative merits of a stylus type and an interferometric type of surface finish measuring instrument?

11. A pneumatic comparator has a linear characteristic given by $P_2/P_1 = -0.5\ A_m/A_c + 1.1$ over the range P_2/P_1 from 0.55 to 0.85 and is supplied with air at 2.5 bar. The control jet is 0.6 mm diameter and the magnification factor of the meter used is 2000.
    (a) What diameter measuring jet should be fitted so that a range of 0.03 mm is just within the linear range stated?
    (b) What will be the mean position of the work from the measuring jet at the middle of the linear range and what will be the overall magnification of the system?

16 Standards of Limits and Fits

16.1 Specification and drawing

The object of manufacturing is to produce saleable goods. To be saleable the goods must function satisfactorily, and this is primarily the concern of the designer; they must further satisfy purchasers by giving value for money. Manufacturing methods need to be considered at the design stage. To give value for money attention must be given, among other things, to:

1. Production costs.
2. Quality which includes reliability, of the resulting product.

The efficiency of, say, a petrol engine can be expressed as the ratio of energy output to energy input. Its efficiency in terms of manufacture is a more complex matter involving the cost of manufacture and the resulting quality of the product measured in such terms as finish, reliability and useful life. Cost and quality are perhaps the two most important criteria for judging a manufacturing method, but production rate and availability as required by the market are important commercial factors.

Manufacturing is based on production drawings and specified details such as quality and delivery date. The production drawing usually incorporates dimensional tolerances and specifies the material surface finishes, any heat treatments, etc. So much information is needed for a complete specification that the drawing cannot possibly include it all; the problem is generally solved by reference to published *standards*. For British industry the standards most widely used are those provided by the British Standards Institution (2 Park Street, London W1A 2BS) and are identified by their BS number.

16.2 Interchangeable manufacture

Present methods of quantity production have arisen because the interchangeable system has certain important economic advantages:

1. Parts can be made in quantity with less demand upon skill and effort if suitable gauges and tooling are employed. The quantity must, however, be sufficient to recover the special tooling costs.

2. Parts can be *assembled* instead of *fitted*, i.e. they will not need final adjustments of a skilled character in order to produce satisfactory assemblies.
3. Assemblies made in this way can be serviced by a simple system of replacement parts drawn from stock. This is convenient for the user and is cheaper than reconditioning involving the manufacture of new parts to special sizes.

Before a system of interchangeable assembly can be operated, certain fundamental conditions must be met:

1. The permissible variation (tolerance) of each dimension must be fixed.
2. The mating condition of each pair of parts assembled must be decided.

These conditions are generally satisfied by use of a system of limits and fits. The object of such a system is to make the decisions required under (1) and (2) conform to a rational pattern. For reasons given above, a *standard* system should be employed.

Interchangeable manufacture requires that the parts made should be as nearly identical as possible. The manufacturing processes to be employed must be chosen accordingly. The procedure will involve the following:

1. *Process planning.* A schedule of the individual operations and their sequence must be drafted.
2. *Jig and tool design.* Special tooling and equipment must be designed in accordance with the process plan. Choice of the datum faces for the consecutive locations of the part while individual operations are performed is very important. Functional datums must be considered.
3. *Limit gauges* and *gauging equipment* must be designed in order to control the accuracy of the work. The accuracy of the gauges and special tooling must be checked to high precision standards.

A component designed and dimensioned by someone aware of the implications of interchangeable manufacture is likely to be produced much more cheaply and accurately than one designed by someone who has a poor appreciation of quantity manufacturing techniques.

16.3 Dimensioning

The metric system uses decimal fractions only of the basic unit (mm up to 1000 mm). The millimetre enables small increments to be expressed using fewer decimal places than the inch system. Frequently open dimensions can be expressed in whole numbers; measurement of these by rule can be made to an accuracy of about ±0.5 mm.

Two quite difference circumstances apply when metric dimensions are being chosen:

1. Conversion from inch dimensions for an old design.
2. Metric dimensioning where no relationship to previously employed inch sizes is involved.

Conversion of inch units

Precise conversion based on 1 inch = 25.4 mm is rarely sensible; a 17/64 inches clearance hole for a $\frac{1}{4}$ inch bolt would become 6.746 875 mm. Consider an open dimension of 6.25 inches. Unless otherwise toleranced an accuracy of ±0.01 inches is implied. Since 6.25 inches = 158.75 mm a conversion to 159 mm is within the implied tolerance but 160 mm, a preferred size from the R5 series, is probably a better conversion.

For closely toleranced dimensions greater accuracy would be necessary. At this point it is worth considering the degree of accuracy likely to be achieved by common workshop methods of measurement (Table 16.1).

Size ranges

The advantages of standardisation are lost if too many gradations of size occur within a specified range. This is not just a matter of linear dimensions; weights, volumes, horsepower, electrical resistances or other physical properties may determine the size of a product.

The sizes within a range are generally required to increase at an approximately constant rate. The manufacture of large quantities of each size in a limited size range will enable motors to be supplied at a lower price than if a wider choice is given. The full economic advantages of quantity manufacturing methods can then be exploited.

Similar reasoning can be applied to the dimensional aspects of many products, e.g. the diameters of bolts, the thicknesses of steel sheet, etc. A rational series of standard sizes will tend to follow a geometric series:

$$a, ar, ar^2, ar^3, \ldots ar^n$$

where r is the constant rate of increase. Due to traditional developments, many of the standard sizes arrived at empirically, such as the BS Whitworth threads, do not follow the geometric rate of increase exactly, but the closeness of agreement is sometimes surprising. The sizes of new products should, however, be related to a suitable series of preferred numbers in order to avoid unnecessary duplication and to obtain the maximum advantages of standardisation.

Preferred numbers

These number series are based on ISO Recommendation R3 which has received international agreement. The series is based on the ideas of the French engineer, Col.

Table 16.1

Workshop instrument used	Smallest increment for which the instrument is reliable
Vernier caliper	0.05 mm
Micrometer	0.01 mm
Sensitive dial indicator	0.0025 mm

Charles Renard and are designated R5, R10, R20 etc. The geometric ratios are shown in Table 16.2.

In an earlier BS, now withdrawn, the late Mr J. E. Spears discussed the technical advantages and practical uses of preferred numbers. Extracts are given in BS 2045 and the main points are summarised below.

> The object of formulating an agreed series of preferred numbers is to provide the designer with a guide which, while not operating to restrict the liberty of his choice, will serve to minimise unnecessary size variations. In a majority of cases the designer, having calculated or selected a preliminary value, would not prejudice his final design by adopting instead the nearest size in a pre-selected (preferred) series. The adoption of preferred numbers in grading a series of articles enables the requisite range to be covered with a minimum of different sizes with resulting economy to both maker and user.

Example 16.1

It is required to standardise parallel keyways ranging from 2 mm to 28 mm. The first seven sizes are to follow the R10 series, the remainder to follow the R20 series. Develop a suitable range of key widths.

Solution

For the R10 series, geometric ratio $r = 10^{0.1} = 1.259$

Calculated values	2	2.52	3.17	3.99	5.02	6.33	7.96	10.02
Nominal (rounded) values	2		3	4	5	6	8	10

For the R20 series, $r = 10^{0.05} = 1.22$

Calculated values	10	11.22*	12.59	14.12	15.84	17.78	19.95	22.38	25.12	28.18
Nominal widths	10		12	14	16	18	20	22	25	28

☐   ☐   ☐

Table 16.2

Series	Ratio	Percentage rate of increase
R5	$\sqrt[5]{10} = 1.58$	58
R10	$\sqrt[10]{10} = 1.26$	26
R20	$\sqrt[20]{10} = 1.12$	12
R40	$\sqrt[40]{10} = 1.06$	6
R80	$\sqrt[80]{10} = 1.03$	3

Additional series can be formed by taking every third term of the basic series; e.g. R40/3, ratio $= \sqrt[40]{10^3} = 1.188$.

* It would be illogical to include an 11 mm size because the last step in the R10 series gives a 2 mm interval.

Summarising, the result shows that the rounded values result in three arithmetic series, i.e.

2 mm to 6 mm in 1 mm steps
6 mm to 22 mm in 2 mm steps
22 mm to 28 mm in 3 mm steps

If plotted the three straight lines of the arithmetic series will approximate to the curves of the geometric series. A common situation where a series of sizes are used for a range of bolts and twist drills.

16.4 Tolerances

Machining results defined

Components are drawn at the design stage with the required dimensions and other properties, such as roundness, straightness and angular precision, established. Since no machining can produce the exact size, the deviations which are permitted are always stated on the drawing. Tolerances are those deviations which can be accepted without jeopardising the function. Tolerances are set in such a way that the component is not manufactured with greater accuracy than necessary, since in most cases close tolerances increase the costs of both the machining and subsequent inspection.

By indicating elements for which the tolerances have been set, information is provided (Figure 16.1). Elements are indicated with an arrow and a line leading to a tolerance rectangle where the following questions are answered:

- For what property is the tolerance intended? Size, shape, direction, position, surface finish or run-out?
- How large may the deviation be?
- Deviation in relation to what?

There are various standards for how tolerances should be shown on a drawing. ISO (the International Organization for Standardization) covers 95 per cent of the world's industrial production. However, in many countries national standards are

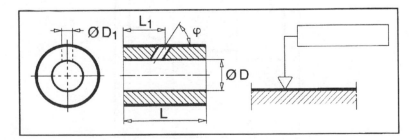

Figure 16.1 Component elements limited by tolerances.

applied which differ from the international ones to a certain extent. Furthermore, since standards are the subject of continual revision, only the principles of tolerances are dealt with here without going more deeply into drawing specifications and terminology. Surface texture is described in detail in Chapter 15.

Reference surfaces

Reference surfaces must be made quite clear in the production drawings and those features which are used as datums during machining must be kept to the specified tolerances. In order to achieve the best machining economy and quality, a sequence for the process should be set up during production planning, with the same reference surface being used where possible. Every time a reference surface that is used is machined, or replaced by another one, new deviations will be added to the final measurements. This can mean that unnecessarily close machining tolerances may need to be achieved. Such is the case in Figure 16.2(1). Instead, a tolerance setting based on a reference surface is recommended as in Figure 16.2(2).

Tolerances for properties

Properties for tolerances can be divided up as follows:

- Size.
- Surface texture.
- Shape.
- Direction.
- Position.
- Run-out.

Figure 16.3 compares theoretical ideal requirements (B) with the actual resulting shapes (A) in various situations: (1) shape; (2) direction; (3) position; (4) run-out.

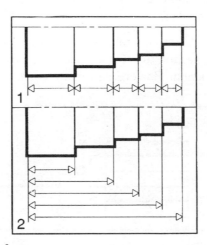

Figure 16.2 Reference surfaces.

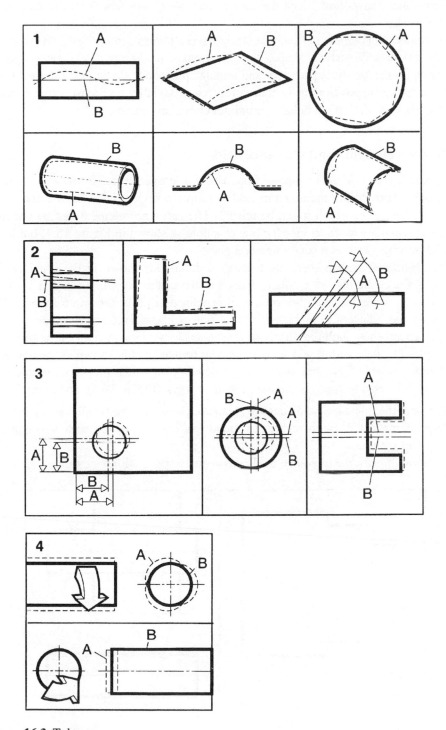

Figure 16.3 Tolerances.

The tolerance shows how much the actual outcome (A) differs from the theoretical exactness (B).

Tolerances for *shape* (1) include straightness, flatness, roundness, cylindricity, profile shape and surface shape. Specifications for *direction* (2) can apply to demands on parallelity, squareness and angular precision. *Positional tolerances* (3) apply to position/positioning in the workpiece, coaxiality and symmetry. *Run-out* (4) applies to deviations in radial or axial directions on rotation.

Reference elements and tolerance fields

For meaningful tolerances on a property, the property itself must be clearly identifiable both during manufacture and subsequent inspection. The tolerance must state a reference to which it can be related. The reference element can be concrete (R), for instance a surface, an edge line or a hole as shown in Figure 16.4, but it can also be abstract, such as a centre line or a plane.

For abstract references there are theoretical definitions shown in the standard. For example, the centre line of a hole is defined as the centre line of the largest inscribed cylinder (1), while the centre line of a shaft is defined as the centre line for the smallest circumscribed cylinder (2). In this way the position of abstract references are determined.

The property to which the reference applies is shown by a symbol in the tolerance rectangle. The field within which deviation can be accepted is known as the tolerance field and consists of an area or a space, depending on which property the tolerance refers to. The area or space is limited in various ways (Figure 16.5), for example:

- For size, the tolerance field consists of an area between two parallel straight lines.
- The tolerance field with respect to roundness consists of an area between two concentric circles.
- For straightness, the tolerance field consists of a space within a cylinder.

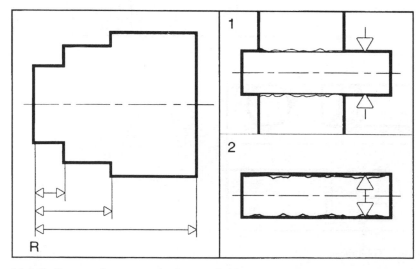

Figure 16.4 Reference elements and tolerance fields.

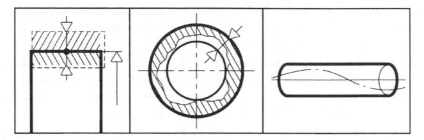

Figure 16.5 References.

Where nothing else is stated, shape, surface finish and direction can vary within the tolerance field without the component being faulty.

When more than one reference is used, it is important that the sequence of priority is given. The sequence is of considerable significance to the results obtained. In Figure 16.6(1) reference A is given as the first reference, while in Figure 16.6(2) reference B is used as the first reference.

Dimensional tolerances

There are three different types of dimensional tolerance:

1. Numerical tolerances.
2. General tolerances.
3. System tolerances.

Numerical tolerances are sometimes also referred to as 'wild tolerances', because they do not follow any system but are added directly to the dimension, for instance 22+0.11/−0.13.

For dimensions where tolerances are not of significance to the function, the designer's drawing work is simplified by using general tolerances. This means that dimensions are given without tolerance information and that a note on the drawing describes what tolerances are applicable for dimensions for which

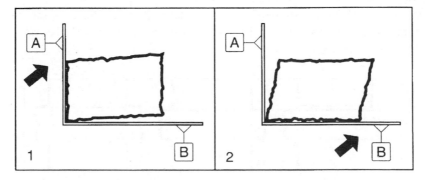

Figure 16.6 Datum position.

tolerances have not been set. The fit gives the dimensional difference between two related parts. This difference will give rise to a positive or negative play between the parts.

To facilitate design work there are ready for use *system* tolerances available. There are three different types of fit (Figure 16.7): play fit (A), transition fit (B) and grip fit (C). The basic measurement gives the exact diameter of the hole or shaft in Figure 16.8. From this basic dimension an upper measurement (D) and a lower one (E) known as limit measurements, are allowed. Together, the limit measurements form the tolerance width (F).

In accordance with the ISO system, the tolerance position of a hole is shown in capitals and the tolerance position of a shaft in lower case letters. J and JS plus j and js tolerances respectively, lie symmetrically around the basic dimension, while other tolerances permit greater or smaller over and under sizes or, alternatively, a tolerance width in one direction only. The type of fit depends not only on the tolerance widths for holes and shafts, but also on the mutual tolerance position of the parts. The tolerance width or degree of tolerance is shown by a figure, where higher figures indicate larger tolerance widths. With the ISO system the fit is indicated by an oblique stroke between the tolerance information relating to the hole and to the shaft, for instance H8/h7.

Measurement adjustment

Sometimes it is possible to exceed the specified tolerance limits when one or more measurement work in conjunction. It is then necessary to ensure that the function requirements are met. When adjustments are made to measurements, subsequent inspections are usually made with function gauges (Figure 16.9).

In Figure 16.9(1), the distance between the centre lines of the dowels is set at 50 ± 0.1 mm. The diameter of the dowels is set at $\varnothing 9.8$ $0/-0.1$ mm. When assembled the component will fit into a disc with two holes. The distance between the centre lines of the holes is set a 50 ± 0.1 mm and the holes have a diameter of $\varnothing 10 + 0.1/0$ mm.

On inspection, it is shown that the diameter dimension of the dowels lies on the top limit ($\varnothing 9.8 - 0.1 = 9.7$ mm) and at a centre distance of 50.2 mm, that is 0.1 mm outside of the tolerance specified. Since the hole diameters lie on their go limits ($\varnothing 10$ mm) and with 49.9 mm between the centre lines, it will be possible to assemble the components.

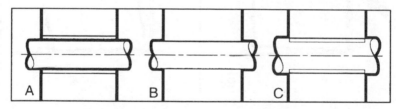

Figure 16.7 (A) Clearance fit; (B) transition fit; (C) interference fit.

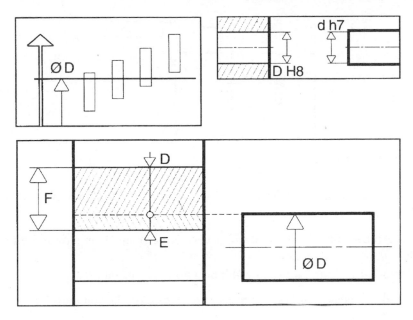

Figure 16.8 Tolerancing on shaft.

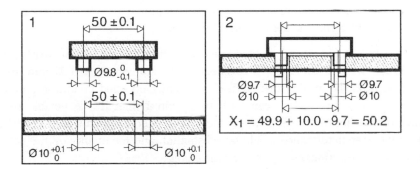

Figure 16.9 Tolerance interference on holes.

The function demand has been met and the component can be approved despite the distance between the centre lines of the dowels not meeting the tolerance specified.

16.5 Economic aspects of tolerancing

Designers and draughtsmen responsible for production drawings are advised to consult BS publication PD 6470: *The Management of Design for Economic Production.*

Consider possible methods of producing the slot shown in Figure 16.10. The tolerance on the width of the slot is too small to be held by production milling,

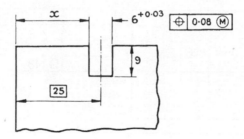

Figure 16.10 Tolerancing of slot.

although the position could be held by this method. The alternative methods of production are:

1. Surface broaching, suitable only if a very large quantity is required due to the high cost of special tooling.
2. Mill to 5.75/5.67 mm, finish to width by a grinding operation.

The second method is expensive because each side of the slot must be finished by a 'cut-and-try' method:

1. Grind to produce dimension x using a limit gap gauge.
2. Grind to produce $6^{+0.03}$ mm using a limit slip gauge. It will not be possible to maintain sharp corners at the bottom of the slot.

Rather than use the second method it is likely that the production workshop will appeal to the drawing office for a revision of the tolerance. A 0.05 mm tolerance could be held on the milling operation if a machine in good condition and a well-ground cutter are used. More care will be required than if the normal 0.07 mm tolerance on the slot were available, but the saving in cost which would result from a 0.02 mm increase in tolerance would be substantial, and the quality of the product is hardly likely to be seriously impaired by the change.

Some examination of the method of dimensioning the position of the slot is perhaps necessary. As dimensioned, x has a tolerance of 0.08 mm minimum, more if the slot exceeds 6 mm in width. Suppose the 25 true position dimension and positional tolerance of 0.08 Ⓜ are omitted, limits must then be given for dimension x. The position of the centre of the slot is now subject to tolerance build-up, and the tolerance for x will be $0.08 - 0.015 = 0.065$. The limits of x are found as follows:

$$x_{max} + 3.015 = 25.04$$
$$x_{max} = 22.025 \text{ mm}$$
$$x_{min} + 3.00 = 24.96$$
$$x_{min} = 21.96 \text{ mm}$$

Limits of $x = 22.025/21.96$; tolerance 0.065 mm. The method of dimensioning given in Figure 16.10 is superior to a method giving limits of x if the work is to be made in quantity and controlled by limit gauging.

Table 16.3 Tolerances associated with manufacturing processes.

Process	Tolerance (mm)	Tolerance grade
Sand casting:		
Small	±0.3	IT16
Large	±1.6	
Forging and drop forging:		
Small	±0.8	IT15–IT16
Large	±1.5	
Die-casting, plastic moulding	General ±0.4	IT14
Precision die-cast zinc alloys	0.05–0.2	IT10–IT12
Press work, tube drawing and extrusion	0.1–1.4	IT10–IT13
Planing and shaping	0.1–0.3	IT10–IT11
Drilling:		
6–12 mm	0.1	IT11–IT12
12–20 mm	0.18	
Reaming:		
6–25 mm	0.02	IT7–IT8
Over 25 mm	0.035 upwards	
Milling:		
Gang milling	0.08–0.12	IT8–IT10
Small slots	0.05–0.08	
Turning: capstan and turret lathes,		
roller box to 18 mm diameter	0.05	
Turning:		IT8–IT10
25–50 mm diameter	0.10	
Over 50 mm diameter	0.12 upwards	
Broaching:		
Up to 25 mm diameter	0.02	IT7–IT8
25–50 mm diameter	0.04	
Honing, up to 50 mm diameter	0.01–0.016	IT6
Grinding:		
Up to 25 mm diameter	0.007–0.012	IT5–IT6
25–50 mm diameter	0.012–0.016	
Lapping, machine	0.002–0.01	IT4–IT5
Lapping, standards, reference gauges etc.	Less than 0.002	IT01–IT3

Economic tolerance

We have shown that production costs are closely linked to the tolerances specified. It may be an invitation to criticism to lay down in very precise terms what tolerances can reasonably be held by the common manufacturing processes. Table 16.3 is included as a guide only, and the actual values which may be worked to in industry will reflect the quality of the equipment and skill of the operators, both of which vary considerably.

16.6 Limit gauging

Consider a bore of 50.046/50.000 mm diameter and 50 mm long. As shown in Figure 16.8 the limits define a zone within which the surface of the bore must lie. It is possible to test the bore, to see if it does comply with the specification, by means of a simple *go* and *not-go* gauging system, Figure 16.11.

To check the work completely by direct measurement would be much more difficult, because, not only has the diameter to be tested, but the roundness and straightness of the cylindrical bore as well. A correctly designed and made limit gauge checks geometrical and dimensional features simultaneously; it can show the work to be correct, but it cannot completely reveal the particular errors which may render it incorrect. Advantages of the limit gauging method are:

1. It is generally much quicker than direct measurement, and may be carried out by less skilled labour.
2. It automatically takes into account combinations of errors such as errors of work diameter and roundness.

   Disadvantages of the method are:

1. Particular sources of error are less easily revealed.
2. Gauges cannot be manufactured entirely without error, and they are subject to loss of accuracy from wear.
3. Practical considerations, such as weight, flexure and manufacturing complexity, restrict gauging to certain gauge sizes and types known to be successful.
4. The manufacturing cost of the gauges may not be recovered from a small quantity of work.

Taylor's principles of gauging

Figure 16.11 illustrates *go* and *not-go* gauges made as closely as possible to these principles. The *go* end of the gauge must be of perfect form made to the maximum metal condition of workpiece. It must completely cover the surface to be tested. A gauge which conforms to these conditions is termed a *full-form gauge*. The *not-go* end of the gauge ideally comprises points, which explore the surface being tested in order to detect any position outside the minimum metal condition. The two points at the ends of the bar gauge shown in Figure 16.11 come as near to these ideal requirements as possible.

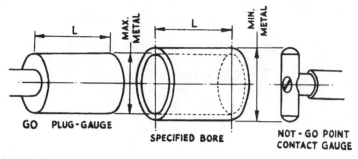

Figure 16.11 Limit gauging a bore.

For practical reasons the *go* gauge may not be full-form, e.g. the weight or length of such a gauge may not be convenient; similarly the *not-go* gauge will depart from the ideal because point contact leads to rapid gauge wear and may also involve a more expensive design of gauge. Generally, where machine-tool alignments and bearings are satisfactory, the geometric errors of machining are relatively small and a common type of gauge, e.g. caliper-type diameter gauge, checks the size only. Where components are easily deflected, e.g. thin-walled tubing, a gauge based on Taylor's principles is desirable; a *go* type caliper gauge might then accept work which would not enter a *go* ring gauge.

Example 16.2

Design gauges suitable for the full inspection of the square end to be straddle milled on the shaft illustrated in Figure 16.12.

Solution

A solution is shown in Figure 16.13.

Note the significance of the Maximum Metal Condition (MMC) specification Ⓜ included in the work limits. If the work is on top limit it must be exactly square, and the full-form *go*

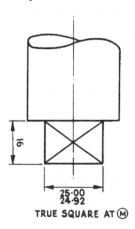

Figure 16.12 Square-ended shaft.

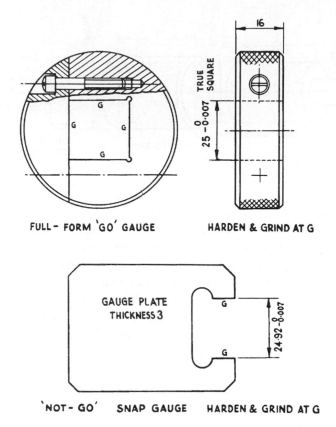

FULL- FORM 'GO' GAUGE   HARDEN & GRIND AT G

'NOT-GO' SNAP GAUGE   HARDEN & GRIND AT G

Figure 16.13 Gauges for inspection of square end.

ring will test this feature. If the work is below top limit errors of squareness are acceptable provided the work is passed by *both* gauges. A serious error of squareness might pass the *go* gauge if the work size were small enough; the *not-go* gauge, having only limited contact on opposite faces, would then reject the work.

The full-form gauge would be expensive to make and would not be justified for a short run of work. Indexing mechanisms are generally reliable, and straddle milling produces faces which are closely parallel, so that a caliper-type limit gauge for the size only could be satisfactory. For 'incoming parts' inspection, where there is no knowledge of the quality of the manufacturing equipment employed, the ring gauge is desirable; also one such gauge might be supplied, along, with several caliper-type limit gauges. in order to control a long run of work in a general machine shop.

□   □   □

Work specification in relation to gauging

Consider the problem of designing a gauging system to control the hole diameter and its position, specified in Figure 16.14(a). The specification requires the centre of the hole to be within limits of 32.1/32.0 mm from the edge, whatever the actual size of the hole. To achieve this, a position gauge designed around some form of expanding location pin is required. Such a gauge would be expensive to make and slow in use.

Alternatively, consider a method of gauging for position based upon dimension A. As the component drawing stands, A is subject to tolerance build up; it can vary by 0.126 mm, and a limit gauge for A based upon this figure could accept work outside the stated conditions.

If the dimensioning is changed to that shown in Figure 16.14(b) this is overcome because the positional tolerance now applies only to a bore at Maximum Metal Condition (MMC). The gauge (Figure 16.15) represents the conditions of the mating part; the increase in the positional tolerance due to the increase in bore size will still allow the parts to assemble.

The MMC specification changes the drawing to make it conform to a simple gauging method. The main reason for doing this is that agreement between the two is exact, whereas for the original dimensions an exact gauging method would be unreasonably expensive in equipment, and in inspection time. As so often in production engineering, the ultimate justification for the suggested change in specification is an economic one.

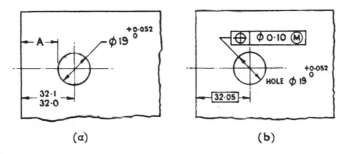

(a)   (b)

Figure 16.14 Work specification.

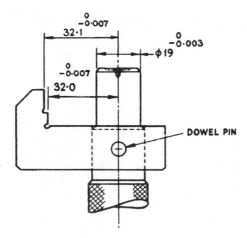

Figure 16.15 Gauge for 32.1/32.0 dimension, bore at MMC.

Standard gauge designs

An excellent series of designs for gauges in common use, such as doubled-ended plug gauges, ring gauges, caliper gauges, etc., is given in BS 1044. There are two principal advantages to be gained by employing standard designs: quantity-produced gauge blanks may be purchased at much lower cost than they can be made in the average tool-room, saving both time and money; and because there is no need to make special drawings for gauges illustrated in the British Standard, considerable time is saved in the tool design office.

Certain standards, such as BS 3550: *Involute Splines*, contain complete information on the gauges required for controlling work to the specification: sizes of the gauge blanks, dimensions of the gauging faces and the appropriate gauge tolerances. Fully worked out systems of gauge inspection, stating roller dimensions required to check the gauges, are given. Very great savings can be made in technical departments, and in tool making and inspection, if full use is made of such information.

Gauge tolerances

BS 969 gives a complete coverage of the recommended tolerances for plain gauges. (Screw thread gauges are discussed in Chapter 14.) The system is illustrated in Figure 16.16.

The magnitudes of the gauge tolerance T, and of the wear allowance for the *go* gauge W, depend upon the workpiece tolerance as shown in BS 969.

BS 969 gives no distinctive tolerances of 'workshop' or 'inspection' gauges. Gauges are made to a common tolerance and, where required, are sorted on final inspection. Those having the largest amount of metal for wear on the *go* dimension are marked *workshop*, the remainder are marked *inspection*.

Certain expensive type receiver gauges, e.g. the *full-form go* gauges for splines and serrations, have rather more generous gauge tolerances and wear allowances than are specified for plain limit gauges. Tolerancing of the more complex type of gauge requires experience and discretion, because special difficulties of tolerance build-up and of manufacturing methods are involved. Examples are given later in this chapter.

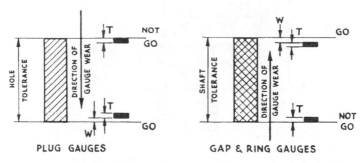

Figure 16.16 Gauge tolerances to BS 969.

Limitations of the gauging method

Rigidity

Caliper gauges are easily 'sprung' over a work diameter; also they vary in size, when held in the hand, due to temperature changes. For these reasons BS 969 suggests that they are only suitable where the work tolerance exceeds 0.018 mm. This is reasonable where small work diameters are involved, but becomes impractical at work diameters of 75 to 100 mm due to flexure and temperature distortion. The alternative method, using a calibrated standard and bench type comparator, is far superior for inspecting external diameters to very close tolerances.

Plug gauges are relatively free from flexure effects, and for this reason may be used on a *go*, *not-go* basis where the work tolerance is as low as IT6.

Weight

Plug gauges for large bores are heavy and awkward to handle, unless attention is paid to weight reduction, Figure 16.17 shows how weight may be reduced. Spherically ended bar gauges are more suitable for bores of 120 mm diameter and above.

Profile gauging

Sighting methods, as shown in Figure 16.18, introduce discretionary judgements into inspection and are not always acceptable on that account. The 'size' of any gap between the component and the gauge, as it appears when held up to the light, depends upon the thickness of the gauging edge, upon surface finish of the work and upon cleanliness, especially freedom from oil films. If the gauge edge is reduced to about 0.8 mm thickness, a gap of about 0.013 mm can be seen. The alternative is to make a more costly form of gauge and to use feeler gauges, or *go* and *not-go* limit plug gauges, to eliminate the discretionary element. A typical gauge of this type is shown in Figure 16.19. Such gauges can precisely control a profile to a tolerance zone specified in accordance with BS 308.

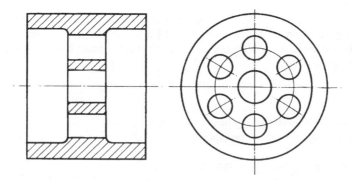

Figure 16.17 Weight-reduced section for plug gauge.

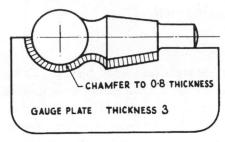

Figure 16.18 'Sighting' method of profile gauging.

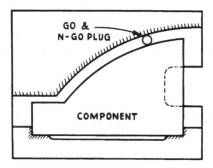

Figure 16.19 Limit gauging of a profile.

16.7 Gauging of tapers

Taper cone gauges are in common use and can effectively control a diameter in a specified plane perpendicular to the work axis on a *go* and *not-go* basis. They are less effective in their control of included angle. 'Blueing' methods may be employed to test the accuracy of the taper angle, but the method is discretionary. The elaborate set of gauges sometimes described for control of taper angle on a *go* and *not-go* basis is rarely employed; the gauges are both costly to make and slow in use.

Example 16.3

Figure 16.20 illustrates a taper fit. Tests show that when the nut is tightened, the 6.35 mm nominal gap closes by an average amount of 0.13 mm. Design a set of limit gauges which will control the assembly to the required limits, given that the taper hole is to be bored and the shaft end to be ground.

Solution

Let d = component diameter at any position on the taper, and
       l = length along the axis of the taper.

For 25° included angle, $\dfrac{\Delta d}{2\Delta l} = \tan 12\tfrac{1}{2}° = 0.222$

i.e. $\Delta d = 0.444\,\Delta l$

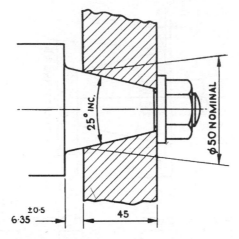

Figure 16.20 Taper assembly gauging.

The tolerance on the 6.35 nominal dimension = ±0.5

$$\Delta d = 0.444 \times 1 = 0.444 \text{ mm}$$

i.e. the total tolerance on taper diameter is 0.44 mm (say)

A suitable division of this tolerance is:

(i) bored hole, say 0.27
(ii) ground shaft, say 0.17

The limit steps required on the taper gauges are:

$$\text{Plug gauge for hole, } \Delta l_h = \frac{0.27}{0.444} = 0.61 \text{ (say)}$$

$$\text{Ring gauge for shaft, } \Delta l_s = \frac{0.17}{0.444} = 0.39 \text{ (say)}$$

If a unilateral hole-based system is employed the nominal diameter of the taper at the large end is 50 mm.

Figure 16.21(a) shows the plug gauge required.
Figure 16.21(b) shows the ring gauge and *go* and *not-go* slip gauge required.
The dimensions of the *go* and *not-go* slip gauge are determined as follows:

Nominal value of the gap, $x = 6.35$ mm.
Allowance for closure on bolting, add 0.13 mm.
Bore on *top* limit absorbs 0.61 mm of the tol on x.
Go size of slip; bore on top limit, shaft on bottom limit

$$x_1 = 6.48 - 0.5 + 0.61 = 6.59 \text{ mm}$$

Not-go size of slip; bore on bottom limit, shaft on top limit

$$x_2 = 6.48 + 0.5 = 6.98 \text{ mm}$$

Check $x_2 - x_1 = 0.39$, which agrees with Δl_s above.

□   □   □

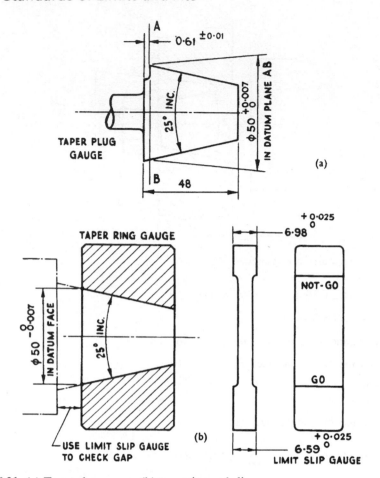

Figure 16.21 (a) Taper plug gauge; (b) taper ring and slip gauge.

Two comments can be made on the above solution:

1. The ±0.5 tolerance on the 6.35 dimension could easily be held under average machine shop conditions. Should the 6.35 dimension have a closer tolerance, say ±0.13, it would be necessary in order to hold the limit, to machine the face of the taper bore after a trial assembly. Production methods are in part determined by the design tolerances.
2. The given solution is based upon algebraic association of tolerances. If the principles of Section 17.10 can be employed because the probability of a very few assemblies lying outside the specified limits is acceptable, somewhat wider tolerances could be permitted.

Limit gauging of taper angle

Taper cone gauges do not give positive control of taper angle. For external cylindrical work a much more effective method is to support the work so that the upper edge lies parallel with a plane surface, and to gauge for parallelism by means of dial gauges as illustrated in Figure 16.22. When grinding 'on-centres' sine centres provide the most suitable means of support. For centreless ground work a special angle block may be made and the work located on this between two guide strips, or alternatively, it can be supported in a vee-block tipped to the appropriate angle. One attraction of the method is that the straightness of the sides of the taper can be checked.

Figure 16.23 illustrates the geometry upon which the relationships between the various angles depend.

Let α = angle of inclination of the centre-line of the cone with respect to the vee;

θ = semi-angle of the cone;

ϕ = semi-angle of the symmetrical vee.

The cone contacts the vee flank along the line OD. The normal AD to the point of contact at D lies in the plane PQ, which is perpendicular to the line of intersection

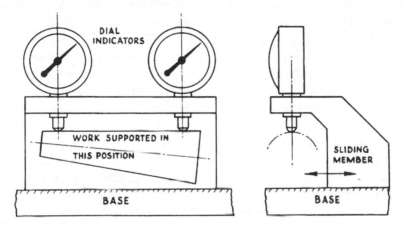

Figure 16.22 Gauging of taper angle.

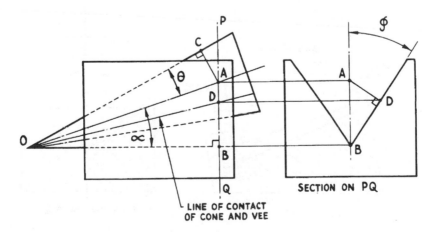

Figure 16.23 Geometry of cone and vee.

of the flanks of the vees. AD is perpendicular to BD, as shown in the sectional view on plane PQ, and is also perpendicular to OD.

$$AD = OA \sin \theta \tag{16.1}$$

(The reader may visualise this relationship more easily by imagining the cone to rotate on its axis until D has the position shown by point C; AD = AC in length.)

It can be seen from the diagram that

$$AB = OA \sin \alpha = \frac{AD}{\sin \phi} \tag{16.2}$$

and by substitution of equation 16.1 in equation 16.2

$$\sin \alpha = \frac{\sin \theta}{\sin \phi} \tag{16.3}$$

To test any particular angle, θ and ϕ will have known values and α can be found from equation 16.3. The vee block can then be inclined to an angle $(\alpha + \theta)$ to bring the top edge of the cone parallel with the surface plate on which the vee block rests. A dial indicator test will reveal any error of parallelism, and this may be restricted to a specific amount, depending upon the accuracy to which θ must be produced.

It would be possible to extend this principle in order to make a semi-automatic gauging system, as illustrated in Figure 16.24, which would control taper-angle and diameter. If automatic loading and ejection were used, and coupled with gauging heads which would sort the work according to signals produced by its accuracy, an automatic method of inspection could be developed. Such methods are necessary in conjunction with high-output automatic machines, e.g. centreless grinder, where the work is required to be 100 per cent within the specification, so that statistical quality control cannot be used. If signals from the automatic gauging equipment are fed

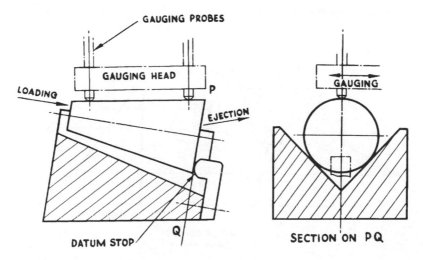

Figure 16.24 Design of semi-automatic gauging system.

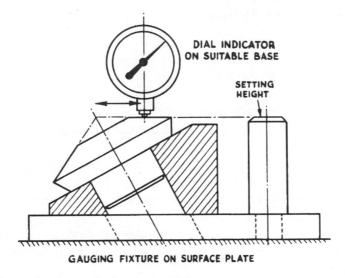

DIAL INDICATOR
ON SUITABLE BASE

SETTING
HEIGHT

GAUGING FIXTURE ON SURFACE PLATE

Figure 16.25 Gauging fixture for bevel gear blanks.

back to the machine, to cause it to make corrective adjustments, the basis of automatic size control is established. Control methods of this kind are now being used for very high-output work of a precision character. They represent the ultimate development in gauging technique, the gauge 'instructs' the machine. Automatic size control of grinding operations is desirable because of the fine tolerances set, and of the steady drift of size due to wheel wear.

Gauging fixture for bevel gears

Figure 16.25 shows a scheme developed for the inspection of the cone angle, and the diameter in a datum face, of a bevel gear blank. The fixture is used on a surface plate, and a dial gauge reading over the setting pin, and over the work, gives an easy control of angle and diameter. Any alternative gauging method would be slow, require much more equipment and might be less reliable. A taper ring gauge would not be effective due to the large included angle of the taper.

16.8 Gauge making materials

The chief properties required are:

1. Hardness, so that the gauging surfaces are not easily damaged.
2. Stability, so that dimensional changes do not occur.
3. Wear resistance, so that the wear life is reasonable.

The material must also take a high-grade surface finish, generally by the processes of grinding and lapping.

The main materials used for gauges are:

1. *Carburised and hardened mild steel*. Available in a wide range of sizes and easily machined; a case depth of about 0.2 mm obtained by carburising is hardened to about 850 HV prior to the final grinding. A relatively cheap material for large gauges.
2. *High-carbon steel*. Generally oil hardening steel of around 0.9 per cent to 1.0 per cent carbon is used for small plugs or gauges made from plate. Only the gauging surfaces should be hardened.
3. *Special wear-resistant materials*. A small diameter screw plug gauge used on a cast iron component will lose its size very rapidly due to abrasive wear. For situations like this improved wear resistance can be obtained from:
   (a) high-speed steel (better wear resistance than carbon steel);
   (b) 'hard' chromium plating on hardened steel, a 'flash' of plating on thread ground gauges, a moderate thickness of plating finally ground to size for plain gauges;
   (c) tungsten carbide, sintered metal powder-mouldings ground with diamond grit abrasive and so very expensive, but having a greatly improved wear life.

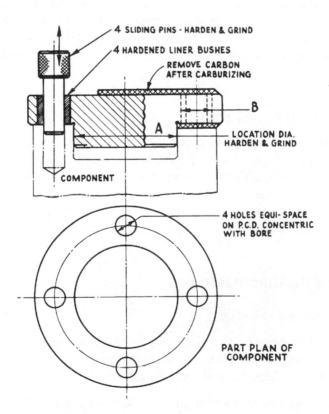

Figure 16.26 Manufacturing details of receiver gauge.

The heat-treatment of all steel gauges should include a stabilising process after hardening to reduce the small dimensional changes which normally occur over a period in the finished gauge. This is very important for gauges of large size or where very high accuracy is involved (e.g. slip gauges).

Figure 16.26 gives details of the construction and manufacturing method for a receiver type gauge for checking the location of four holes relative to a bored diameter.

16.9 Component tolerancing and gauge design

Many of the principles given earlier in this chapter are linked with problems of maintaining accuracy during manufacture. It is difficult to treat all the practical aspects of this which may arise in process planning or gauge design, in a textbook. An attempt is made in the following worked examples to show how such problems can be tackled. There are, of course, many possible alternatives to the solutions presented, which are only intended to give an insight into some aspects of tolerancing and gauge design.

Example 16.4

Figure 16.27 gives details of an assembly. The functional requirements are:

1. Pin (3) to be an interference fit in shaft (1).
2. Sleeve (2) to be a slide fit on shaft (1).
3. The slot of (2) to engage pin (3) in either possible position with a minimum of 'play'.

Tolerance the functional dimensions of the assembly in accordance with BS 4500, as far as possible, and on the assumption that the manufacturing costs are to be typical of average machine-shop work.

Design suitable gauges for the control of the limits set, assuming that there is to be a fairly long production run.

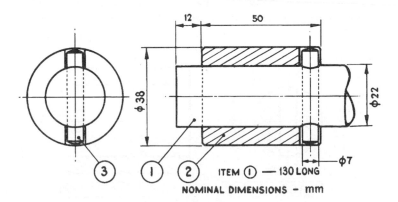

Figure 16.27 Assembly showing nominal dimensions.

Solution

Selection of the main fits

1. Pin (3) in shaft (1), H7/p6 is a suitable interference fit.
   Reamed hole in shaft (1), $\phi\ 7.015/7.000$
   Centreless ground pin (3), $\phi\ 17.025/7.015$
2. Sleeve (2) on shaft (1) (it is assumed that the shaft will be centreless ground), H7/g6 is a suitable close-clearance fit.
   Reamed hole in sleeve (2), $\phi\ 122.021/22.000$
   Ground shaft (1), $\phi\ 121.993/21.972$
3. Position of hole in shaft (I) and of milled slot in sleeve (2). The positional tolerances are shown in Figure 16.28.
4. Limits for the milled slot of sleeve (2). If the slot is to engage the pin in both possible positions its minimum width will depend upon the positional tolerances given in Figure 16.28.

   Let w = minimum width of the slot

   $$\text{then} \qquad \frac{w}{2} = \text{rad. of pin at MMC} + \tfrac{1}{2}(t_1 + t_2)$$

   $$w = 7.025 + 0.200 = 7.225 \text{ mm}$$

   A suitable tolerance for the milled slot is 0.08 and the limits are 7.305/7.225.

   Since BS4500 gives no suitable limits for a basic size of 7 mm and there is no good reason to change this basic size, it seems best for the *special* limits to remain. The slot can be produced by means of a cutter of 8 mm (standard) thickness, specially reduced in width to cut the required size, i.e. cutter grind to $7.230^{+0.02}_{0}$ mm thickness.
5. Figure 16.29 shows a component drawing for the sleeve (2). Provision has been made to accommodate the burr and swollen edge which will arise when the pin is driven into the shaft.
6. The standard type gauges required are as follows:
   (a) 7H7, D/E plain plug (D/E = double ended);
   (b) 22H7, D/E plain plug.
   The 7p6, and 22g6, limits, will be controlled by means of slip gauges and a measuring comparator, because caliper type gauges are unsuitable for such close tolerances, see p. 517, *Rigidity*.
7. Special gauges:
   (a) Figure 16.30 shows the special double ended slip gauge required for the width of the milled slot.
   (b) Figure 16.31 shows a receiver-type gauge for the position of the hole in the shaft. While such a gauge would be relatively expensive, it would be desirable for a long

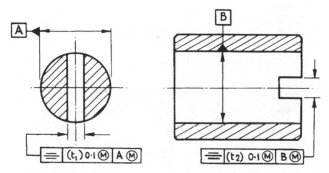

Figure 16.28 Tolerances controlling the relationships of the pin and slot.

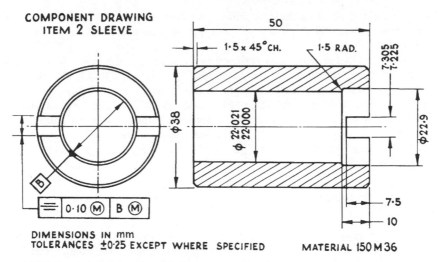

Figure 16.29 Production drawing of sleeve.

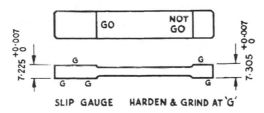

Figure 16.30 Double-ended slip gauge for milled slot.

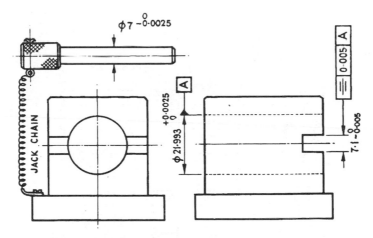

Figure 16.31 Position gauge for reamed hole.

production run. The position of the hole would depend upon the accuracy of the jig provided for drilling it. The jig may function satisfactorily for a time, but wear of the drill bush will occur, and the position of the hole drilled will vary from this cause. The position of the hole must be checked at regular intervals. It would be possible to dispense with the gauge, provided a 'first-off' inspection of each batch was made to check the position of the hole. A skilled inspector would be required for this work; if a 'viewer' or unskilled operator is to do the inspection the receiver gauge must be provided.

The width of the gauge slots is shown as $7.1_{-0.005}^{0}$ mm, a figure arrived at as follows. The positional tolerance of the hole, item (1), is (0.05×2) wide, the nominal diameter 7 mm: the nominal width of the slot is therefore 7.100 mm. The gauge has to control the *position* of the hole to a tolerance of (0.05×2), and a reasonable gauge tolerance is 10 per cent of this, or 0.01, opposite to the direction of gauge wear (i.e. minus). The gauge maker needs a positional tolerance and a width tolerance, so the 0.01 has been split as shown in Figure 16.31. As arranged, it is possible for one side of the slot to lie at

$$\left(\frac{7.1}{2}\right) + 0.0025$$

from the centre-line of the datum hole, i.e. to be just outside the specified work tolerance.

(c) Figure 16.32 shows a *full-form* receiver gauge, designed in accordance with Taylor's principles, for checking the position of the milled slot. If this gauge accepts the work, and if separate *not-go* gauges for the bore and for the width of the slot also pass the work, the component inspected has been proved to lie within the tolerance zones specified. The $7.125_{0}^{+0.005}$ width of the tongue is determined as follows:

Nominal width of the tongue = Minimum width of the slot
  − Positional tolerance of the slot = $7.225 - 0.1 = 7.125$ mm

The 10 per cent gauge-making tolerance has again been divided between the positional and width dimensions of the tongues.

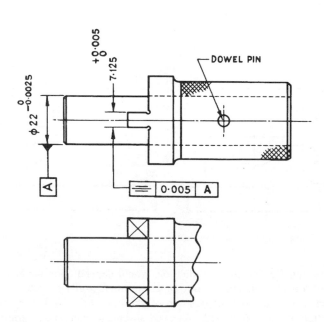

Figure 16.32 Receiver gauge for position of milled slot.

Example 16.5

A taper cone, nominal included angle 25°, rests in a symmetrical vee of included angle 90°. To what angle must the vee be inclined for the upper edge of the taper cone to be parallel with the surface plate upon which the vee-block rests?

If dial indicator readings taken over the taper cone show a difference of 0.03 mm over a length of 50 mm, lower at the large end of the taper, what error of included angle is indicated?

Solution

By equation 16.3

$$\sin \alpha = \frac{\sin \theta}{\sin \phi}$$

From the given conditions $2\theta = 25°$, $2\phi = 90°$ and α is required.

$$\sin \alpha = \frac{\sin 12.5°}{\sin 45°}$$

$$= 0.306\ 09$$
$$\alpha = 17° 49.5'$$

Angle of inclination required $= 17° 49v5' + 12° 30'$
$$= 30° 19.5'$$

Angular error of $(\alpha + \theta)$ indicated $= \dfrac{0.03}{50}$ radians

$$- 0.0006 \text{ radians or } 2.06'$$

α is however a function of θ. By differentiation of equation 16.3, ϕ being regarded as a constant,

$$\cos \alpha.\Delta\alpha = \frac{\cos \theta}{\sin \phi}.\Delta\theta \text{ (very nearly)}$$

$$\Delta\alpha = \frac{\cos 12.5°}{\cos 17°50' \times \sin 45°}$$

$$= \frac{0.9763}{0.952 \times 0.7071} \Delta\theta$$

$$= 1.45\ \Delta\theta$$

But $(\Delta\alpha + \Delta\theta) = 2.06'$
$$\therefore 2.45\ \Delta\theta = 2.06'$$
$$\Delta\theta = 0.84' \text{ or } 50 \text{ sec of arc}$$
Error of included angle $= -0° 1'40''$

□   □   □

16.10 Alternatives to limit gauging

Refinements in machine tool control (see Chapter 7) and better instrumentation, mainly due to electronics have had the following impact on the machining of components:

1. Output rates are raised.
2. Dimensional and geometric accuracy has become more reliable.

Due to (1) manual inspection by limit gauging becomes a disproportionately large element of the production cost. Due to (2) dimensional control is more easily integrated with the machining process so that inspection can be limited to 'first off' checks supplemented by sampling methods for important dimensions (see Chapter 17).

Such technical improvements have almost eliminated the need for expensive receiver gauges to check dimensional and geometric features as required by Taylor's principles of gauging. However, limit gauging is still the cheapest and most convenient method of control for such things as small plain bores and both internal and external screw threads. Full sets of gauges based on Taylor's principles remain an economic form of dimensional control for such items as splines and serrations, where the alternative of 100 per cent inspection by measurement requires standards room equipment and highly skilled labour.

16.11 Multi-gauging based on comparators

A pre-set comparator has advantages over a caliper-type limit gauge; one instrument can cover a wide range of sizes and the actual departures from the ideal size can be detected. By building a number of comparator units into a type of gauging fixture rapid inspection of a number of dimensional features can be performed simultaneously.

Figure 16.33 illustrates schematically how a back pressure air gauging system operating as explained on p. 488 can be developed for multi-gauging. Back pressure P_b determines displacement h of the manometer associated with each measuring

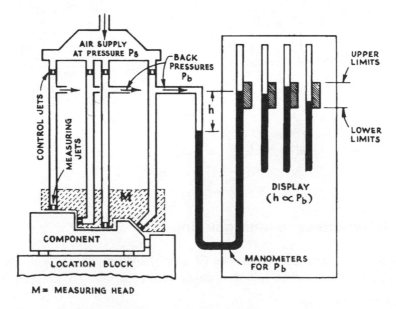

Figure 16.33 Multi-gauging fixture, based on air gauging.

element. If the control and measuring jets are suitably chosen and a small adjustable air bleed is incorporated in each line, it is possible to have common limits for h on the display, even where different limits of size apply to the workpiece. The system involves certain specially manufactured elements, namely:

1. The measuring head.
2. The component locating block.
3. A 'master' component for setting up and subsequent checking.

These parts need not be manufactured to the same high precision as gauges because adjustments can be made, via the air bleeds, during setting up. It is very easy to see at a glance if all the columns of coloured liquid terminate within the limits for h, an arrangement ergonomically superior to that of reading separate dials. Electronically operated comparators are now available which employ a column of light (neon discharge) in place of the manometer of the air gauging system. By further processing the output signals from electronic comparators, it is possible to operate just one indication lamp at each pass: green light – accept, red light – reject.

16.12 In-process measurement

A machine tool fitted with a control system may be suitably adapted to measure and control a particular dimension during the machining. Figure 16.34 illustrates such a system developed by Ferranti for control of an external diameter during a grinding process. This system incorporates post-process measurement with feedback to give greater reliability.

Item A is the primary control unit receiving a signal from the caliper-type measuring head B which it compares with a pre-set value in A. When 'size' is

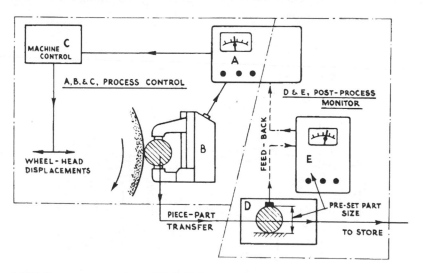

Figure 16.34 In-process gauging with feed-back.

signalled, the information passes to the machine control C, the grinding wheel head retracts and the ground part is transferred to a post-process measuring unit D operating in conjunction with E, where the diameter required is pre-set. Any error detected at E is fed back to A to adjust the size setting. The post-process monitor is away from disturbance factors of the grinding and gives a more reliable reading from which to control the size. Both B and D are adjustable for diameter, as also the size target analogues of A and E, making the equipment adaptable to a range of work provided parts are compatible with the transfer mechanism B to D.

16.13 Co-ordinate measuring machines

Co-ordinate measuring machines (Figure 16.35) are a natural development of CNC machine tools. They were considerably enhanced by the invention of touch trigger probes by Renishaw, and these types of probes are now universally used. They provide very low contact forces and an extremely high degree of repeatability (around 0.0003 mm) they thus allow extremely complex measurements to be performed, either manually or automatically. In manual mode they will provide arc and circle measuring and calculation facilities, as well as pitch circles, tapers, planes, etc., they will also locate the workpiece origin on the machine table, and avoid the need for a precise location system for each part. In automatic mode all these facilities can be performed very rapidly while the operator is doing other work.

There is a wide range of styli (Figure 16.36) available to reach into complex workpieces, and the provision of a motorised head and automatic probe changing system can enhance the flexibility of the CMM considerably. Scanning probes – either contact or non-contact are used for digitising models for program generation for die making purposes. The laser probe shown in Figure 16.37 allows soft workpieces to be measured at high speed, avoiding the problems of workpiece deformation in these situations.

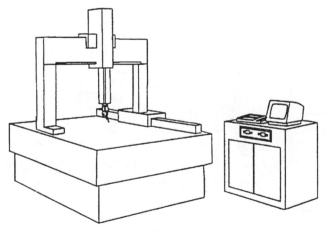

Figure 16.35 Computer controlled co-ordinate machine.

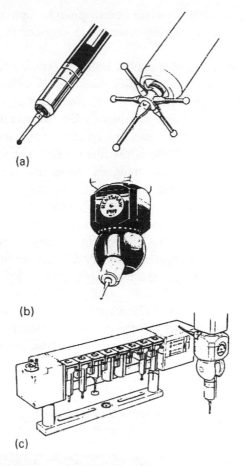

Figure 16.36 (a) Probes and styli; (b) motorised heads; (c) autochange system.

Figure 16.37 Laser scanning probe.

The machines usually have a granite bed to provide high accuracy and a stable base for the machine, and air bearing systems are very common to provide friction free motion for the axes.

A CMM should be recalibrated at regular intervals (say, every six months) (Figure 16.38) to ensure that its accuracy is being maintained. A quick check on its reliability can be performed by remeasuring a standard sample to ensure that repeatable results are being obtained, but a full calibration involves the use of a laser interferometer to ensure that all axes are square and true, and that the measuring system is within tolerance.

An obvious advantage of a CMM is that the CAD/CAM system can be used to generate the CNC program for machining the parts, and then the same database information can be used to generate the CMM inspection routine to ensure that the part conforms to the design specification (Figure 16.39).

Their principal merit is that they can provide the independent assessment of dimensional features produced on CNC machine tools, profiles, hole positions, angular displacements etc. The reliability of CNC machine tools is such that once the program is proved in this way, the major geometrical and dimensional features are reproduced consistently. The main cause of dimensional variation when machining is tool wear, and frequent checks are necessary to monitor this, for which reason simple limit gauges for hole diameters, screw threads, width of milled slots etc., are likely to remain as the economic method for these features (Figure 16.39).

Exercises – 16
1. Make a drawing of a limited plug gauge which would be suitable for inspecting the 25.025/25 mm bore of a bush, 63 mm long, in accordance with the Taylor principle.

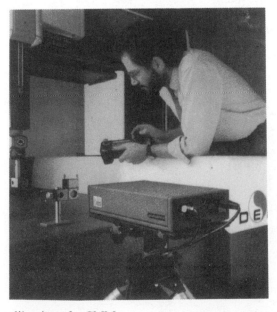

Figure 16.38 Laser calibration of a CMM.

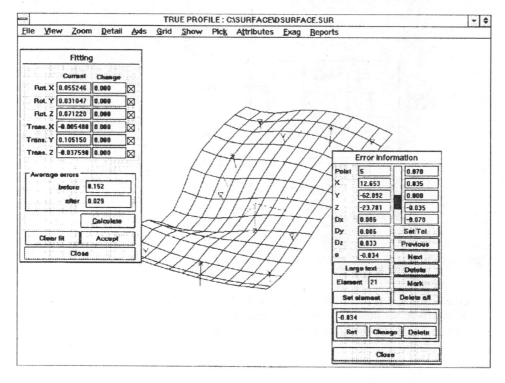

Figure 16.39 3D surface measuring system on a CMM.

The drawing should indicate:
(a) the tolerance to which the *go* and *not-go* ends of the plug should be manufactured;
(b) the tolerance to which the measuring anvils will be subsequently inspected to control the degree of permissible wear;
(c) the material for which the measuring anvils of the gauge will be made;
(d) the sequence of operations used in making the measuring anvils.

2. State the principles which govern the design of limit gauges giving examples of such gauges to check:
(a) length;
(b) diameter;
(c) depth;
(d) distance between an internal and an external face.

3. The part shown in Figure 16.40 is to be inspected on all linear dimensions by limit gauging. The angle of the taper bore is not important, but the distance from the ϕ 20 mm ball must be held.
   Specify all the gauges you would require. Make dimensioned sketches of the gauges you would use for:
(a) the 38/37.5 mm shoulder length;
(b) the 25.0/24.9 mm square projection;
(c) control of the 51.20/50.08 mm dimension.
   Show the gauge tolerances on these sketches.

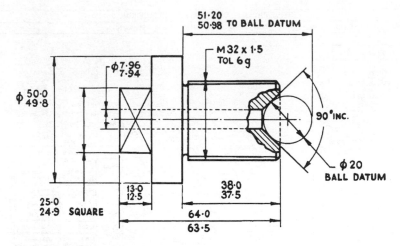

Figure 16.40 Gauge design example.

4. A static coarse clearance fit for ease of assembly, is required between a shaft and hole of 100 mm nominal diameter. The fit is designated H8-c9 and the hole limits are $^{+54}_{+0}$ in 0.001 mm units. The extremes of fit are $^{321}_{180}$ clearance.

   (a) Determine the limits of the shaft, and design and dimension a gauge to check it.

   (b) Draw up a detailed manufacturing sequence for the gauge.

5. Figure 16.41 shows a bush in which a slot has to be milled. For convenience in manufacture the slot position must be redimensioned, working from the new datum face shown, without causing any departure from the previous specification. Give the required new dimensions.

   Make dimensioned sketches of the gauges required to control the overall length of the component, and the width and position of the slot (as re-dimensioned). Show the gauge-making tolerances.

6. A sectional view through a valve block is shown in Figure 16.42. Design gauges on the flush-pin principle suitable for checking the following:

   (a) the $33^{+0.25}$ mm depth of the $45^{+0.13}$ mm diameter hole;

   (b) the $87.6^{-0.38}$ mm dimension of the taper cone seating.

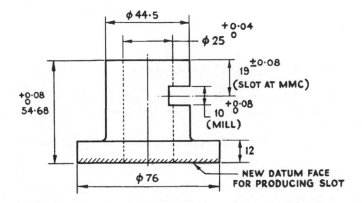

Figure 16.41 Gauge design example for slot position.

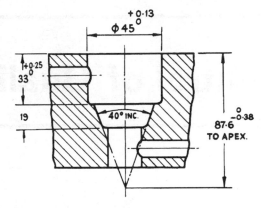

Figure 16.42 Flush pin gauge design example.

7. A taper shaft, basic taper 0.25:1 is 63.5 mm long and the diameter at the large end is $44.5^{-0.076}$ mm. If the shaft rests in a 60° included-angle vee block, to what angle must this be tipped for the upper edge of a shaft to be horizontal?

   If the taper angle is inspected by means of two dial gauges placed 50 mm apart, what difference in their readings will occur for the maximum permissible error of the shaft taper? (Assume parallel tolerance zones.)

8. Figure 16.43 gives the principal dimensions of a bevel gear blank.

   Design a gauging fixture which checks the faces marked A and B by means of a sliding element carrying a profile plate gauge. Also dimension the fixture so that the cone apex position is tested by taking a dial-gauge reading over face A, and over a suitably positioned height-setting pin. What is the maximum acceptable variation of reading of the dial gauge when checking the cone apex position?

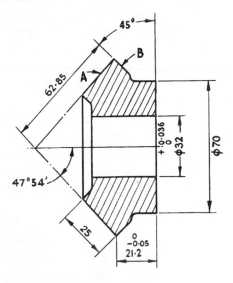

Figure 16.43 Gear gauging example.

17 Control of Quality

17.1 Variability in manufacturing processes

It is a common misconception that an automatic machine will produce identical components. Unfortunately, real life considerations interfere with this theoretical ideal; the properties of the workpiece material vary along the length of the bar, the machine tool slideways must have clearances to allow them to move, and lubrication conditions will constantly be changing, and such random variations will mean that the actual sizes of the parts produced will vary, distributed closely around the target sizes. When the quantity involved is large the pattern of variation can be studied on a statistical basis; it then becomes possible to assess the quality achieved by the process without testing every piece produced. A statistical method which reveals the pattern of variation in a product provides a more certain basis for the assessment of the quality of a large volume of work than would be provided by a detailed inspection of some parts made without reference to the pattern of variability present.

Consider a 200 lot of 12.5 mm diameter pins taken at random from a large batch. If these are measured on the shank diameter they will be found to vary either side of a mean-size. There will be similar variations for each of the remaining dimensions; also there will be variations of the mechanical properties of the material as between one pin and another. For every manufacturing process it is accepted that such variations in quality will occur, and *tolerances* are generally introduced into the specification in order to define the acceptable amount of variation.

If we wish to test the quality of the pins in respect of shank diameter the 200 items can be limit gauged (Figure 17.1). This is a *sorting* process; any work varying more than the permitted amount is separated from the acceptable work. Limit gauging isolates the 'defectives', it gives no information about the pattern of variability of the parts produced.

Alternatively, the shank diameter of the 200 pins could be individually measured and their sizes recorded. This procedure will yield a large amount of data which, when suitably arranged, can reveal the pattern of variability. The arrangement and analysis of such data, in order to obtain a clear picture of the pattern of variability, is based upon statistics, a mathematical science for the study of variability.

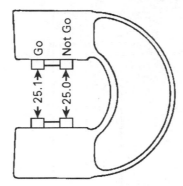

Figure 17.1 A go/not go gauge for 25.0/25.1 tolerance.

17.2 Statistical concepts and variability

For an introductory treatment of the mathematics employed in this section see *Modern Applied Mathematics* by J. C. Turner (Hodder and Stoughton, 1970).

Suppose the measured shank diameters of the 200 rivets to be as given in Table 17.1.

The groupings of Table 17.1 are made so that a part measuring 12.2 exactly would be in the first group, a part measuring 12.3 in the second group, etc.

If the results shown in Table 17.1 are plotted as a bar-chart (histogram), Figure 17.2, a picture of the variability is obtained. If a smooth curve is now drawn through the highest point of the mid-ordinate of each strip a graphical representation of the variability is obtained. The graph, called a frequency distribution curve, would tend to have the same general shape if the number of pieces measured was increased, or if the groupings were varied in some small degree.

Suppose now that a very much larger number of pieces was measured, and the results arranged in at least twice as many groups so that the size-range of the groups became smaller; a frequency distribution curve of more closely defined shape would emerge and would reveal the variability pattern of the production process with greater certainty.

Table 17.1

Diameter (mm)	Mid-size (mm)	Frequency
12.1 up to 12.2	12.15	4
12.2 up to 12.3	12.25	21
12.3 up to 12.4	12.35	50
12.4 up to 12.5	12.45	72
12.5 up to 12.6	12.55	41
12.6 up to 12.7	12.65	12

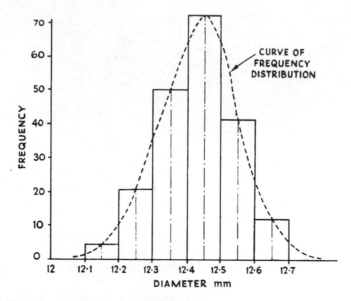

Figure 17.2 Histogram and frequency distribution curve.

The shape of a frequency distribution curve (or simply 'distribution curve' as it is generally called) is influenced by two important features of variability:

1. The frequency with which parts close to the mean-size occur.
2. The 'spread' either side of the mean-size, called the range.

Mean-size (the arithmetical mean of all the sizes)

The mean-size can be conveniently calculated from a set of values, such as Table 17.1, by the following method.

> Let x_1, x_2, etc., be the mid-size of the groups.
> Let f_1, f_2, etc., be the frequency of pieces in the groups.
> Let N be the total number of pieces.

Select some value of the size near to the anticipated mean-size, the fictitious mean to be represented by A. Subtract A from the mid-size of a group and multiply the result by the group frequency, i.e. $f_1(x_1 - A)$, $f_2(x_2 - A)$, etc. Take the sum of all these values, paying due regard to the sign of each; divide this sum by N. The arithmetic mean, represented by $\bar{x}$, is now given by the expression:

$$\bar{x} = \frac{f_1(x_1 - A) + f_2(x_2 - A) + f_3(x_3 - A) + \cdots}{N} + A \qquad (17.1)$$

A convenient tabular method of working is shown in Table 17.2, columns 3 and 4.

Table 17.2

1	2	3	4	5	6
Mid-value x	Frequency f	$(A = 124.5)$ $x - A$	$f(x - A)$	$(x - A)^2$	$f(x - A)^2$
121.5	4	−3	−12	9	36
122.5	21	−2	−42	4	84
123.5	50	−1	−50	1	50
124.5	72	0	0	0	0
125.5	41	1	41	1	41
126.5	12	2	24	4	48
	$N = 200$		$65 - 104$		259

(*Note:* the errors have been multiplied by 10 to simplify the working.)

Range

This is the difference between the largest and the smallest part, e.g. 0.6 mm (or slightly less) for the measurements given in Table 17.1.

Standard deviation (σ)

Variability is considered to be the amount of the departure from the mean-size. It is roughly indicated by the range, but since range is a comparison between the sizes of two pieces only, it cannot give an accurate measure of the variability of *all* the pieces. The standard deviation is a more accurate measure of variability because the size of every part produced is taken into account.

If we are to take a particular size, and the frequency with which it occurs, into account in measuring variability, an expression of the form

$$f_1(x_1 - \bar{x}) + f_2(x_2 - \bar{x}) + \ldots$$

might conceivably give what is wanted. Due to the fact that $\bar{x}$ is the arithmetic mean, however, the positive and negative terms of this expression will be equal, and the sum of the terms zero. The difficulty is overcome by squaring the values $(x_1 - \bar{x})$, $(x_2 - \bar{x})$, etc., so that there are no negative signs. The resulting value is called the *variance*.

$$Variance = \frac{f_1(x_1 - \bar{x})^2 + f_2(x_2 - \bar{x})^2 + f_3(x_3 - \bar{x})^2 + \cdots}{N} \tag{17.2}$$

Standard deviation $= \sqrt{Variance}$

Both quantities are important measures of variability. The standard deviation is of particular importance because of its relationship to the *normal* distribution curve.

The standard deviation can be calculated from data such as given in Table 17.1, by an extension of the methods already given on p. 540 for finding $\bar{x}$.

The fictitious mean-size (A) is first employed to find a fictitious value (S) for the standard deviation.

$$S^2 = \frac{f_1(x_1 - A)^2 + f_2(x_2 - A)^2 + f_3(x_3 - A)^2 + \cdots}{N} \tag{17.3}$$

The process is conveniently carried out in tabular form, as shown in Table 17.2, columns 5 and 6. It can be shown mathematically that

$$\sigma^2 = S^2 - (\bar{x} - A)^2 \tag{17.4}$$

where σ represents the standard deviation.

Tabular method for finding $\bar{x}$ and σ

The data given in Table 17.1 is used to illustrate a convenient method of finding the required values.

$$\bar{x} = \left[\frac{65 - 104}{200} + 124.5\right]10^{-1} \text{ by equation 17.1}$$

$$= 12.43 \text{ mm (to the nearest 0.01)}$$

$$S^2 = \tfrac{259}{200} \times 10^{-2} = 1.295 \times 10^{-2} \text{ by equation 17.3}$$

$$\sigma^2 = S^2 - (\bar{x} - A)^2 \text{ by equation 17.4}$$
$$= 1.295 \times 10^{-2} - (0.2 \times 10^{-1})^2$$
$$= 1.295 - 0.04)10^{-2}$$
$$= 1.255 \times 10^{-2}$$
$$\sigma = 1.12 \times 10^{-1}$$
$$= 0.11 \text{ mm (to the nearest 0.01)}$$

The diameters of the pins can now be represented numerically in a way which indicates both the size and variability resulting from the manufacturing process.

Mean diameter   = 12.43 mm
Range of diameters = 0.6 mm
Standard deviation = 0.11 mm

17.3 Normal curve of distribution

The 200 pins considered would have been taken from a much larger quantity manufactured. If very many more pins were measured and then grouped into smaller intervals, say 0.02 mm, it would be possible to draw a histogram from which a much smoother curve of frequency distribution would result, because by greatly increasing the number of pieces considered the influence of any exceptional sizes which have occurred will be reduced. At very large quantities, and where the causes of variation are strictly random, the distribution curve will tend to be *normal*.

It is possible to arrive at this *normal* form of distribution by mathematical reasoning; the resulting equation of the normal curve is

$$y = \frac{1}{\sigma\sqrt{(2\pi)}}\, e^{-(x-\bar{x})^2/2\sigma^2}$$

where σ is the standard deviation.

The equation can be simplified by putting $t = (x - \bar{x})/\sigma$. For the condition where $\bar{x} = 0$ the curve will have its axis of symmetry on the Y axis instead of at the mean value, and for this condition $t = x/\sigma$, so that t now represents x to some scale depending on the value of σ. If, in order to represent the curve as simply as possible, we now put $\sigma = 1$, the equation to the curve becomes

$$y = \frac{1}{\sigma\sqrt{(2\pi)}}\, e^{-t^2/2}$$

The term $1/\sqrt{(2\pi)}$ is included to make the area under the curve equal unity. The areas bounded by the curve and by different values of t, expressed as a fraction of unity, then represent the probability of parts lying between those values of t. Table 1 of BS 2564, p. 59, gives the area relationships at 0.2 intervals of the value of t.

The term $1/\sqrt{(2\pi)}$ can be ignored in plotting a curve of normal distribution, since it only causes a change of scale on the Y axis. The shape of the curve may be obtained quite accurately by plotting the equation $y = e^{-t^2/2}$, and the area properties referred to above can be estimated by counting squares. Mathematical tables which give values of e^{-x} over a reasonable range of x, facilitate the plotting of such curves.

Example 17.1

Plot a curve of normal distribution and use this to estimate the number of pins. Table 17.1 having diameters outside the limits 12.43 ± 0.1 mm.

Solution

Equation to be plotted, $y = e^{-1^2/2}$

t	$t^2/2$	$e^{-t^2/2}$
±0	0	1
±0.6	0.18	0.835
±1	0.50	0.606
±1.6	1.28	0.278
±2	2.00	0.135
±2.6	3.38	0.034
±3	4.50	0.011

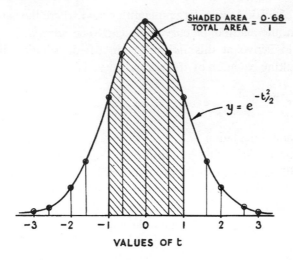

Figure 17.3 Curve showing normal distribution.

Figure 17.3 shows the normal curve. This should be plotted on squared paper as accurately as possible. Estimate by counting squares, or by means of a planimeter, the area bounded by the curve and by the ordinates of $t = \pm 3$.

For the equation used to plot the curve, $\bar{x} = 0$ and $t = \bar{x}/\sigma$. It has been shown on p. 542 that σ for the pins is approximately 0.11 mm; we also know that the limits given in the question are ± 0.1 either side of the mean-size ($\bar{x}$).

Hence, for the range of diameter expressed in the limits, $t = \pm 0.1/0.11$, i.e. $t = \pm 1$ roughly. Estimate the area between the curve and the ordinates $t = \pm 1$, and express this as

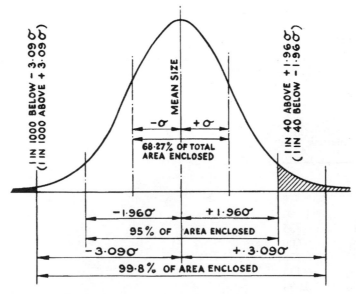

Figure 17.4 Area properties of the normal distribution curve in terms of the standard deviation (σ).

a ratio of the area between $t = \pm 3$. As shown in Figure 17.3, approximately 68 per cent of the area is enclosed between $t = \pm 1$. From the area, property of the distribution curve it can now be estimated that 0.68×200 pins lie between the limits stated, i.e. 136 out of 200.

□   □   □

The significance of the area property of the normal distribution curve is revealed by the above example, and Figure 17.4 illustrates this property in terms of σ. It should be noted that ordinates of $\pm 3.09\sigma$ do not enclose the entire area beneath the curve, because the 'tails' of the curve extend to infinity. Since production patterns do not follow the normal distribution exactly, the approximations that $\pm 3\sigma$ enclose 100 per cent of the area, and $\pm 2\sigma$ enclose 95 per cent of the area, are often made.

17.4 Causes of variation

A manufacturing process may be analysed in order to identify the causes of the variations which occur. Consider a steel bar turned to a given diameter on a CNC lathe. Causes of variation of diameter may be:

1. Fluctuations of temperature.
2. Variations in material properties.
3. Variations in the coolant supply.
4. Variable amounts of eccentric running of the bar.
5. Vibrations caused by the cutting.
6. Errors of tool setting.
7. Tool wear.

Items (1) to (5) are unlikely to follow any regular pattern and may be regarded as the causes of the *inherent process variation*. It will not be possible to relate these causes to specific changes in component diameter because their operation is of a random character.

Items (6) and (7) are *assignable causes* of variation. An error of tool setting will give rise to a discrepancy between the mean-size achieved and the mean-size of the tolerance band. Tool wear will cause a gradual drift of the mean-size of consecutive samples across the tolerance band.

Sampling methods have the advantage that, to a considerable extent, they separate variation due to assignable causes from the inherent process variation. It is possible to make reasonably accurate predictions of the distribution pattern of the bulk from information about the mean-size and variability of samples, and this is the basis of the technique of quality control by measurement of samples.

17.5 Relationships between bulk and sample parameters

Suppose the diameter we have considered as being turned on a CNC lathe is to be controlled for size by the measurement of samples of four consecutive pieces, taken at half-hourly intervals.

If the sizes of the parts measured are plotted as shown in Figure 17.5 a comparison between the samples and the distribution pattern of the bulk can be made. Figure 17.5 shows that the sample range is smaller than the bulk range, and that the mean-sizes of samples do not spread so far either side of the bulk average as do the sizes of individual pieces.

The most significant parameters of variability are average value and standard deviation. In order to show the relationships of these values as between the samples, and the bulk of which they form a part, the following symbols will be used.

	Bulk	*Sample*
Average value	$\overline{X}$	$\bar{x}$
Standard deviation	σ	s
Number of parts	N	n
Range of sample	—	w
Average of sample ranges	—	$\overline{w}$

The following mathematical relationships link the above parameters:

1. Mean value of sample averages,

   Mean of $\bar{x}$ values $= \overline{X}$

2. Standard deviation of $\bar{x} = \dfrac{\sigma}{\sqrt{n}}$ (17.5)

   (often called the standard error of $\bar{x}$)

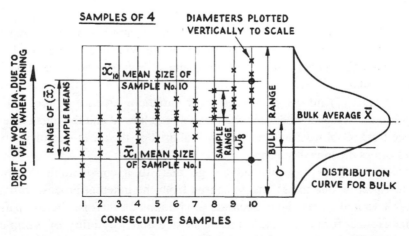

Figure 17.5 Relationship between sizes of parts in a sample and the bulk distribution curve, CNC turning.

3. Mean value of the square of sample ranges,

$$\text{Mean of } s^2 \text{ values} = \frac{n-1}{n} \times \sigma^2 \tag{17.6}$$

4. Standard error (deviation) of $s = \dfrac{\sigma}{\sqrt{(2n)}}$ (17.7)

If $\bar{X}$, σ and n are known it is possible to predict the spread of values of $\bar{x}$ to be expected from the measurement of samples, provided there has been no change in the inherent process variation. Should the values of $\bar{x}$ for the samples exceed the predicted spread, some assignable cause of variation, such as tool wear or tool-setting error has occurred, and this must be rectified in order to bring the process under control again; i.e. to restore the quality of production to the conditions represented by $\bar{X}$ and σ.

17.6 Control chart for sample average

It is usual to set *warning* limits for $\bar{x}$ based upon the probability of one value in 40 lying outside each limit ($p = 0.025$, where p represents the probability) and *action* limits based upon a probability of one value in 1000 lying outside each limit ($p = 0.001$). Reference to Figure 17.4 shows that these conditions are achieved at ordinates of $\pm 1.96\sigma$ and $\pm 3.09\sigma$.

The control limits of $\bar{x}$ are given by:

$$\text{Action limit} = \bar{X} \pm 3.09 \, \frac{\sigma}{\sqrt{n}} \tag{17.8}$$

since σ/n is the standard deviation for $\bar{x}$ by equation 17.5.

$$\text{Warning limit} = \bar{X} \pm 1.96 \, \frac{\sigma}{\sqrt{n}} \tag{17.9}$$

Example 17.2
Use the information on p. 542 to construct a control chart for sample averages suitable for a sample size of 6 pieces.

Solution
From

$\bar{x} = 12.43, \sigma = 0.11$

$$\text{Action limits} = 12.43 \pm \frac{3.09}{\sqrt{6}} \times 0.11 \text{ by equation 17.8}$$

$$= 12.43 \pm 0.14 \text{ mm}$$

$$\text{Warning limits} = 12.43 \pm \frac{1.69}{\sqrt{6}} \times 0.11 \text{ by equation 17.9}$$

$$= 12.43 \pm 0.09 \text{ mm}$$

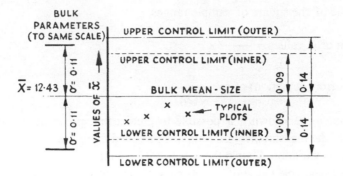

Figure 17.6 Control chart for sample average.

□   □   □

The chart is shown in Figure 17.6. The operator should be warned when a value of $\bar{x}$ lies outside the inner control limits, but a result outside the action limits, means that the machine must be stopped immediately and reset.

It is often more convenient to set the limits for $\bar{x}$ from a knowledge of the average range of samples ($\bar{w}$). The relationship between σ and $\bar{w}$ depends upon the number in the sample, and σ may be reasonably estimated from $\bar{w}$ if n is small, by means of the expression:

$$\sigma = \frac{1}{d^n} \times \bar{w} \tag{17.10}$$

where d_n has a value depending on the value of n. Values of d_n, given in BS 2564, Table 3, are as shown (Table 17.3).

The control limits are then given by:

$$\bar{X} \pm \frac{3.09\,\bar{w}}{d_n\sqrt{n}} \quad \text{and} \quad \bar{X} \pm \frac{1.96\,\bar{w}}{d_n\sqrt{n}}$$

Values of $3.09/d_n\sqrt{n}$ and $1.96/d_n\sqrt{n}$ are given in BS 2564, Table 2, under the symbols $A'_{0.001}$ and $A'_{0.025}$, the suffixes indicating the probability of sample averages outside these limits (see Figure 17.4).

Control chart for range

The control chart for sample average is a sufficient check on the quality of the work only if it is certain that the inherent process variation remains constant. It is desirable to test whether the work still being produced has the same value of σ as was used to set the limits for $\bar{x}$. The standard error of the standard deviation of the samples (s) is given by $\sigma/\sqrt{(2n)}$, equation 17.7. A control chart for s can thus be drawn in much the same way as a control chart for $\bar{x}$. This control chart will

Table 17.3

n	2	3	4	5	6
d_n	1.13	1.69	2.06	2.33	2.53

indicate any exceptional values of s and thus show if σ for the process is changing. One serious disadvantage of a control chart based upon s is the arithmetical work necessary to determine its value for each sample taken. When n is small ($n < 10$) there is a predictable relationship between s and the sample range (w) of sufficient accuracy to enable a chart based upon values of w to be used to monitor σ. The control chart limits for range are:

> Upper control limits:
> > Outer limit $= D'_{0.999} \times \bar{w}$;
> > Inner limit $= D'_{0.975} \times \bar{w}$.
> Lower control limits:
> > Outer limit $= D'_{0.001} \times \bar{w}$;
> > Inner limit $= D'_{0.025} \times \bar{w}$.

The suffixes relate to the probability of values lying outside the limits of 1 in 1000 and 1 in 40 as previously discussed. Lower control limits for range are rarely charted because the purpose is to detect any increase in the inherent process variation, as indicated by an increase in σ.

BS 2564, Table 3, gives the required values of D', reproduced as Table 17.4 here.

It can now be seen that control charts for *average* and for *range* depend upon the value of σ for the process, or upon an estimate of σ based upon $\bar{w}$. To start a control chart, values of $\bar{x}$ and w should be plotted for about 10 samples (not less than 40 separate measurements) without making any attempt to determine the control limits. When this stage has been reached an initial estimate of $\bar{w}$ can be made, and limits set on a provisional basis. After a further 10 values have been plotted a better estimate for the value of $\bar{w}$ can be made, and new and more accurate values of the control limits set. Eventually, sufficient data will be available for an accurate assessment of $\bar{X}$ and σ to be made. Normal inspection methods should be continued until sufficient data has been collected from which reliable control limits can be set.

Table 17.4

n	2	3	4	5	6
$D'_{0.999}$	4.12	2.98	2.57	2.34	2.21
$D'_{0.975}$	2.81	2.17	1.93	1.81	1.72
$D'_{0.025}$	0.04	0.18	0.29	0.37	0.42
$D'_{0.001}$	0.00	0.04	0.10	0.16	0.21

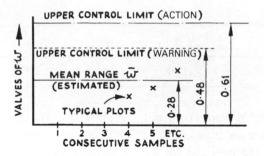

Figure 17.7 Control chart for sample range.

A chart for the sample range may be constructed as follows:
By equation 17.10, $\sigma = \bar{w}/d_n$ and a probable value of $\bar{w}$ can be estimated from this.

$$\bar{w} = 0.11 \times 2.53 = 0.278$$

The upper control limits are:

Action limit $= D'_{0.999} \times \bar{w} = 2.21 \times 0.278 = 0.61$ mm
Worming limit $= D'_{0.975} \times \bar{w} = 1.72 \times 0.278 = 0.48$ mm

The control chart for range is shown in Figure 17.7.

It can be seen that the upper limit for range just exceeds the bulk range of 0.6 mm. This arises because the initial data of Table 17.1 does not conform strictly to normal distribution, and because the strict meaning of 'sample range' is involved.

For the range chart drawn 'sample range' means the difference between the largest and smallest diameter of any six pieces selected at random from the 200. However many times six pieces are selected from a much larger bulk having the same inherent process variation, the range should not exceed 0.61 mm more than once in 1000 times!

More generally, samples collected during a production run are taken consecutively. This tends to give smaller values of sample range than for samples selected at random from the bulk. The drift of sample average due to tool wear, is the main reason why samples taken from the bulk will have a greater range than samples collected consecutively during the production run. Figure 17.5 illustrates this point.

17.7 Control to a specification

These statistical measurements allow accurate prediction of the best possible tolerance that can be produced on the particular dimension being examined. It is important to realise that the various random factors discussed above refer to the particular workpiece on *that* particular machine. A different machine may be able to produce the same parts to a better (or worse) degree of accuracy, depending on the condition of its slideways, bearings, etc. This has lead to major companies (Ford, Rover, Nissan, etc.) insisting that the supplier systematically audits the process used to make parts to ensure that, firstly, the parts can be manufactured to consistent quality levels, and secondly that the machine remains in a satisfactory condition. It is thus necessary to relate the quality produced to the specification given by the customer.

For present purposes a tolerance band may be regarded as fixing the permissible variation on either side of a mean-size. Let $\bar{X}$ be the bulk average required, and a tolerance (T) be the bulk range; the dimension to be controlled is then specified in terms we have already employed.

Figure 17.8 relates a process distribution curve to three tolerance bands of common mean-size, but varying range. From the area property of the distribution curve, it is clear that the process represented by the curve can produce to tolerance band F with certainty, to tolerance band E provided there is no drift of the mean-size, but that for tolerance band G some scrap is bound to occur. It is useless to try and produce to tolerance band G unless a proportion of scrap is permitted; the inherent process variation must be reduced by an improvement of the equipment, or of the method, before scrap can be eliminated. It is one advantage of quality control methods that any inadequacy of the equipment or process is revealed at an early stage of production.

Let T = tolerance magnitude. The maximum value of σ for which the probability of work outside the limits is 2 per 1000, is given by $\sigma = T/2 \times 3.09$. A specification and manufacturing process are reasonably matched if $T \geqslant 6\sigma$.

In BS 2564 a relationship of similar importance between T and $\bar{w}$ is called the relative precision index (RPI).

RPI = $T/\bar{w}$, and BS 2564, Table 4, gives values related to the number of pieces in the samples and to the degree of precision, classified as *low*, *medium* and *high*.

Medium relative precision means that the quality of work produced is just good enough to meet the quality specified.

Since by equation 17.10 $\bar{w} = d_n \times \sigma$, it is possible to show from the values given in BS 2564, Table 4, that in round figures:

$T < 6\sigma$ represents *low relative precision.*
T between 6σ and 8σ represents *medium relative precision.*
$T > 8\sigma$ represents *high relative precision.*

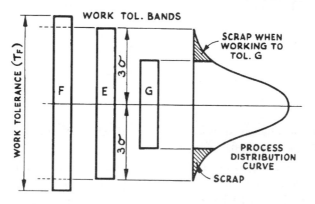

Figure 17.8 Relationship of process variation to work tolerance.

Process Capability Index

The Process Capability Index is an alternative measure of process capability. It is defined as

$$C_p = \frac{\text{tolerance allowed}}{6\sigma}$$

and $C_p = 1$ when a process matches the required tolerance. It closely matches the RPI described above, e.g. consider two CNC lathes A and B producing components with $\sigma_a = 0.02$ mm and $\sigma_b = 0.015$ mm when the specified diameter is 20 +0/−0.1 mm. The tolerance is therefore 0.1 mm. Which machine should be used to produce the parts?

	By RPI	*By C_p*
Lathe A	RPI = 0.1/0.02 = 5	$C_p = 0.1/(6 \times 0.02) = 0.866$
	Indicates low relative precision	This is less than 1, thus
	thus machine is not suitable	machine is not suitable
Lathe B	RPI = 0.1/0.015 = 6.66	$C_p = 0.1/(6 \times 0.015) = 1.11$
	Larger than 6 thus machine	Larger than 1, thus
	is suitable	machine is suitable

It can be seen that these are different ways of measuring the suitability of a process. The comparison is:

	RPI	*C_p*
High relative precision	$>8\sigma$	>1.33
Medium relative precision	$6-8\sigma$	1–1.33
Low relative precision	$<6\sigma$	<1

The methods of Example 17.3 are suitable for the construction of control charts for sample average when the relative precision index is *medium* or *low*. For low relative precision, however, defective work is bound to occur. An estimate of the proportion of the work likely to be defective can be made as follows:

Estimate σ from $\bar{w}$; i.e. $\sigma = \dfrac{\bar{w}}{d_n}$

Let $\dfrac{T}{2\sigma} = k$; find k

Now if $k = 3.09$ the chance of rejects is 1 per 1000 either side of tolerance band, provided that $\bar{X}$ is at the middle of the tolerance band.

Suppose k is found to be 1.8. From the area properties of the normal distribution curve, Figure 17.3, when $t = \pm 1.8$ there are 3.6 per cent of the parts above and 3.6 per cent below the limits set by t. Figure 17.9 illustrates the relationships and shows the percentage scrap to be expected.

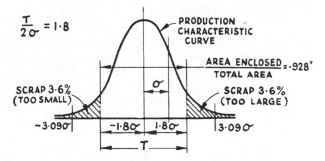

Figure 17.9 Rejects when relative precision is 'low'.

When the RPI is *high* the bulk of work produced, if it is controlled by means of a chart for sample average constructed according to Figure 17.10, will be well inside the tolerance band. Some drift of size, as revealed by a change in sample average, might then be allowed without serious risk of producing scrap; the resultant longer run between tool adjustments would increase output. In order to do this it is necessary to set the control limits by a different method.

Figure 17.10 shows the relationships for $T > 8\sigma$ between the bulk distribution curve, the distribution curve of samples and the tolerance band. The new principle used in setting the control limits is examined in terms of the outer control limit on one side of $\bar{X}$ only; the same principle will apply for the other control limits.

If the work tolerance $= \pm 3.09\sigma$ the control limits for sample average would normally be set at $\pm 3.09\sigma/\sqrt{n}$. The difference between 3.09σ and $3.09\sigma/\sqrt{n}$ is represented by dimension 'a' in Figure 17.10. Now supposing the work tolerance is 8.5σ, the outer control limit for $\bar{x}$ could be set by measuring dimension 'a' from the drawing limit towards $\bar{X}$. The inner control limits can be similarly set by measuring dimension 'b' from the drawing limits towards $\bar{X}$.

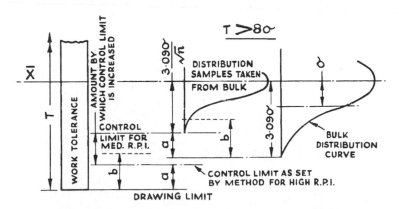

Figure 17.10 Method of setting control limits for 'high' relative precision.

From Figure 17.10 it can be seen that

$$a = 3.09\sigma\left[1 - \frac{1}{\sqrt{n}}\right] \qquad (17.11)$$

and

$$b = \sigma\left[3.09 - \frac{1.96}{\sqrt{n}}\right] \qquad (17.12)$$

Values of 'a' and 'b' can be found from a known value of $\bar{w}$ by substituting $\bar{w}/d_n$ for σ, equation 17.10.

BS 2564 Table 5 gives values from which 'a' and 'b' can be determined directly:

$$a = A''_{0.001} \times \bar{w}; \; b = A''_{0.025} \times \bar{w}$$

The suffixes indicate probability as previously explained.

Example 17.3

·Sample ranges for the first 12 samples of 5 pieces, taken at half-hourly intervals, during the production of a turned diameter 48 ± 0.1 mm on an automatic lathe, are given.

Unit of range = 0.1 mm

Sample no.	1	2	3	4	5	6	7	8	9	10	11	12
Range (w)	0.2	0.4	0.2	0.5	0.7	0.2	0.5	0.6	1	0.5	0.2	0.3

Determine the relative precision index and construct a control chart for sample averages.

Solution

$$\bar{w} = \frac{\Sigma w}{12} = \frac{5.3}{12} = 0.44 \qquad \text{RPI} = \frac{T}{\bar{w}} = \frac{2}{0.44} = 4.6$$

Reference to BS 2564 shows this to be high relative precision. Alternatively,

$$\sigma = \frac{\bar{w}}{d_n} = \frac{0.44}{2.33} = 0.19$$

by estimation from $\bar{w}$, since $d_n = 2.33$ when $n = 5$,

Hence $\quad \dfrac{T}{\sigma} = \dfrac{2}{0.19} = 10.5 \; (T > 8\sigma)$

To construct a control chart for $\bar{x}$:

$$a = 3.09 \times 0.19\left(1 - \frac{1}{\sqrt{5}}\right) = 0.32 \qquad \text{by equation 17.11}$$

$$b = \left[3.09 - \frac{1.96}{\sqrt{5}}\right] \times 0.19 = 0.42 \qquad \text{by equation 17.12}$$

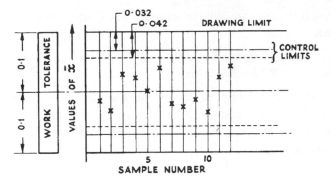

Figure 17.11 Control chart for 'high' relative precision.

The chart is illustrated in Figure 17.11, where a series of possible values of $\bar{x}$ have also been plotted.

□   □   □

17.8 Control chart for attributes

Limit gauging sorts work into two classes – 'good' and 'bad'. A similar classification may arise where practical conditions do not permit precise measurement, e.g. blow holes in castings or bubbles in glass. Provided a criterion is set distinguishing the lowest acceptable quality, it becomes possible to plot the proportion of defectives in batches on a control chart and to use the chart for detecting a decline in the general quality being produced, before a large quantity of scrap has occurred. The method is of course equally applicable where a direct measurement, e.g. linear size or electrical resistance, determines the outer limits of acceptable quality.

The number of defectives found in random samples of n pieces enables the 'fraction defective' to be calculated.

$$\text{Fraction defective } (P) = \frac{\text{number defective per sample}}{n}$$

When a reasonable number of samples have been taken it is possible to obtain the mean value for P.

$$\text{Mean value of fraction defective } (\bar{P}) = \frac{\text{total defective found}}{\Sigma n}$$

Since random factors underlie the variation in quality, the value of P varies from sample to sample and for such conditions it can be shown that P tends to follow a binomial distribution and have a standard deviation given by

$$s = \sqrt{\left(\frac{\bar{P}(1-\bar{P})}{n}\right)}$$

Having found s, it becomes possible to set limits for P which can then be marked on the control chart.

> Outer limits (1 in 1000) are $\bar{P} \pm 3.09s$
> Inner limits (1 in 40) are $\bar{P} \pm 1.96s$

It can be seen that these limits are set on the assumption that the distribution is 'normal' while s has been calculated on the assumption that the distribution is 'binomial'. This is allowable because, for conditions where n is large and $nP > 5$ the two distributions are similar.

Example 17.4

The following data relates to the daily output of a foundry producing small castings. The sample size is 150.

Day	1	2	3	4	5	6	7	8	9	10	11	12	13	14	15
Defectives in sample	4	3	1	4	5	2	3	3	4	3	3	2	3	2	4

Draw up a control chart for the fraction defective. What number of defectives could be expected to occur once in 40 samples?

Solution

Number of defectives	1	2	3	4	5	Sample size
Fraction defective	0.007	0.013	0.020	0.027	0.033	150

$$P = \frac{46}{15 \times 150} = 0.02044$$

$$s = \frac{0.02044(1 - 0.02044)}{150} = 0.01155$$

Upper limits, (1 in 1000)$P + 3s = 0.055$
(1 in 40)$P + 2s = 0.044$
(lower limits not applicable).

Possible defectives per sample occurring once in 40 times

$$= 0.044 \times 150 = 6.6$$

That is, 7 defectives should not occur as frequently as once in 40 samples.

Note: Sometimes a specification gives the allowable 'average' percentage defectives, e.g. 2 per cent. If so, this should be used for calculating the standard deviation to be met, i.e. $\bar{P}$ would take the value 0.02.

□   □   □

17.9 Sampling of incoming goods

When regular deliveries of goods are received in large batches and the supplier is using a reasonable system of quality control, it is wasteful to submit the work to 100 per cent inspection unless there are very special requirements. Sampling methods enable a check on the incoming quality to be made. The principles are similar to those detailed in Section 17.8.

A brief consideration of the statistical basis of sampling will indicate what is involved.

Let p be the probability of a defective, then $1 - p = q$ is the probability of a 'good' part. If a sample of n pieces is taken at random from the batch, the probability of there being 0, 1, 2, 3 or more defective depends upon:

1. The general level of quality being supplied.
2. Successive terms of the binomial expansion of $(q + p)^n$.

Example 17.5

A large batch of components contains 8 per cent defectives. If a sample of 50 pieces is taken, what will be the probabilities of finding 0, 1, 2, 3, 4 and 5 defectives in the sample? What is the probability of there being less than 4 defectives in a sample of 50 pieces?

Solution

$$p = 0.08 \qquad q = 1 - 0.08 = 0.92$$

$$(0.92 + 0.08)^{50} = 0.92^{50} + 50(0.92)^{49}(0.08) + \frac{50 \times 49}{2!}(0.92)^{48}(0.08)^2 + \cdots$$

$$= 0.016 + 0.067 + 0.144 + 0.199 + 0.204 + \ldots$$

It follows from successive terms of the expansion that the probability of there being less than 4 defectives is given by addition.

Defectives	Probability
0	0.016
1	0.067
2	0.144
3	0.199
Less than 4	0.426

Less than 4 defectives should appear in about 43 per cent of samples taken.

□   □   □

Example 17.5 shows that the binomial expansion is not very convenient for calculation. However, when n is large and p small, a Poisson distribution gives a reasonably close approximation to the binomial and is much more convenient to calculate. The Poisson distribution is based on the exponential functions e^x and e^{-x}

$$e^{-x} \times e^x = 1 = e^{-x}\left(1 + x + \frac{x^2}{2!} + \frac{x^3}{3!} + \cdots\right)$$

$$= e^{-x} + xe^{-x} + \frac{x^2 e^{-x}}{2!} + \frac{x^3 e^{-x}}{3!} + \cdots$$

If x is given the value of the expected average number of defectives per sample, the successive terms of the Poisson series will be found to approximate quite closely to those of the binomial series. Considering Example 17.5

$$x = 0.08 \times 50 = 4$$

The probabilities given by this series are;

$$e^{-4} + 4e^{-4} + \frac{4^2 e^{-4}}{2!} + \frac{4^3 e^{-4}}{3!} + \cdots$$

giving $0.018 + 0.073 + 0.147 + 0.195 + 0.195 + \ldots$

By this series the probability of there being less than 4 defectives per sample is seen to be 0.433 (the sum of the first four terms as before).

For sampling purposes there is no serious difference between the result obtained from the Poisson distribution and the binomial distribution.

Operating characteristics of sampling systems

A commercial arrangement between a supplier and a purchaser can be operated as follows. Both agree on a permitted proportion of defectives and presumably the supplier will have some quality control system designed to maintain this quality. The purchaser operates as follows. From each incoming batch a random sample of n parts is taken. If this sample contains less than k defectives, accept the batch; if it contains more than k defectives submit the batch to 100 per cent inspection and return defective items; if more than k defectives appear in successive batches reject and return all the work to the supplier.

To successfully achieve its objective of minimising the purchaser's inspection costs while ensuring a reasonable standard of quality the system must:

1. Give a high chance of acceptance at the agreed proportion of defectives.
2. Give a high chance of rejection if the proportion of defectives exceeds the agreed figure.

The Poisson distribution (p. 562) shows that, for 8 per cent defectives, chances of acceptance at this quality when $n = 2$, are $0.018 + 0.073 + 0.147 = 0.238$ (about 24 per cent).

This is low, if only 24 per cent of batches are likely to satisfy the customer a great deal of work would be returned to the supplier. However, suppose the agreed proportion defective was set at 2 per cent.

The Poisson distribution then becomes: $x = 0.02 \times 50 = 1$ and the first three terms of expansion are $0.368 + 0.368 + 0.184 = 0.92$. For 2 per cent proportion defective, where acceptance depends upon there being no more than 2 defectives in a batch of 50, the probability of acceptance is 92 per cent. Summarising the above considerations:

Sample number $(n) = 50$; maximum permitted defectives $(k) = 2$.

Proportion defective	Probability of acceptance
2% (0.02)	92% (0.92)
8% (0.08)	24% (0.24)

This is described as a $50_{2/3}$ sampling plan where 2 is the acceptance number, and 3 is the rejection number. If these principles are followed it is possible to develop a graph to show the probability of acceptance for differing quality levels for specified values of n and k. Figure 17.12 shows such a graph, commonly called an *operating characteristic*. It follows from the above considerations and from Figure 17.12 that:

1. The producer's risk of having work containing no more than 2 per cent defectives rejected is $1 - 0.92 = 0.08$ (8 per cent).
2. The consumer's risk of accepting work containing no more than 6 per cent defectives is 0.42 (42 per cent).

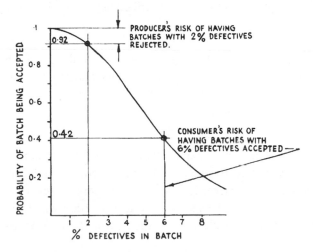

Figure 17.12 Operating characteristic for $50_{2/3}$ sampling plan.

Economics of single sampling

Since it would be normal practice to give 100 per cent inspection to a batch rejected by sampling as on p. 558, and since the reliability of the sampling method increases as the sample size is increased, it is possible to decide upon a sample number (n) such that the total amount of inspection is near the minimum. Clearly, as n is increased, k can be increased for the agreed values of producer and consumer risk. For any sample size the total amount of inspection I depends upon n and the producer's risk P_1.

$$I = n + (N - n)(1 - P_1),$$

where N is the total number in the batch from which sample n is taken.

The matter has been investigated by Dodge and Romig and their *Sampling and Inspection Tables* (Wiley) enable economic sampling sizes to be found for specified conditions.

Alternative sampling methods

Basically the criterion in sampling is to increase the discriminating power of the sampling without greatly increasing the amount of inspection involved. Two developments in this direction are discussed in the next two sections.

Double sampling

Proceed as in single sampling, sampling size n_1, acceptable defectives per sample k_1. Whenever k_1 is exceeded take a further sample of n_2 pieces. Should the cumulative result, sample size $(n_1 + n_2)$, have total defects of less than k_3 $(= k_1 + k_2)$ accept the batch; if k_3 is exceeded reject the batch. (See Dodge and Romig – Double sampling.)

A double sampling plan could thus be described as:

100, 100 1/4

where the first figure is the size of the first sample (S_1), the second figure is the size of the second sample (S_2), the third figure is the acceptance number on S_1, and the fourth figure is the rejection number on S_1 and S_2.

The decision rules for this plan are:

1. Take a first sample of 100 parts:
   (a) If the number of defective parts is 0 or 1 – accept the batch.
   (b) If the number of defectives is 2 or 3 – take second sample.
   (c) If the number of defectives is 4 or more – reject the batch.
2. If (b) above was the case, take a second sample of 100 parts then:
   (a) If the *total number* of defectives on S_1 *and* S_2 is 3 or less – accept the batch.
   (b) If the *total number* of defectives is 4 or more – reject the batch.

The probability of accepting a batch of these parts can be calculated by using the table of cumulative probabilities of the Poisson distribution shown on Table 17.5.

Example 17.6

Calculate the probability of acceptance of a batch of parts that is 2 per cent defective when sampled using the 100, 100, 1/4 double sampling scheme.
There are two possible ways in which a batch of parts can be accepted using this plan:

1. If 0 or 1 defective parts are found in the first sample S_1.
   The probability of this is given by reading $p_{(0 \text{ or } 1)}$ from the chart at $n = 2$ (the expected number of defectives $n = 2$)

   $$p_{(0 \text{ or } 1)} = 0.406, \text{ thus } P_a = 0.406$$

2. If a total of 3 or less defectives are found on S_1 and S_2.
   This is actually two separate cases:
   (a) If 2 defectives in S_1 were found, there must be 0 or 1 in S_2.
       The probability of this calculated thus:

       $$p_2 \text{ on } S_1 = p_{(0,1,2)} - p_{(0,1)} = (0.677 - 0.406) = 0.271$$
       $$p_{0,1} \text{ on } S_2 = 0.406$$

       These are linked events – both must happen.
       Thus probability of acceptance for this case is:

       $$p_{(bi)} = 0.271 \times 0.406, \text{ i.e. } 0.110$$

   (b) If 3 defectives in S_1 were found there must be 0 defectives in S_2.
       The probability of this is calculated thus:

       $$p_3 \text{ on } S_1 = p_{(0,1,2,3)} - p_{(0,1,2)} = (0.857 - 0.677) = 0.180$$
       $$p_0 \text{ on } S_2 = 0.135$$

       Thus $p_{bii} = 0.135 \times 0.180 = 0.0243$

The total probability of acceptance is therefore the sum of these three probabilities.

$$p_{accept} = 0.406 + 0.110 + 0.024 = 0.540$$

The idea behind double sampling is to reduce the amount of inspection work that has to be done. Good quality work will be accepted immediately and low quality work rejected immediately on the first sample. Borderline work will be assessed by the use of the second sample.

The above double sampling plan could be used in place of a $200_{3/4}$ single sampling plan, and would give much the same level of quality control with less inspection. We can examine the extent of this by calculating the average sample number (ASN). This is done by assuming the batches of work are consistently delivered with a particular percentage of defects.

Example 17.7

Calculate the average sample number for the 100,100 1/4 double sampling plan if the parts 2 per cent defective are sampled.

$$ASN = \text{size of } S_1 + (\text{size of } S_2) \times (\text{probability of taking } S_2)$$
$$= 100 + 100 p_{(2 \text{ or } 3)} = 100 + 100 \ (0.857 - 0.406)$$
$$= 145 \text{ parts}$$

This can be calculated for a number of percentages defective and can be plotted against the percentage defective to give a comparison between single and double sampling (Figure 17.13).

Table 17.5 Cumulative Poisson distribution, probability of *r* or less events.

n	0	1	2	3	4	5	6	7	8	9	10	11	12	13	14	15	16	17
0.25	0.779	0.974	0.998	1.0														
0.30	0.741	0.963	0.996	1.0														
0.35	0.705	0.951	0.994	1.0														
0.40	0.670	0.938	0.992	0.999	1.0													
0.45	0.638	0.925	0.989	0.999	1.0													
0.50	0.607	0.910	0.986	0.998	1.0													
0.55	0.577	0.894	0.982	0.998	1.0													
0.60	0.549	0.878	0.977	0.997	1.0													
0.65	0.522	0.861	0.972	0.996	0.999	1.0												
0.70	0.497	0.844	0.966	0.994	0.999	1.0												
0.75	0.472	0.827	0.959	0.993	0.999	1.0												
0.80	0.449	0.809	0.953	0.991	0.999	1.0												
0.85	0.427	0.791	0.945	0.989	0.998	1.0												
0.90	0.407	0.772	0.937	0.987	0.998	1.0												
0.95	0.387	0.754	0.929	0.984	0.997	1.0												
1.0	0.368	0.736	0.920	0.981	0.996	0.999	1.0											
1.1	0.333	0.699	0.900	0.974	0.995	0.999	1.0											
1.2	0.301	0.663	0.879	0.966	0.992	0.998	1.0											
1.3	0.273	0.627	0.857	0.957	0.989	0.998	1.0											
1.4	0.247	0.592	0.833	0.946	0.986	0.997	0.999	1.0										
1.5	0.223	0.558	0.809	0.934	0.981	0.996	0.999	1.0										
1.6	0.202	0.525	0.783	0.921	0.976	0.994	0.999	1.0										
1.7	0.183	0.493	0.757	0.907	0.970	0.992	0.998	1.0										
1.8	0.165	0.463	0.731	0.891	0.964	0.990	0.997	0.999	1.0									
1.9	0.150	0.434	0.704	0.875	0.956	0.987	0.997	0.999	1.0									
2.0	0.135	0.406	0.677	0.857	0.947	0.983	0.995	0.999	1.0									
2.2	0.111	0.355	0.623	0.819	0.928	0.975	0.993	0.998	1.0									
2.4	0.091	0.308	0.570	0.779	0.904	0.964	0.988	0.997	0.999	1.0								
2.6	0.074	0.267	0.518	0.736	0.877	0.951	0.983	0.995	0.999	1.0								

How to use this chart

e.g. sample plan 80 2/3: what is the probability of acceptance for parts that are 2% defective?

Expected no. of defectives

$n = 2\%$ of $80 = 1.6$

Read across from $n = 1.6$ to column 2 – read the probability value: 0.783

(*Continued*)

n	0	1	2	3	4	5	6	7	8	9	10	11	12	13	14	15	16	17
2.8	0.061	0.231	0.469	0.692	0.848	0.935	0.976	0.992	0.998	0.999								
3.0	0.050	0.199	0.423	0.647	0.815	0.916	0.966	0.988	0.996	0.999	1.0							
3.2	0.041	0.171	0.380	0.603	0.781	0.895	0.955	0.983	0.994	0.998	1.0							
3.4	0.033	0.147	0.340	0.558	0.744	0.871	0.942	0.977	0.992	0.997	0.999	1.0						
3.6	0.027	0.126	0.303	0.515	0.706	0.844	0.927	0.969	0.988	0.996	0.999	1.0						
3.8	0.022	0.107	0.269	0.473	0.668	0.816	0.909	0.960	0.984	0.994	0.998	0.999						
4.0	0.018	0.092	0.238	0.433	0.629	0.785	0.889	0.949	0.979	0.992	0.997	0.999	1.0					
4.2	0.015	0.078	0.210	0.395	0.590	0.753	0.867	0.936	0.972	0.989	0.996	0.999	1.0					
4.4	0.012	0.066	0.185	0.359	0.551	0.720	0.844	0.921	0.964	0.985	0.994	0.998	0.999	1.0				
4.6	0.010	0.056	0.163	0.326	0.513	0.686	0.818	0.905	0.955	0.980	0.992	0.997	0.999	1.0				
4.8	0.008	0.048	0.143	0.294	0.476	0.651	0.791	0.887	0.944	0.975	0.990	0.996	0.999	1.0				
5.0	0.007	0.040	0.125	0.265	0.440	0.616	0.762	0.867	0.932	0.968	0.986	0.995	0.998	0.999	1.0			
5.2	0.006	0.034	0.109	0.238	0.406	0.581	0.732	0.845	0.918	0.960	0.982	0.993	0.997	0.999	1.0			
5.4	0.005	0.029	0.095	0.213	0.373	0.546	0.702	0.822	0.903	0.951	0.977	0.990	0.996	0.999	1.0			
5.6	0.004	0.024	0.082	0.191	0.342	0.512	0.670	0.797	0.886	0.941	0.972	0.988	0.995	0.998	0.999			
5.8	0.003	0.021	0.072	0.170	0.313	0.478	0.638	0.771	0.867	0.929	0.965	0.984	0.993	0.997	0.999	1.0		
6.0	0.002	0.017	0.062	0.151	0.285	0.446	0.606	0.744	0.847	0.916	0.957	0.980	0.991	0.996	0.999	1.0		
6.2	0.002	0.015	0.054	0.134	0.259	0.414	0.574	0.716	0.826	0.902	0.949	0.975	0.989	0.995	0.998	0.999	1.0	
6.4	0.002	0.012	0.046	0.119	0.235	0.384	0.542	0.687	0.803	0.886	0.939	0.969	0.986	0.994	0.997	0.999	1.0	1.0
6.6	0.001	0.010	0.040	0.105	0.213	0.355	0.511	0.658	0.780	0.869	0.927	0.963	0.982	0.992	0.997	0.999	1.0	1.0
6.8	0.001	0.009	0.034	0.093	0.192	0.327	0.480	0.628	0.755	0.850	0.915	0.955	0.978	0.990	0.996	0.998	0.999	1.0
7.0	0.001	0.007	0.030	0.082	0.173	0.301	0.450	0.599	0.729	0.830	0.901	0.947	0.973	0.987	0.994	0.998	0.999	0.999
7.2	0.001	0.006	0.025	0.072	0.156	0.276	0.420	0.569	0.703	0.810	0.887	0.937	0.967	0.984	0.993	0.997	0.999	0.999
7.4	0.001	0.005	0.022	0.063	0.140	0.253	0.392	0.539	0.676	0.788	0.871	0.926	0.961	0.980	0.991	0.996	0.998	0.999
7.6	0.001	0.004	0.019	0.055	0.125	0.231	0.365	0.510	0.648	0.765	0.854	0.915	0.954	0.976	0.989	0.995	0.998	0.999
7.8	0.000	0.004	0.016	0.048	0.112	0.210	0.338	0.481	0.620	0.741	0.835	0.902	0.945	0.971	0.986	0.993	0.997	0.999
8.0	0.000	0.003	0.014	0.042	0.100	0.191	0.313	0.453	0.593	0.717	0.816	0.888	0.936	0.966	0.983	0.992	0.996	0.998
8.5	0.000	0.002	0.009	0.030	0.074	0.150	0.256	0.386	0.523	0.653	0.763	0.849	0.909	0.949	0.973	0.986	0.993	0.997
9.0	0.000	0.001	0.006	0.021	0.055	0.116	0.207	0.324	0.456	0.587	0.706	0.803	0.876	0.926	0.959	0.978	0.989	0.995
9.5	0.000	0.001	0.004	0.015	0.040	0.089	0.165	0.269	0.392	0.522	0.645	0.752	0.836	0.898	0.940	0.967	0.982	0.991
10.0	0.000	0.000	0.003	0.010	0.029	0.067	0.130	0.220	0.333	0.458	0.583	0.697	0.792	0.864	0.917	0.951	0.973	0.986

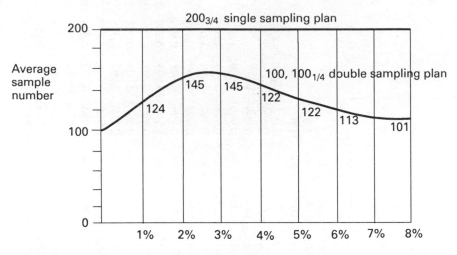

Figure 17.13 Graph of average sample numbers/per cent defective.

Sequential sampling

The procedure is based on a chart such as Figure 17.14, drawn on squared paper. Values of h and slope θ are available from tables or may be calculated as shown by Moroney, *Facts from Figures* (Pelican). As the inspection proceeds the cumulative result is plotted on the squared paper until eventually a point is obtained outside the middle band, above for 'good' work, below for 'bad'. The system is said to give a decision with the minimum amount of inspection but the statistical concepts involved are rather complicated.

While both these systems have attractions, their greater sophistication means that a higher level of supervisory work is involved and single sampling is more satisfactory where those operating the control system have only a modest appreciation of the statistical basis of their work.

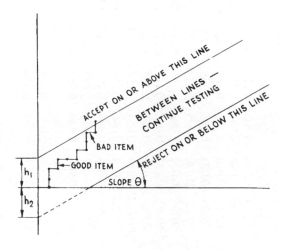

Figure 17.14 Sequential sampling.

17.10 Tolerance 'build-up' in assemblies

Figure 17.15 shows the tolerance build-up which may occur if two bars of length $l_1 \pm \frac{1}{2} t_1$ and $l_2 \pm \frac{1}{2} t_2$ are placed end to end, where t_1 and t_2 represent the tolerances. The overall length is seen to be $(l_1 + l_2) \pm \frac{1}{2}(t_1 + t_2)$, the tolerance on the assembly being the sum of the tolerances of the separate parts.

Suppose, however, that the two bars are taken at random from two large groups for which the distribution curve is somewhere near *normal*. The chance of two bars, one at the maximum-metal condition and one at the minimum-metal condition, being selected for the assembly is very remote, more so than the chance of selecting one bar at the extreme tolerance condition. The frequency curves sketched in relation to the tolerance bands, Figure 17.15, illustrate this fact. Notice the long 'tails' of the assembly distribution curve in relation to the algebraic tolerance of $\pm\frac{1}{2}(t_1 + t_2)$.

The important thing about the relationships illustrated, is the magnitude of σ_a, and its relationship to the magnitudes of σ_1, and σ_2. It can be shown mathematically that provided a random selection of the parts is made from bulks having normal distributions of standard deviations σ_1 and σ_2, the assemblies will also follow a normal distribution of standard deviation σ_a where

$$\sigma_a = \sqrt{\sigma_1^2 + \sigma_2^2}$$

The principle may be extended to any number of separate items of an assembly such that:

$$\sigma_a = \sqrt{(\sigma_1^2 + \sigma_2^2 + \sigma_3^2 \ldots \sigma_n^2)} \tag{17.13}$$

Since, for the production characteristic curve discussed, the tolerances are approximately $\pm 3\sigma$, a similar rule applies to the tolerances, viz:

$$t_a = \sqrt{(t_1^2 + t_2^2 + t_3^2 \ldots t_n^2)} \tag{17.14}$$

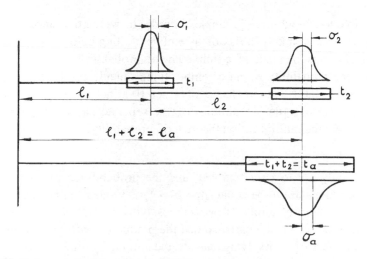

Figure 17.15 Tolerance 'build-up' in assemblies.

The value of t_a, obtained by the statistical method of associating tolerance, is seen to be lower than the value obtained by algebraic association. An important corollary for quantity production is that the assembly tolerance t_a could still be met if the component tolerances t_1 and t_2 based upon algebraic association were somewhat increased.

Example 17.8

Five components of equal tolerance are associated in an assembly, to build up a length for which the maximum permissible variation is 0.42 mm. What increase in component tolerance is possible when statistical concepts of tolerance build-up are used to replace the normal algebraic considerations?

Solution

Let t = tolerance on components.
For tolerance build-up considered algebraically:

$$t_1 = \frac{0.42}{5} = 0.084 \text{ mm}$$

For tolerance build-up considered statistically:

$$t_a = \sqrt{5 t_2^2} \qquad \therefore \quad t_2 = \frac{0.42}{\sqrt{5}} = 0.188 \text{ mm}$$

$$\text{Permissible increase of tolerance} = 0.188 - 0.084$$
$$= 0.104 \text{ mm}$$

The increase is about 120 per cent. The very big increase is due to the large number of items in the assembly.

$\square$ $\square$ $\square$

It is possible to extend these principles in order to widen tolerances on individual components, provided a known risk of assemblies falling below a specified standard is accepted. Use must be made of a simple law of probability.

Let p_a be the probability of an assembly lying outside certain limits. A suitable probability is $p_a = 0.001$, i.e., a chance of 1 per 1000 not meeting the specification. Let p_1, p_2, p_3, etc., be the probabilities of the separate parts lying outside certain limits. Then from the multiplication theorem of probability,

$$p_a = p_1 \times p_2 \times p_3, \text{ etc.}$$

If two parts only are to be assembled, and the probabilities of these parts lying outside certain limits are to be equal, $p_1 = p_2 = \sqrt{p_a}$. Values can be found for p_1 and p_2, in terms of the probability of assemblies lying outside a specified condition. Limits can then be set for the parts, so that the production will have probabilities of p_1 and p_2 amounts of work lying outside the limits, by making use of the area property of the normal curve.

Example 17.9

A hole-and-shaft assembly is required to provide a close running fit, nominal diameter 32 mm, minimum clearance 0.015 mm, maximum clearance 0.05 mm. The hole tolerance is to be 1.5 times the shaft tolerance. A risk of 1 in 1000 is to be taken of assemblies lying outside the specified clearances, and it may be assumed that the parts, produced in quantity, follow a normal curve of distribution. The assemblies are built by random selection of the mating parts. Find suitable hole and shaft limits and compare them with limits set to give 100 per cent perfect assembles. (Use a unilateral hole-base system.)

Solution

$$p_a = 0.001 \therefore p_h = p_s = \sqrt{0.001} = 0.0316$$

The hole and shaft tolerances must be set so that approximately 3 per cent of the parts lie outside the limits required for 100 per cent perfect assemblies.

For 100 per cent perfect assemblies:

$$t_h + t_s = 0.05 - 0.015 = 0.035$$
$$t_h = 1.5t_s \quad \therefore \quad 2.5t_s = 0.035 \quad t_s = 0.014 \text{ mm}$$
$$t_h = 0.035 - 0.014 = 0.021 \text{ mm}$$

Reference to Table 22 of BS 600 shows, that for a risk of 3 per cent of the work outside the ideal limits (97 per cent 'good' work), the required limits are $\pm 2.2\sigma$. It is usual to assume that $\pm 3.09\sigma$ will just contain all the work produced, hence the component tolerances representing assemblies having the risk of defectives as stated will be those for 100 per cent good assemblies increased in the ratio of 3.09/2.2.

The required tolerances are:

$$t_h = 0.021 \times \frac{3.09}{2.2} = 0.03$$

$$t_s = 0.014 \times \frac{3.09}{2.2} = 0.02$$

The limits of size must now be set.

For 100 per cent good assemblies these are:

Hole   32.021/32.000 mm
Shaft   31.985/31.971 mm

Limits for 99.9 per cent good assemblies (1 per 1000 defective) must be set so that the same mean-fit is achieved.

Hole   32.030/32.000 mm

The mean-fit required is 0.0325 clearance.
The mean-size of the shaft is $32.015 - 0.0325 = 31.9825$ (say 31.982)

Shaft   31.992/31.972 mm

The 'fits' are illustrated to scale in Figure 17.16.

□   □   □

The clearances possible for the second set of tolerances have extreme values (on an algebraic basis) of 0.008 mm minimum and 0.058 mm maximum. However, it is

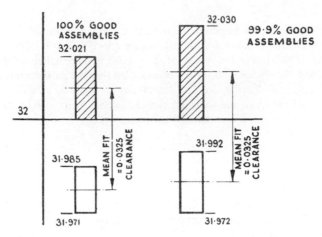

Figure 17.16 Diagram of fits showing increase of work tolerances for 1 per 1000 assemblies outside the desired condition.

not likely that the specified assembly condition will be exceeded more frequently than 1 in 1000 assemblies. For this small risk, a 40 per cent increase of component tolerance has been gained ('rounding' the figures has made this appear to be somewhat larger).

It is possible to test the above solution by applying equation 17.14.

If the sizes are controlled by the quality control chart technique a watch can be kept on the distribution pattern, which must, of course, be something approaching 'normal' for the consideration made above to apply.

It is possible to test the above solution by applying equation 17.14.

$$t_a = \sqrt{(0.03^2 + 0.02^2)} = 0.036 \text{ mm}$$

The limits set for the hole and shaft could be expected to give an assembly tolerance of 0.036 mm which would be exceeded not more than 1 in 1000 times. The result compares favourably with the required tolerance on the assembly of 0.035 mm.

The production engineer is always faced with the problem of controlling the quality of the goods produced, although the time spent in doing this does not contribute directly to the output. Skilled inspection is slow and expensive; semi-skilled viewing by means of limit gauges tends to fall in effectiveness as quantities are increased, due to the monotonous nature of the task and the unavoidable wear of equipment. Process control, which neither wastes skill nor fails in its objective as quantities are increased, is best effected by either the employment of automatic gauging equipment or by quality control based on statistical methods. The latter method has the advantage that a large expenditure on special equipment is avoided.

Exercises – 17

1. Plot a normal distribution curve and use it to estimate the percentage of the total area under the curve lying between the following limits:

   $\pm 0.8\sigma$, $\pm 1.28\sigma$, $\pm 1.64\sigma$, $\pm 1.96\sigma$.

2. From the information given in Figure 17.7 determine, for samples of 5 pieces, the values of $A_{0.001}$, $A'_{0.001}$, $A_{0.025}$ and $A'_{0.025}$.

   On a particular control chart it is required to draw control limits for average which are likely to be exceeded 1 in every 20 times. If samples of 4 are taken, find the constants required to set these limits: (a) assuming $\bar{w}$ is known; (b) assuming σ is known. (i.e. find $A'_{0.05}$ and $A_{0.05}$.)

3. After a machining operation on the diameter of a component specified as 47.500 ± 0.025 mm, a sample of 300 components was inspected the dimension being measured to the nearest 0.001 mm. The readings have been grouped into discrete classes having equal intervals and the frequencies of occurrence are tabulated below:

Diameter of components (mm)	Frequency
47.480 – 84	8
85 – 89	21
90 – 94	38
95 – 99	54
47.500 – 04	66
05 – 09	52
10 – 14	34
15 – 19	20
20 – 24	7

Calculate (a) the arithmetic mean, (b) the standard deviation, of the diameter for the batch and draw a histogram of the distribution. (c) What degree of relative precision is indicated?

4. During the grinding of a large batch of components three modifications were made to the operating conditions with the object of improving the surface finish produced.

   Measurements were made on components processed by the original method and after each modification.

| Surface finish CLA | Number of components | | | |
	Method 1	Method 2	Method 3	Method 4
1μ	90	70	116	110
2μ	124	164	190	240
3μ	178	94	64	92
4μ	240	100	130	100

Determine if the changes in method produced any improvement in surface finish.

5. (a) To what kind of manufacturing process is statistical quality control specially suited? Describe the steps which should be taken in applying statistical quality control to a typical manufacturing process.
   (b) Write down the formula for each standard deviation of a group of parts when:
      (i) the whole of the parts are measured;
      (ii) a sample batch is inspected.
   (c) Make a drawing of a normal distribution curve and show the percentage of parts included in variations from the average of $\pm\sigma$, $\pm 2\sigma$, and $\pm 3\sigma$.

6. A certain dimension of a component produced in quantity on an automatic lathe is specified as 84.60 ± 0.05 mm. A 5 per cent inspection check resulted in the following variation of the dimensions measured to the nearest 0.01 mm.

Dimension (mm)	84.56	84.57	84.58	84.59	84.60	84.61	84.62	84.63	84.64
Frequency	1	8	54	123	248	115	44	6	1

Calculate the mean value and standard deviation for the sample check and draw the frequency curve. Show that for practical purposes the variation of the dimension for the whole batch would be expected to fall within the prescribed limits.

7. (a) What is the significance of the RPI when considering the suitability of a process to produce to a particular specification?
(b) Give the relationship between σ and the process tolerance for low and high relative precision.
(c) When producing under low RPI conditions, show how the percentage of scrap may be estimated.
(d) Why can the control limits for average be widened when the RPI is known to be high?

8. For a controlled operation on a lathe, a particular dimension of a part is specified as 55 ± 0.25 mm. Samples of five components were each measured to the nearest 0.001 in at 10 equal time intervals and the following readings obtained:

Sample No.	1	2	3	4	5	6	7	8	9	10
Dimension (in)	55.21	55.13	55.11	54.93	54.88	55.22	55.11	55.02	55.12	55.02
	54.99	55.03	55.01	55.12	55.14	55.15	55.07	55.06	55.03	54.97
	55.10	54.98	54.99	55.04	55.05	54.97	54.99	54.97	54.95	55.06
	55.02	54.96	54.97	54.98	55.11	54.95	54.93	54.99	54.98	54.99
	54.95	55.04	54.98	54.91	54.97	55.02	55.00	54.01	54.93	55.01

Calculate: (a) the relative precision index (RPI); (b) the control limits on means and ranges, given that the RPI corresponds to medium relative precision.
From the given data and calculated limits construct control charts for means and ranges of operation. What do you deduce from your charts?
For samples of 5; $A'_{0.025} = 0.38$ $D'_{0.975} = 1.81$
$A'_{0.001} = 0.59$ $D'_{0.999} = 2.21$

9. The table gives the number of defectives found in 30 consecutive samples.

10	7	9	12	8	11	10	17	13	6
8	11	8	10	9	6	12	14	7	3
9	11	4	15	11	17	5	9	11	8

(a) Using information from the first 20 samples, draw a control chart for fraction defective.
(b) Plot the remainder of the results and comment upon the quality of work which they indicate.

10. A sampling scheme is operated from the following instructions:
    'From incoming batches take samples of 50 and inspect. If the sample contains no more than 3 defectives accept the batch; if it contains more than 3 defectives reject the batch.'
    Using the Poisson distribution, plot the operating characteristic for up to 10 per cent defectives in a batch. State the producer's risk of having batches containing 2 per cent defectives rejected. and the consumer risk of having batches containing 8 per cent defectives accepted.

11. A sampling inspection plan uses a single sample of size 200 and an acceptance number of 5. It is used for lots which are large in relation to the sample size. Using the Poisson distribution, determine the approximate probabilities of acceptance for lots which are 1, 2, 3 and 5 per cent defective.

12. An acceptance sampling plan uses a single sample of 150 items and an acceptance number of 3. Using the Poisson distribution to determine approximate probabilities of acceptance, construct an operating-characteristic curve.

13. A single-sampling acceptance inspection plan is required for a purchased product. The probability of acceptance must be 0.95 or more if the per cent defective is 0.5 per cent or less. Find three different combinations of sample size and acceptance number to meet this condition. The sample size must not be greater than 400 items. The Poisson distribution may be used as an approximation.

14. The following is a double-sampling procedure: (a) draw a first sample of 200 items, (b) if 1 or less defectives are found, accept the lot at once, (c) if 4 or more are found, reject the lot at once, (d) otherwise draw a second sample of 200 items, (e) if the total number of defectives found in the combined samples is 4 or less, accept the lot, and (f) if the total number is 5 or more reject it.
    (a) What will be the probability of acceptance for a lot that is 1 per cent defective.
    (b) What will be the average or expected amount of inspection if this plan is applied to a number of 1 per cent defective lots?
    (The Poisson distribution may be used).

15. A single-sample acceptance plan is required to provide the following operating characteristics; (a) if the lot per cent defective is 0.5 per cent or less, the probability of acceptance should be 0.95 or more, and (b) if the lot per cent defective is 3 per cent or greater, the probability of acceptance should be 0.10 or less. Using the Poisson distribution as an approximation, determine suitable values for the sample size and acceptance number.

16. Consider the double sampling plan $N_1 = 50$, $N_2 = 100$, $C_1 = 2$, $C_2 = 5$.
    (a) Determine the probability of acceptance for a lot 6 per cent defective on the first and second samples, hence find the probability of acceptance total for this lot. Also find probability of rejection for the first and second samples.
    (b) Determine the probability of rejection on first and second samples for lots 3 per cent and 8 per cent defective, then comment on the principle of double sampling.

17. The required quality of a mass produced assembly is obtained by using the following limits because the permitted extremes of fit are just acceptable: bore 25.020/25.000 mm diameter shaft 24.992/24.980 mm diameter.

During manufacture it is found that excessive scrap is produced because of the very close limits required, and in order to widen these, it is decided to accept a chance of 1 assembly per 1000 falling below the desired standard.

Given that the range for 99.8 per cent good work is $\pm 3.09\sigma$ and for 97 per cent good work is $\pm 2.2\sigma$, estimate suitable new limits on a statistical basis, and illustrate both sets of conditions in a conventional diagram.

Why is it desirable to use quality control methods when the new limits are introduced?

18 Part Handling and Location

18.1 Introduction

Whether the requirement is the machining of a part or the assembly of the part into product, the engineer is faced with a three-fold problem.

1. The presentation of the part separate from other parts and correctly oriented for the subsequent operation.
2. The picking up of the part and transportation of the part to the subsequent operation.
3. The location and holding of the part.

In a production environment these requirements must be satisfied with minimum loss of operational time with maximum safety to operator and machine and with consistent quality, all within tight cost constraints.

18.2 Presentation

Whether the parts are handled by manual means or by mechanical devices, there is an initial requirement that the parts should be positioned within reach of the operator or device in such a manner that they can be taken up individually. Each part will present its own problems, and the solution to those problems may involve redesign of the part as well as provision of suitable feeding devices.

In its simplest form the problem may only involve care in positioning the parts relative to the equipment, so that the operator has the parts within easy reach when required. The principles of ergonomic workplace layout are largely common sense, the object being to minimise non-productive movement. Thus in the manual loading/ unloading of a machine the containers for raw and finished work are placed on the same side of the operator to cut out the need for the operator to turn from one to the other, and are placed within the operator's easy reach.

If the parts are jumbled in the container, time will be lost in locating and removing the part, particularly if the parts are such that they can become tangled together, and a significant saving of time will be made if the parts are symmetrically set out. Even more time will be saved if the parts are arranged to arrive at the same position every time, so that the operator no longer has to seek the part. Important for manual operations, this becomes essential for mechanical operations.

Large parts will be presented individually, or on trays. Small parts supplied in bulk will require some form of feeding device. A wide range of part feeding devices are available, the selection of the device depending on the nature of the part to be handled. A representative selection of these will be considered.

Bowl feeder

Possibly the most ubiquitous feeding device for small parts is the bowl feeder (Figure 18.1). A bowl with a domed base is mounted on leaf springs radially disposed about its circumference. Beneath the bowl is an electromagnet fed with an intermittent or oscillating current which causes the bowl to vibrate radially. In the interior of the bowl is a helical track rising up the side of the bowl in a spiral.

A mass of components placed in the bowl are shaken apart and ride down the domed base to the bottom of this track, and the rotational vibration causes them to climb the track. Since the track is only of sufficient width to take a single component, a single row of components emerges from the mass and passes on to a feed track.

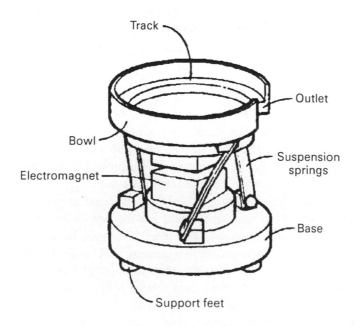

Figure 18.1 Bowl feeder.

In a bowl feeder, orientation can take place within the bowl or outside, whichever is the most suitable. The method whereby incorrect parts are rejected back into the bowl is known as 'positive orientation'. Where the bowl feeder is used only as a feeding device, and orientation takes place outside the bowl, the term used is 'active orientation'.

In bowl orientation

This is very common and various methods are used.

1. Wiper blade which will reject parts standing on end.
2. Pressure break to allow only one part to pass at a time – rejecting the others back into the bowl.
3. Tilted track for chamfered parts – this only accepts parts with the correct orientation and rejects others.
4. Slotted track to accommodate headed parts such as screws.
5. Cut-outs which reject cup-shaped parts which are the wrong way up.
6. Rails to carry 'U'-shaped parts.
7. Steps in the track to cater for parts which are heavier one end than the other.

Figure 18.2 shows a typical set up feeding cap screws.

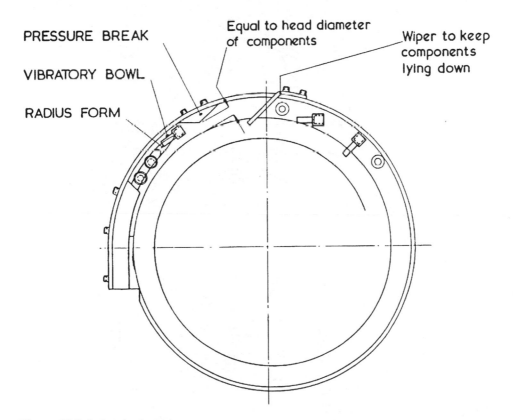

Figure 18.2 In bowl orientation.

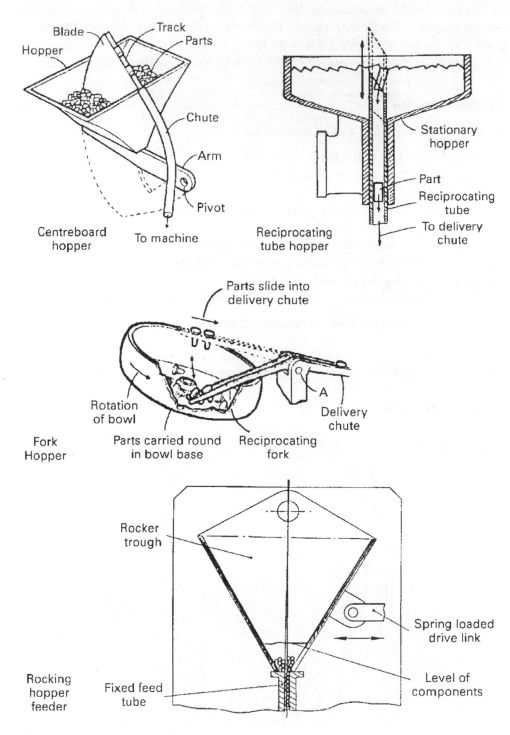

Figure 18.3 Hopper feed devices.

Out of bowl orientation often uses the shape of the part to orientate them which can involve the sensing of different shape or weight.

Hopper feeds

Other types of feed mechanism based on hoppers are also commonly found, and a few of the most common are shown (Figure 18.3). Again, subsequent orientation may be necessary.

The alternative

The above assumes that the parts are received in random orientation. Before setting out to design devices to handle these, it is useful for the manufacturing engineer to ask whether it is necessary to receive the parts in this way. In most cases the primary process producing the part will have discharged the part individually in a spaced and oriented manner, and it is only subsequently that the parts have been 'jumbled up'. Co-operation between supplier and user can ensure that the parts are packed and transported in a form which eliminates or minimises the need for complex feeding mechanisms. This may have an additional advantage in reducing the bulk of the shipments, since, generally, parts oriented and stacked take up less space than random parts.

Parts received in such packaging can often be dispensed by simple slide mechanisms where a slide carries the component from a stack while retaining the next component, and the return of the slide allows the next component to fall into position, to be carried forward on the next slide movement (Figure 18.4).

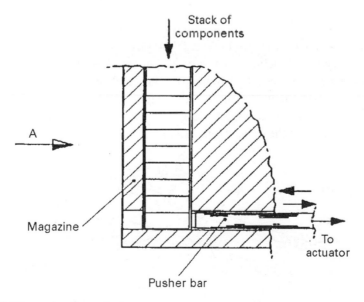

Figure 18.4 Dispensing from stack.

18.3 Reorientation of parts

It may be convenient on occasions to feed a component with a suitable orientation for pick-up, but to reorientate prior to placement.

18.4 Transfer

Conveying

The transfer of a component from feeding device must be carried out in a controlled manner, maintaining the orientation imparted by the feed device. Commonly used transfer techniques are conveyors, chutes, or mechanical pick-and-place devices (Figure 18.5). Robots are a complex version of the latter (Figure 18.6).

The basic requirement of a transfer device is to take the component from the feeding device, and place it, correctly oriented, in its required location.

Separation

Most feeding devices produce a continuous flow of oriented parts. Usage of parts, however, is almost invariably one at a time, so it is necessary to provide some form of escapement to control the flow of the parts. Such escapements will be designed to suit the parts and the method of placement of the part. A few of these are illustrated (Figure 18.7).

Placement

The last item of the transfer mechanism is the placement device itself. This will be synchronised with the machinery of the subsequent operation, and can take many forms, from a simple push rod pushing a part into a chuck or fixture to the autofeed screwdriver head shown in the illustration (Figure 18.8), or a secondary transfer device of the pick-and-place variety (Figure 18.9). On occasion the jig or fixture itself may form part of the escapement and placement mechanism.

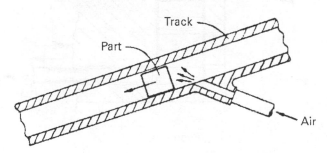

Figure 18.5 Air powered track.

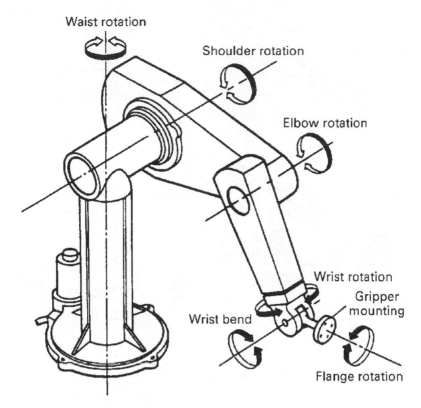

Waist rotation

Shoulder rotation

Elbow rotation

Wrist rotation

Gripper
mounting

Wrist bend

Flange rotation

Figure 18.6 PUMA robot.

18.5 Location

Basically, a jig locates and a fixture holds, but this is of only academic interest, since in practice the terms are used almost synonymously. There is a popular conception that the need for jigs vanished with the coming of computerised machines: this is far from the truth, and a knowledge of the principles of fixturing is as important now as ever.

Before the manufacturer can effectively produce a product in quantity, it is necessary to understand a few basic principles of design and location.

Firstly, the jig/fixture must locate the part accurately. Then it must hold it firmly. At the same time, however, the part held must be capable of rapid loading and release, since expensive machines and equipment are not earning money except when they are working. There must therefore be adequate clearance between jig and component to prevent jamming, especially where the component is subject to variations within wide tolerances. Clearances must also allow for dirt or swarf removal, and for the flow of coolant or lubricant where this is required by the operation. If a fixture, it must be firmly mounted, or if a jig, must sit squarely and positively without rocking.

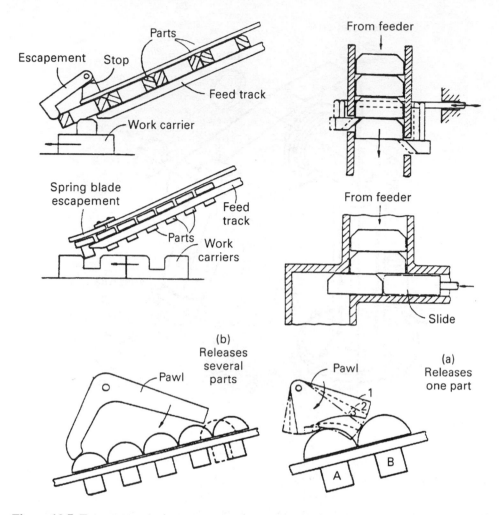

Figure 18.7 Escapement devices.

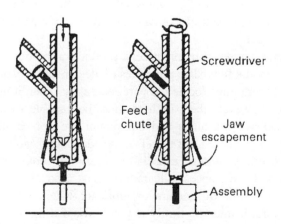

Figure 18.8 Autofeed screwdriver.

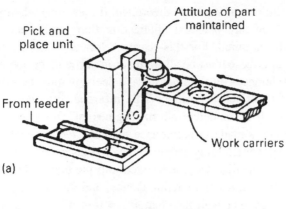

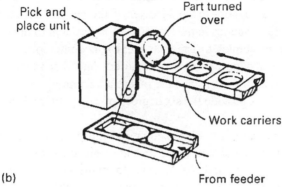

Figure 18.9 Pick and place transfer.

The principles of location

In Chapter 4 the fundamental aspects of movement were considered, with respect to the kinematics of machine tools. These same fundamentals apply to location, which is the prevention of movement, and it is appropriate to repeat these here. A body freely suspended in space will have three possibilities of linear movement:

1. Movement along the X axis.
2. Movement along the Y axis.
3. Movement along the Z axis.

It will also have three possibilities of rotational movement

1. Rotation about the X axis.
2. Rotation about the Y axis.
3. Rotation about the Z axis.

All movements of this body can be achieved by the combination of these basic movements. It therefore follows that to fully inhibit the movement of the body it is necessary to inhibit each of these basic movements.

All automated equipment works to theoretical dimensions when positioning work or tools. However, in any practical application there will be dimensional error, departures from the nominal. For this reason tolerances, i.e. limits of acceptable deviation, are set, and these must be allowed for in subsequent operations.

Since there are tolerances on all dimensions, a decision must be made as to the point of reference used to establish position. This reference, if the item is to be truly located, must be such as to define all six degrees of freedom outlined above. In practice this selection is made in accordance with the requirements of the operation being performed, e.g. for a turning operation the obvious parameters are the centre line of the workpiece and the distance it is inserted into the chuck. The former is implied by gripping the exterior in a self-centering device, and the latter by a stop acting on a datum face. These, together with the clamping action of the chuck, provide the six degrees of restraint. Variations in the diameter of the bar will not affect the position of the centre line, while variations in the length are not pertinent to the operation.

Extending these principles to irregular parts it becomes obvious that in order to locate such a part six datum points are required, and that more than six points of location will lead to ambiguity, since the inherent variation in the part will cause it to locate whichever points provide the six degrees of restraint, and these may not be the same six points each time.

The practical application of this is shown in Figure 18.10. Here the component is placed on three points of contact. These provide restraint in the Z axis, and rotational restraint about the X axis and the Y axis. In plain language, the part sits down without rocking. However, so far it is not restrained in the X axis or the Y axis, and can rotate about the Z axis. The addition of two further contact points 4 and 5 provide restraint along the first two of these, but it is not until a contact point 6 is provided that the part is totally restrained.

It should be noted that this locates the component only with respect to chosen points on its perimeter. In practice this may not be what is required. In an assembly operation, for example, the requirement may be to locate the part with respect to some feature. The most common application of this is relative to an existing group of holes, and the application above must be modified to take account of this. Figure 18.11 shows such an application, where the part is to be located with reference to a group of four holes.

Consideration of the principles indicates that location cannot be with respect to all four holes, as this would provide conflicting restraints due to tolerances. One hole is therefore chosen as the primary location, with a second to provide the necessary further restraint. A fitting dowel in the chosen datum hole provides the linear restraint in the X and Y axes. Rotation about the Z axis is prevented by a peg in the second hole. Notice that this peg is not round. If a round peg were used, a second set of X and Y constraints would be introduced, and tolerances on the hole centres would prevent location. This is overcome by providing a peg which is flatted at right angles to the line between hole centres so that it provides only rotational constraint. The part is placed on three pads as previously.

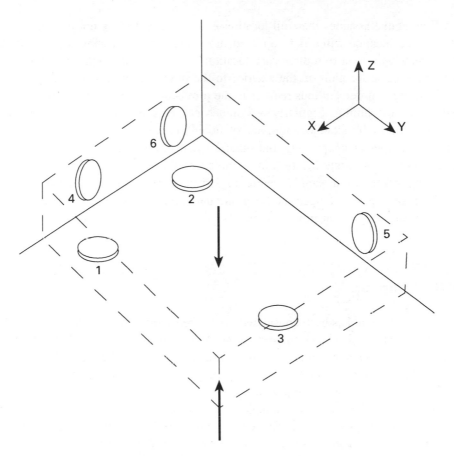

Figure 18.10 Degrees of restraint.

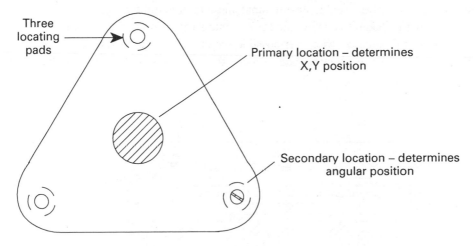

Figure 18.11 Location on holes.

All the above assumes that full location is necessary. This is not always so, and often partial location will satisfy operational requirements. A casing with a complex profile may only require rudimentary location for the initial machining operations, since the precise location of the exterior to the machined surfaces is unimportant. The fixturing requirement thus reduces to the provision of a nest to take the part and hold it for machining. Similarly a symmetrical part in an assembly may have requirement only for axial location, the rotational location being unimportant.

The development of probing and optical location devices has cut down on the requirement for accurate location of complex parts, as it is now possible for the machining centre or assembly centre to locate a randomly-positioned part and modify its program to compensate for positional variation. Such ability is costly both in equipment cost and in process time, and should be used only where the cost is justified.

18.6 Clamping

So far only restraint has been considered. It will be obvious that for accurate location some force must hold the component in position against the restraints provided. In some cases gravity may suffice, in other cases the operation being performed may provide these forces, but in many cases, particularly in machining operations, some form of applied clamping is required.

The design of the clamping is a major factor in achieving a successful production jig or fixture (Figure 18.12). The clamp must hold the part firmly against its locations, but at the same time must not interfere with the easy removal of the part. For this reason the majority of clamps are designed to swivel out of the way, and the motion of clamping is split into two elements: the bringing of the clamp into position, and the actual tightening of the clamp. Cam mechanisms or toggle mechanisms are often used to ensure rapid and effective tightening.

Care must be taken to ensure that the clamping forces do not distort the part being clamped. For this reason wherever possible the clamping action should be positioned directly opposite the locating pad. Where this is not possible, and a danger of

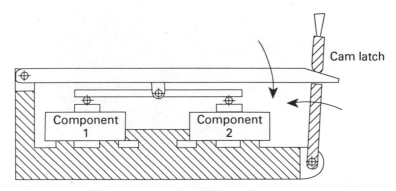

Figure 18.12 Clamping of two components with cam latch.

distortion exists, auxiliary support may have to be provided within the fixture, but care should be taken to ensure that this does not interfere with the location of the part.

Where multiple clamps are required on a fixture, or where rapid clamping or controlled pressure are required, pneumatic or hydraulic operation is often utilised. This may be combined with pneumatic or hydraulic ejection and some form of mechanised component feeding to give automatic or remotely controlled loading or unloading of parts.

Quite often in industrial applications it is necessary to locate and clamp a number of components at the same time, and it is convenient to arrange for a single clamping motion to hold them all. In this case care must be taken to incorporate some device which will distribute the clamping equally over the parts, and which will allow for minor variations from part to part. This may be done by a system of rocking beams, or by utilising individual hydraulic or pneumatic clamping.

18.7 Foolproofing

This entails designing the jig or fixture in such a manner that it is impossible for an operator to insert the component in any way other than the correct way, or to insert a defective or wrong component.

This is normally achieved by having a pin or something similar placed in a position where it will clear the correct and correctly-oriented component but will foul a defective or wrongly-positioned component. Some thought in product design may be required to make foolproofing effective.

18.8 Service features

For the purposes of orienting, holding or foolproofing it is sometimes convenient to provide holes, bosses, or milled surfaces which serve no purpose on the finished part, but which make the manufacturing engineer's job much easier. This is particularly true in the area of orientation for assembly, where the feature, particularly if it is an internal feature, which determines the required orientation of

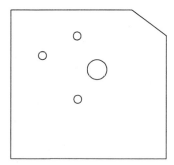

Figure 18.13 Corner removed to simplify orientation.

the part may be difficult to locate. In such cases the provision of an external feature, produced at the same time as the critical feature, simplifies the orientation, sometimes even making the apparently impossible possible. This is illustrated in Figure 18.13, where a corner has been cut off an otherwise symmetrical exterior to simplify orientation of hole group.

19 Assembly Technology

19.1 Introduction

A product is seldom a single part, but is more usually an assembly of a number of parts. Traditionally, assembly has been the least mechanised part of manufacturing, but with improving living standards and increasing disposable income there has been an increasing necessity to produce goods in larger quantities with improved quality at lower cost. The rising cost of manual labour has accentuated the demand for increased assembly efficiency.

The demand for efficient assembly is not new, however. Henry Ford is generally credited with the concept of the assembly line, but he was only putting into operation concepts developed in other areas. Following the ideas set out by Taylor and Gilbreth, the work of assembly was broken down into a series of short operations carried out by a number of operators, the work passing from one to another until the total assembly was complete. Each operator was provided with the tools required to allow him to carry out his particular operation, and all parts for that operation were placed conveniently to hand.

Mechanisation of the assembly process evolved gradually, largely in the form of the provision of aids to what was seen as essentially a manual process (Figure 19.1). Powered transfer of the product from station to station by means of conveyor belts became the norm, each operator taking a unit from the belt, carrying out his operation, and replacing it on the belt for transfer to the next operator. Powered devices were provided for the operator to facilitate his work. Fixtures were engineered to make the operators work easier by holding and locating the parts. Sometimes these were unique to the workplace, sometimes they passed from operator to operator locating the product until it was fully assembled.

From this mechanisation, however, emerged the basis of the automated assembly as it exists today. Requirement for greater accuracy and speed resulted in the development of mechanised stations using pick-and-place devices originally developed for machining automation, and the robot.

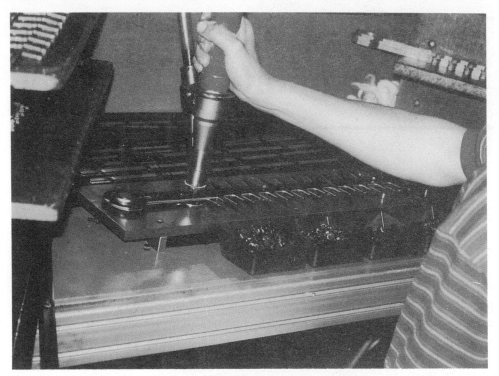

Figure 19.1 Assembly of hedge trimmer blades. Note use of autofeed screwdriver.

19.2 Automatic assembly

When automating an assembly operation there are a number of things which must be considered. The first step is to orientate the parts so that a piece of machinery can be used to access them. This is followed by some means of picking up the part, transporting it to fixture, locating it accurately relative to other parts, and then perhaps fixing it in position. This is often a simple human task which can be performed with very little training and tooling, often requiring only a fixture and manual dexterity. Transfer the same operation to an automatic machine process and it becomes difficult (Figure 19.2). Also, the human operator can carry out selective assembly whereas a machine system cannot, and this leads to the necessity of increased component accuracy, which is an expensive requirement. The maintenance of consistent high-quality has been one of the major impediments, if not the greatest, to the development of automatic assembly processes.

To give some idea of the magnitude of the assembly operations undertaken in the manufacturing industry, it is considered that some 40 per cent of the labour force in the engineering industry is engaged in assembly work of some kind and that assembly costs often account for more than 50 per cent of total manufacturing costs. In other industries, such as electronics, it is known that the cost of assembly is higher.

Figure 19.2 Automatic assembly of hedge trimmer handle.

19.3 Factors to consider for automation

The following factors must be taken into account when automation is considered.

1. Assembly costs.
2. Production rate.
3. Labour availability.
4. Product market life.
5. Cost of automation.

With regard to product market life, it is worth noting that the total costs of the automation undertaken must be paid for in the life of the product. When the product is no longer required there is no further need for the automatic assembly machine.

As far as labour availability is concerned, the lack of suitable labour may lead to mechanisation or automation, even if the economics do not.

Advantages of automatic assembly

1. Reduction of assembly costs – or no point in doing it.
2. Increased productivity – reduction of labour force.
3. Increased consistency – unsuitable parts rejected, fewer faults produced.
4. Removal of operative(s) from hazardous area(s).

Disadvantages of automatic assembly

1. Machine breakdown due to faulty parts – an operator can easily reject faulty parts or overcome small errors.
2. The cost.

An automatic assembly system consists of:

1. A transfer system for moving components, or assemblies, from station to station.
2. Automatic workheads which perform simple tasks.
3. Vacant work stations where inspection can take place and where operators can sometimes carry out difficult tasks.
4. Feeding devices for the workheads in the form of hoppers, magazines, vibratory feeders, etc.

These feeding devices often give part orientation and the controlled feeding of the parts along tracks with some type of escapement control.

19.4 Feasibility study

The first step in an automatic assembly project is a feasibility study which should indicate the practicality of proceeding, the machine performance and the economics of the scheme. As this is often a unique study there are few practical guidelines, and the degree of error can be high, even with experience.

As much information as possible must be readily available when undertaking this work. It is often suggested that automatic assembly is only viable where there is a steady, high volume demand, and where labour costs are high. A study of the various ways in which the assembly being studied can be carried out must be done and all the alternatives considered. The use of sub-assemblies should be taken into consideration.

A network analysis can then be carried out to determine the best theoretical assembly sequence. However, the nature of the product may often dictate the order in which parts must be added to the assembly and a network analysis is therefore unnecessary.

19.5 Quality

Where it is intended to automate an existing system, which is often the requirement, it is relatively easy to determine quality levels of the parts from data at hand. For a new product the only guide is existing knowledge of similar parts. A lengthy study of an existing system would give some indication of what faulty parts could be expected, and in what quantity, and these can be divided into two separate classes.

1. Parts which cannot be assembled – this will give an indication of the downtime of a machine.
2. Parts which can be assembled but which would normally be rejected – this will give an indication of the number of faulty assemblies which would be produced.

19.6 Feeding and assembly

A study should be made of the methods of parts feeding and an estimate made of the rates of parts feeding required to meet the required assembly output. Following this an estimate can be made of the rate of assembly. This can often be determined from past experience or by experiment.

19.7 Machine layout

The general layout of the machine can be determined, following the principles set out later in this chapter. The choice will depend upon several factors such as the complexity of assembly, the methods of loading and unloading, overall costs, the floorspace required, machine output, product life, etc.

Breakdown, downtime and maintenance must be considered and then an outline design can be prepared and a budget price obtained.

19.8 Economic assessment

The final and most important stage is a full economic assessment of the overall scheme, which should, as far as possible, take all factors into consideration, and must be as accurate as possible if a sensible decision is to be made.

For example, for each type of machine consider such factors as:

- *Automated*
    1. Cost of transfer system and work carriers.
    2. Cost of feeding and placement devices.
    3. Total cost of machine (1 + 2).
    4. Effective production rate.
- *Manual*
    5. Number of operators.
    6. Number of assemblies per shift.
    7. Number of operators to achieve this assembly rate.
    8. Cost of equipment for manual assembly.
- *Comparing automated and manual*
    9. Effective cost of machine (8 − 3).
    10. Number of operators saved.
    11. Capital outlay per operator saved (9/10).

It will then be necessary to consider the availability of labour, incentive schemes, the reaction of unions, etc., when making a decision.

19.9 Automatic assembly techniques

Transfer systems

The purpose of a transfer system is to transport a fixture holding assemblies at various stages of assembly, from one station to another. Whilst this is taking place it is essential that the various placement devices do not operate and that positive indexing and accurate location takes place.

Although it is possible to synchronise placement devices and transfer mechanisms, it is not usual for the transfer of assemblies to be continuous and thus be assembled on the move. Normally transfer is intermittent with the assembly stationary when parts are added.

The assembly systems may be rotary or in-line, and in both cases the assembly is completed in one revolution or pass of the machine. In the case of the rotary machine the finished article is returned to the starting station, whilst in the case of the in-line machine the finished article is removed at a distant station at the end of the line. In this case the return of the assembly fixture to the loading station must be achieved by some method. A modified layout for the in-line machine can achieve the same benefits as the rotary machine: the modification required is for the machine to be built in a rectangular layout.

Whilst rotary machines (Figure 19.3) often take up less floorspace they are not as versatile as the in-line machines (Figure 19.4) as they can only allow placement devices around the outside of the assembly fixture. The machine design does not allow for the positioning of devices in the centre of the machine. In the case of in-line machines devices can be located virtually all around the assembly fixture.

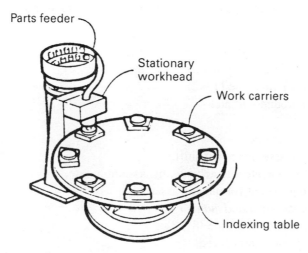

Figure 19.3 Rotary transfer.

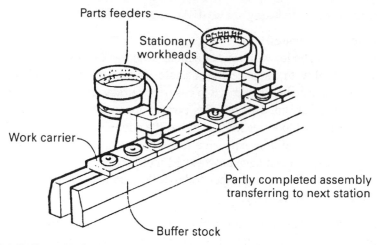

Figure 19.4 In-line transfer.

Another factor in favour of the in-line machine is that stations can be added without too much difficulty. Stations cannot be added to a rotary machine quite so easily, and only if space is available.

A variation on the in-line system is to make the normal straight line machine into a rectangle pattern (Figure 19.5). This simulates the rotary design and give the advantages of that type of machine where the finished assembly is returned to the start position, whilst having all the advantages of the in-line machine. A further advantage is that it is necessary to turn corners and this has the effect of rotating the assembly fixture, which enables different faces to be presented to the outside of the machine.

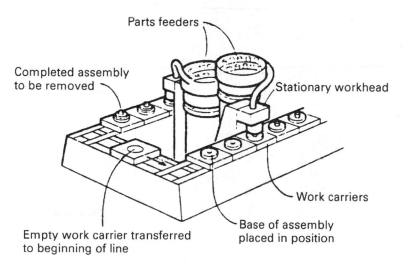

Figure 19.5 Rectangular transfer.

Various methods of indexing are used. Among the commonest are:

1. Rotary table (see Figure 19.6):
   (a) Pneumatic cylinder with ratchet action. This has the advantage of cheapness and the ability to vary the index angle by varying the cylinder stroke, but needs to have a shot pin device to ensure accuracy and positional locking.
   (b) Geneva mechanism. This requires careful construction if it is to be accurate, and is restricted to a sine motion. It has the advantage of being self locking.
   (c) Crossover cam. While expensive to produce, this is capable of indexing at high speeds, and can have any desired motion characteristic designed in. It is self locking. The use of taper roller followers allows adjustment for wear, giving long life accuracy.

Both the Geneva and the crossover cam mechanisms can be continuously driven, or may be timed by the incorporation of a clutch/brake into the drive system.

2. In-line transfer:
   (a) Walking beam (see Figure 19.7). This has the ability to move individual components or fixtures, but has no ability to provide between-station buffering.

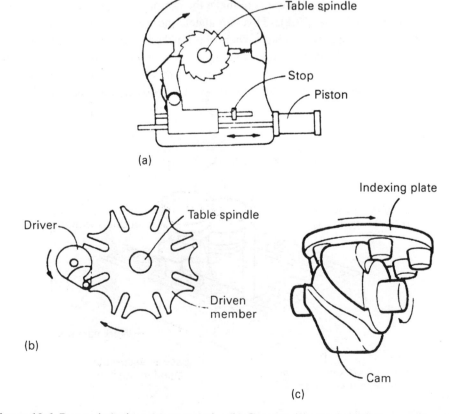

Figure 19.6 Rotary indexing: (a) pneumatic; (b) Geneva; (c) crossover cam.

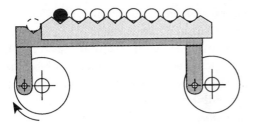

(a) Component rests against fixed index plates

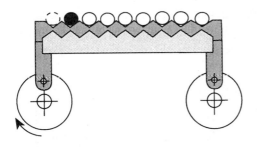

(b) As eccentric drive wheels rotate, 'walking' beams lift component clear of fixed plates

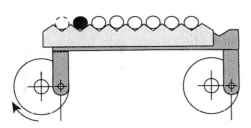

(c) Continued rotation of eccentric wheels causes walking beams to carry component forward one pitch and deposit it on next station of fixed plates

Figure 19.7 Walking beam transfer.

(b) Shunted fixture. Probably the simplest technique. An air cylinder pushes forward a string of carriers. This technique is much used for transfer between lines, e.g. across the end of a rectangular set-up.

(c) Conveyor, either belt or chain. Another simple technique. Stops at each station either raise the fixture from the conveyor, or restrain the fixture, causing it to slip on the belt.

(d) Powered rollers. Detectors at each station power or free rollers as required. This system allows for easy buffering between stations.

In most cases of in-line transfer the transfer device does not in itself provide positional accuracy. This is provided by individual location devices accurately and separately locating each carrier at each station, thus overcoming any danger of cumulative positional errors.

Many methods of transfer have been designed and used although most are variations on the above methods. The choice of transfer method is often determined by the desired accuracy of transfer and the economics of the system.

Assembly machines do not necessarily have automatic transfer at fixed time intervals. It is sometimes better to arrange transfer on a sequential basis when each placement device has completed its assembly work. Transfer then only takes place when all devices have returned to their start positions and it is safe to transfer. This is the most secure method and often the quickest, as there is no need to build in a safety time delay to allow for all systems to be in the correct place. Time delays, for reasons of safety, must cater for extreme conditions and so time is wasted. Another method is 'free transfer' where the assembly fixture is transferred when the operation is finished, if there is space for this to happen.

A major disadvantage with these systems is that a fault at any placement device results in the whole machine stopping. Production will cease until the fault has been located and rectified. An attempt at overcoming this problem would be to consider a system where there is a buffer stock between each station. It is then possible to overcame a problem before the buffer stock is exhausted.

Parts handling in automated assembly

The techniques for feeding parts have already been dealt with, and are as applicable to assembly machines as they were to machining operations. There is an additional problem on an assembly machine, however, as there is likely to be a requirement for a number of parts all requiring to be placed within the one small area. The transfer and placement mechanisms thus become much more complex.

A robot is often specified for assembly operations. Parts are fed into the reach of an industrial robot, which picks out and transfers each part in turn into the assembly fixture and secures it in position. This means that the robot must be able to grip each part, and the 'hand' must not interfere with the assembly operation. The assembly fixture must be designed to hold the part while the robot goes for the next part, or the part must be self-holding on the assembly when released. Since the robot is a programmable device, it has the advantage of being able to put parts in a number of locations within the work area, and the additional advantage of being able to follow a controlled path when carrying out operations. It has the disadvantage of only being able to put one part in at a time.

It follows therefore that the robot is best applied where the parts being added are the same or similar, and where selectability of location and path are required. Thus we find robots used with power fed screwdrivers screwing down assemblies, robots with simple grippers inserting components into PC boards, and robots carrying spot welding equipment and applying spot welds around complex perimeters.

In a large number of assembly applications such flexibility is not required, and it is convenient to bring the component into position above the work by means of a chute, or similar method, then to carry out the actual insertion with a linear motion. It may even be possible just to release the component down a chute and allow it to fall into position.

Many assemblies are co-axial, where a number of parts fit on a shaft or into a recess. In such cases it is more efficient to arrange the feed mechanisms to bring the components in line with one another, and use a linear motion to pick them up and locate them all together onto the protrusion or into the hole.

Figure 19.8 AGV taking hedge trimmers from assembly line to despatch bay.

The simple pick-and-place device, either linear or rotary, forms the basis of many assembly machines, allowing components to be accurately located at low cost. One application of this is the use with a turn-over device to transfer partly assembled product to allow further assembly to be carried out on the other side of the product.

It should be kept in mind that an automated assembly line may well incorporate a variety of devices, each selected to fill a particular purpose.

19.10 Servicing the assembly line

The preceding text has dealt with the actual operations carried out on the product. However, there is another aspect of assembly work, and indeed of manufacturing in general, which should be mentioned briefly, that of *moving* parts in bulk, and finished product, to and from machines and production lines.

This may vary from the box of parts carried by operator or servicer, through pallet systems, moved by pallet truck or fork-lift, to computer scheduled robotic truck systems, according to the requirements of the product and the degree of mechanisation within the company. Figure 19.8 shows an AGV in operation, in this case guided electromagnetically by a cable buried in the floor. Such techniques extend the principles of automation throughout the factory.

20 Set-up Time Reduction

20.1 Introduction

A major pressure on industry is the requirement to reduce stocks, and to reduce lead times to respond to customers. This has lead to the widespread adoption of Just-in-Time techniques, which call for close links between supply companies and their customers. Traditionally suppliers would make a monthly delivery of components to their customers, and would deliver batches of several thousand parts. These occupied valuable floorspace and also needed to be ordered several months in advance, see Figure 20.1. Industry must continually evolve to make what the customer wants, when the customer wants it, and with the options required. The customer will not wait weeks or months for their particular product, and industry must change its whole working practises to provide very short lead times for products. This means that suppliers must be able to respond to customers' orders in a few hours, and to deliver the products that are required, in the order that they are required to be used on the individual customer's production line.

For example Nissan will be making a range of cars an their assembly line, (1600 cc, 2000 cc, 2000 cc Turbo, 2000 cc diesel) in any order, and each of these different models requires a different exhaust pipe. Their local supplier of exhaust systems, Calsonic, is about ten miles away and will receive a computer message giving authorisation to make and deliver, and will then take the relevant exhaust pipe from the assembly cell and load it into a small van ready for delivery. Every two hours the van will deliver the next two hours requirement of exhausts loaded into the van in the order that they will be used on the line. The operator at Nissan merely reaches for the next exhaust, knowing that it will be the one required for the next car. Similarly, Ikeda Hoover will be making and delivering seats in the correct colour and sequence for each car, again with two-hourly deliveries (Figure 20.2).

This greatly increased speed of response to customer's needs demands totally new organisational structures to enable companies to react so quickly. The suppliers are required to make parts in very small batches, and to respond hour by hour to changing customer requirements. It is not the purpose of this book to describe the JIT manufacturing system, our focus here is on the technology of manufacture.

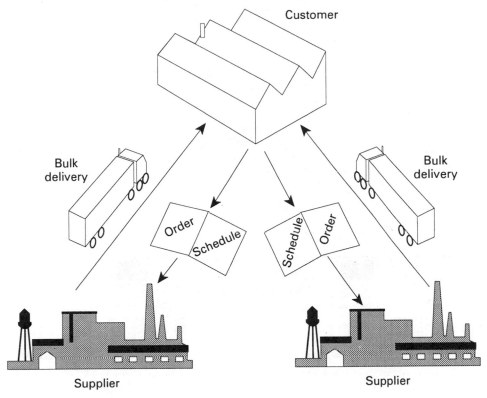

Figure 20.1 Traditional supply chain (large and infrequent deliveries).

There are many excellent books which the reader can refer to for information about JIT. The key to this responsive industrial situation is to reduce batch sizes to the minimum using set-up time reduction techniques to enhance the flexibility of production lines.

Successful set-up time reduction requires a close knowledge of the processes being used, as well as how they can be modified to permit rapid changeovers between batches of different parts. This chapter thus focuses on the technology principles which permit JIT to operate. Many related points have been covered throughout this book, and these strands are drawn together here. Many engineering details can be improved upon to reduce set-up times, but the principles often demand a revised specification for the performance of the machines being used. Two examples are quoted to illustrate the type of changes required:

1. Use of light operated guards for power presses to reduce tool change times will usually require a new press mechanism. The clutch/brake system must be able to respond to a danger signal io a few microseconds and stop the press before an accident occurs. This advanced drive system is only likely to be fitted as original equipment when a new press is purchased.
2. Pallet changers on machining centres are usually built-in features of the machine, and they probably cannot be retrofitted.

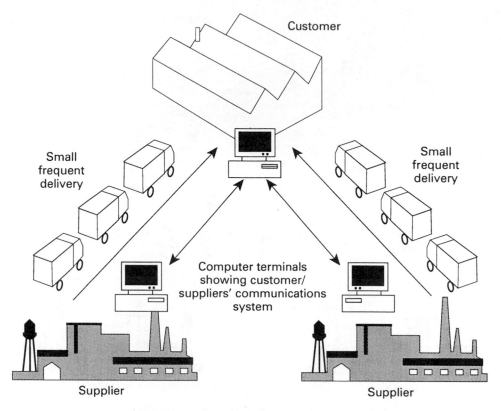

Figure 20.2 JIT supply chain.

These examples show that major advances in set-up time reduction *must* be considered when a new line is being specified. Set-up time reduction *demands* a new approach to analysing the performance requirements and hence the specification of a new line.

This is not to say that major improvements cannot be obtained from re-engineering existing equipment, and both approaches are relevant in different situations.

As mechanisation increases, so also does the investment in machinery, which is only earning the money to pay for the investment while it is actually producing. It is therefore essential that down time for set-up or maintenance should be cut to a minimum. In addition, there is pressure for stockholdings to be cut to release money tied up in stock.

The reasons for holding stock have been twofold; firstly it was necessary to hold stock to allow for rapid fulfilment of customer orders, and secondly large batches were required so that the set-up costs could be spread over a large number of components. It follows therefore that if set-up time could be reduced to zero there would be no need for expensive stocks, and full utilisation of expensive machinery.

Modern CNC machinery and flexible automation techniques offer the possibilities for reduction, perhaps not to zero, but at least by a substantial measure. In this chapter the application of set-up reduction techniques to machining, presswork, and assembly operations is studied.

If a changeover process is examined in detail it will be found that the actual time to change the tools themselves is probably only 10 per cent of the total machine stoppage time. Tool changing systems have been like this for years, with no attempt made to seek improvements. Review of such a system shows that the machine has been idle for an hour while the next set of tools and clamps were prepared, and scrap material removed, guards adjusted, etc. A variety of different clamps, tools, spanners, etc. may be required, different perhaps from those required to dismantle the previous set-up, and this variety slows down the changeover.

When the new tools are installed, there will probably be a need for trial cuts, and adjustment of the tools, before correct parts are produced. All this cuts into the productive machine time.

It is necessary to challenge such established practice. The first challenge should be to ask whether it is necessary for the part to undergo sequential operations on a number of different machines. Many modern CNC machines are capable of carrying out both turning and milling operations, and selection of such machines will not only cut down on set-up time, but will also save on component handling.

To minimise machine downtime the following guidelines are followed:

1. Externalise the setting:
   (a) Prepare the tools while the machine is still running on the previous job.
   (b) Get the next batch of materials ready for the machine *before* it stops.
   (c) Ensure tools are overhauled immediately after use so that they are ready when next required.
2. Minimise the change time:
   (a) Standardise on fixings.
   (b) Use powered clamping where appropriate.
   (c) Minimise range of tools required, e.g. one size spanner for all nuts.
   (d) Use quick change and non-obstructive guarding systems, e.g. light beams, which do not need to be dismantled when changing tools.
3. Eliminate on-machine adjustments:
   (a) Use precision tips and standard sizes.
   (b) Preset tooling on off-machine presetters to standard dimensions.
   (c) Use self-correcting systems e.g. in-cycle probing, automatic wear compensation, CNC adjustment.

20.2 The power press

A good example of this outlook is the the modern power press, where hydraulic die clamping, light beam guards, powered rollers to move the heavy tools, and CNC press stroke adjustment together with memory chips in each tool, all combine to give tool changeover in 2 to 3 minutes instead of the 2 to 3 *hours* by traditional methods.

On the press illustrated automatic tool clamping, in conjunction with an associated tool changing carriage, enables large die sets to be changed in a matter of minutes (Figure 20.3). This carriage, which moves on rails attached to the press, consists of two stations, either of which can be brought in line with the platens. To effect a change, the die set in the press is unclamped and moved to the empty station. The carriage is then traversed, and the new die set moved into position on the press. Spring loaded ball inserts raise the tool above the surface of the bed to allow quick and easy positioning. Hydraulic quick tool change clamps are mounted on the press bed to secure the dieset (Figure 20.4).

The diesets themselves have been standardised. Standard bolster sizes, standard shut height or pre-programmed shut height information, standard fixing positions all render on-press adjustments unnecessary. Proper maintenance of diesets, and

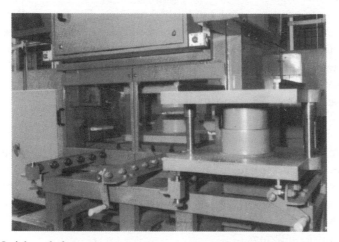

Figure 20.3 Quick tool change on power press.

Figure 20.4 Hydraulic tool clamping located in tee slots.

provision for storage and transportation of these, make it possible to have the next die in position ready for change while the press is running, and changeover time is therefore only minutes, making it economic to produce small batches on demand.

20.3 The lathe

Block tooling systems

Tooling for turning operations normally involves the clamping of a series of tools into a tool post turret or tailstock turret (Figure 20.5). The time taken for this can be minimised if each tool has a standard clamping system. One such system is the block tooling system (BTS) shown in Figure 20.6.

Adjustable centre height of the cutting edge and an internal coolant supply are important features of the system.

Cutting units fit into clamping units, which have an established datum position in the machine. They are available in various mounting types, such as shank, VDI, etc. which means that BTS provides interchangeability between tool positions and machines. The BTS clamping unit becomes the universal adaptor for BTS cutting units to be used in any predetermined position.

Very high stability is achieved through axial support of the cutting unit. The centre line position of the drawbar ensures that the clamping pressure directly opposes the cutting tool forces. Low clamping torque and short drawbar stroke allow simple and efficient tool changing.

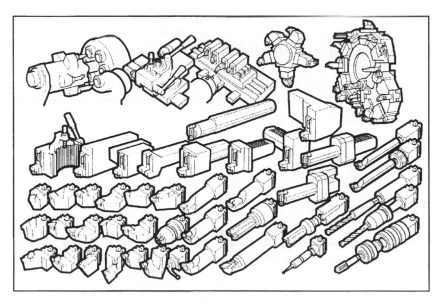

Figure 20.5 Block tools will improve utilisation of any lathe.

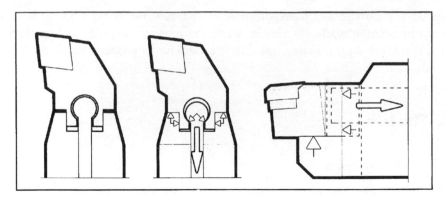

Figure 20.6 The block tool clamping system.

The BTS coupling offers no play in any direction when in the clamped position. The force on the drawbar makes the block tools as rigid as a solid tool. The cutting unit is supported from underneath to the extent that in many cases, stability is even better than that of solid tools. Plain contact faces and high precision between the unit and holder help to maintain the stability of the coupling.

The accuracy of the coupling gives excellent repeatability. When the same cutting unit is clamped and unclamped, it will repeat within ±2 microns in the X axis and ±5 microns in the Z axis. When changing from one unit to another the cutting unit tolerance of ±0.15 mm and the insert tolerance must be taken into account.

This accuracy of coupling allows the tools to be preset on a device distant from the machine, so that no machine time is lost through the need to determine accurately the position of the cutting edge, and to programme in the necessary offset.

20.4 Advanced machining facilities

As the difference between the lathe and the milling machine grows increasingly blurred with the fitting of milling facilities to the lathe to give the turning centre, it is necessary to redefine the machine classifications. It has now been generally accepted that a machine in which the workpiece is rotated is a *turning centre*, while one where the workpiece is basically held in a fixed position is a *machining centre*.

20.5 Tool changing

Tool holding systems have evolved to satisfy the needs of these advanced machines, allowing the investment in a system which can provide common tool holding across the machine shop.

The need is for a precision tool holding system providing rigid support and accurate location for both turning tools and milling cutters, capable of transmitting high torsion forces. One such system is the tapered polygon system, where the tapered polygonal shank on the toolholder is used both to locate the tool and transmit the cutting forces.

The tapered polygon (Figure 20.7), which is tensioned in the clamping unit with a preload force of several tons, produces an extremely stable joint. At the same time, the tool can be quickly and easily removed, either manually or automatically (Figure 20.8). The large contact surface on the three flanks of the taper provides non-slip transmission of the torque. The self-centering and self-aligning effect of the coupling ensures that the constituent parts are never displaced in the radial direction. Moreover, the polygon coupling functions equally well in both directions of rotation. The length of the taper and the precision contact surfaces counteract bending of the cutting-tool holder, and at the same time contribute to the capacity of the coupling to handle large torques (Figure 20.9).

The high precision of the coupling ensures a repeatable accuracy of ±2 microns in the X and Z axes. This repeatable accuracy applies for one and the same edge of an insert in the same clamping unit. When different cutting units are used, the tolerance of the insert seat in relation to the coupling, plus the tolerance of the insert, must also be taken into consideration.

For changing tools, presetting can be carried out to measure deviations from nominal values. These deviations can then be compensated for via the tool offset function of the machine control system.

The system can be used on most machine tools and for many types of machining. It forms the cornerstone for Just-in-Time production. It is the viable solution for turning centres and FMS, but can also be used advantageously on conventional machines. It permits tool system standardisation and reduces inventory costs as well as simplifying administrative and handling of tools. Tools are prepared for through-tool coolant supply as standard. It is equally suitable for manual or automatic installations.

Driven tools for rotating tool applications transform the scope and the efficiency of workpiece production in turning machines. They combine turning, milling,

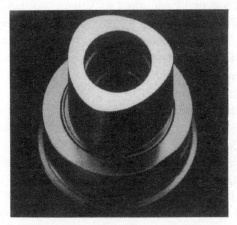

Figure 20.7 Tapered polygon tooling clamp system.

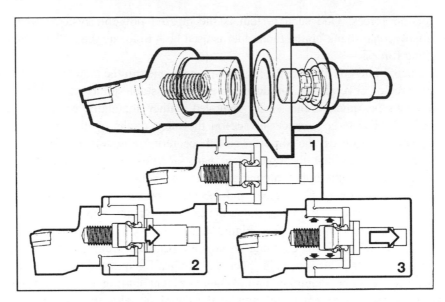

Figure 20.8 Clamping and unclamping: (1) the cutting unit released; (2) pulled into position; (3) locked firmly.

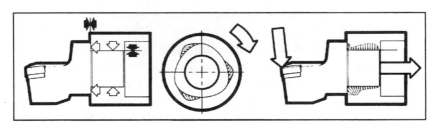

Figure 20.9 Stability, torque, and bending strength of coupling.

tapping, reaming and drilling operations in a single set-up. The adaptable range of driven toolholders can be fitted in almost any type of standard or special turning machines. The system has two basic types of driven tool holders: axial and radial. Each unit consists of standard modules for easy adaptation to different machines and turrets (Figure 20.10). To extend their application range each one can be modified.

Careful planning of the turret layout is essential when integrating driven tool holders into a turret, especially to avoid any risk of collision. The driven tool holders can be integrated into most turret layouts.

Many other proprietary systems exist, all having their strengths and weaknesses. The checklist in Figure 20.11 is useful in assessing which of the available systems is most suited to a specific workplace.

The provision of standard tool clamping systems has led to the development of automatic tool changing systems with tool magazines which can store a substantial number of tools. This is illustrated in Figure 20.12. These can be a range of tools covering several different set-ups or can be duplicate tools if the same workpiece is being run for long periods on the same machine, allowing sets

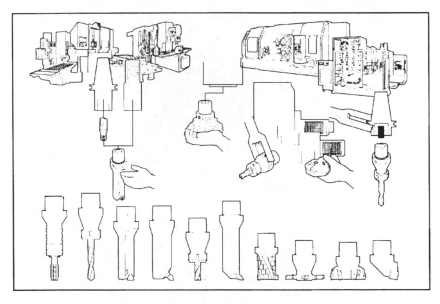

Figure 20.10 Universality in the machine shop with the modular tool system.

Checklist for adopting modular tools

Make sure that real universality is provided for today and tomorrow, to utilise flexibility and rationalising potential in handling and inventory:

- Invest in a system that, without any compromises, can be used for any type of machining operation.
- Ensure that the system is suitable for all foreseeable machine tool types.

Make sure of the best stability and repeatability, so that tool selection is not limited by any operational demands:

- Check for best function as regards stationary and rotating tools.
- Establish values for the built-in safety margins, especially for heavy-duty applications.
- Test for effect on workpiece quality.
- Determine elimination of measuring cuts.

Have quick changing facility of tools as high priority for turning operations. Modular systems vary in speed and complexity, which will affect the practical utilisation of the system and the ability to reap benefits of higher productivity and rationalisation.

   Choose a system that provides coolant through the tools as the standard method. This is often a crucial requirement in those operations needing swarf removal and cutting fluid to be directed to the right spot on the tooling.

Figure 20.11 Checklist for the adoption of modular tooling.

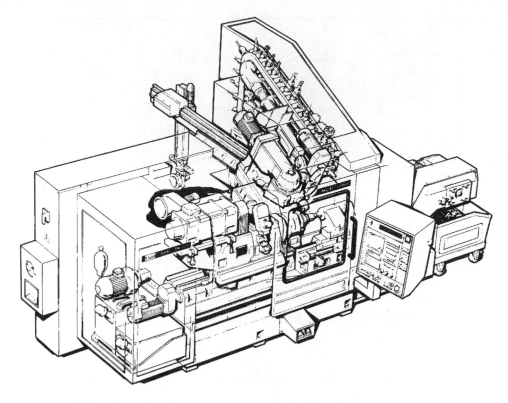

Figure 20.12 Turning cell with tool magazine and robotic loading.

of tools to be changed automatically if a new job is called up on the machine schedule, or for provision of new tools when a particular tool reaches the end of its planned life.

The tools in such systems must either be accurately preset in an off-line presetting device, or must have had any offset accurately noted to allow it to be entered into the machine computer. Since tools may be resharpened several times, and thus have their effective length vary, the latter is the most common technique. Entry of the tool offset is a manual operation, and is a possible area of error; attempts have been made to automate this.

20.6 Tool code tagging

Identification of tools by means of magnetically readable code tags is now becoming common, particularly for large machines. Tool identity and information on length, etc., is recorded at the presetting station, and entered into the machine controller when required, avoiding keyboard errors.

A typical system of this type is shown in Figure 20.13. The system requires a code tag installed in each tool, and a reading head and interface unit at each reading

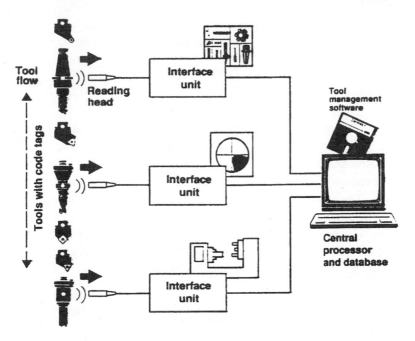

Figure 20.13 Tool code tagging system.

position connected to a central processor. The code tag supplies the central processor with the tool identity, and the processor then retrieves the relevant data from its database and transmits this to the machine controller.

Alternatively, the tool can carry a read/write data tag capable of holding all the relevant information, and can supply this directly through a reading head incorporated in the machine.

Tool code tagging is also applicable to press tools.

20.7 Probing

If the machine itself could measure and compensate for any discrepancies in tool tip position it would be unnecessary to have accurate compensation figures fed in, saving time in presetting and avoiding risk of error. There would also be an ability to compensate for wear taking place in the tool. This is possible using active probing. Probes mounted on the machine itself measure the results of the first cut of the tool, and provide the necessary compensation automatically to ensure that the next cut is accurate (see Section 7.15).

The same probing techniques can be used to determine the location of the workpiece, allowing the workpiece to be mounted in an approximate position, with the machine determining the actual location and compensating accordingly. This eliminates the need for (costly) accurate jigs, and the time necessary to mount the workpiece accurately.

The application of these techniques to machining operations is best illustrated by considering actual examples from industry in the following sections.

20.8 The turning centre

Turned parts are rarely finished components. This is the reason why more and more workpieces are not only turned but also drilled and milled on the turning centre, during the same chucking operation (see Figures 20.14 and 20.15). The workpieces are clamped securely, and shaft-type components can, in addition, be supported by the tailstock. The advantages: increased accuracy, shorter work-in-process times, lower manufacturing cost.

Using the C axis, any desired angular position can be reached rapidly and precisely to carry out drilling and milling operations on the stationary workpiece. A brake unit holds the spindle in the programmed position.

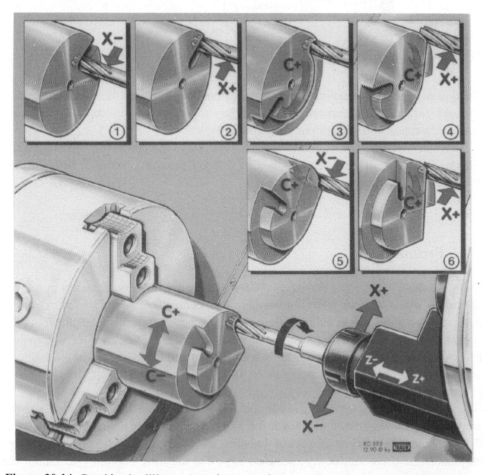

Figure 20.14 Combined milling and turning operations.

Figure 20.15 Milling on side of turned component.

The main spindle can also rotate during a milling operation. Contours can be milled if the rotary motion of the workpiece is superimposed by a linear or circular movement of the milling tool.

Live tooling. (Figure 20.16) applications include:

- Simple programming even of complex 3D milling contours.
- Full geometry support.
- Input in the usual Cartesian coordinate system XYZ.
- Milling on faces, chordal flats and circumferences.
- Milling with cutter radius compensation.

The drilling and milling tools are driven via an external gear. This enables high torques and outputs to be transmitted, without weakening the toolholder shank.

The manufacture of a measuring probe directly from stock material on a single machine is an example of this, although not in fact incorporating any turning as such (Figure 20.17).

The operation sequence for this part is then as follows:

1. Mount billet in chuck (this can be done by a robot arm if required). Probe billet to check for correct loading and to set Z origin at end of billet. Reject billet if sizes are outside the 1 mm initial position tolerance.
2. Mill top, bottom and side faces. Drill and tap holes in top face.
3. Circular mill around 'bore' of component and around outside of body.
4. Probe body to check critical dimensions.
5. Rotate 90° and mill, drill and tap side face.

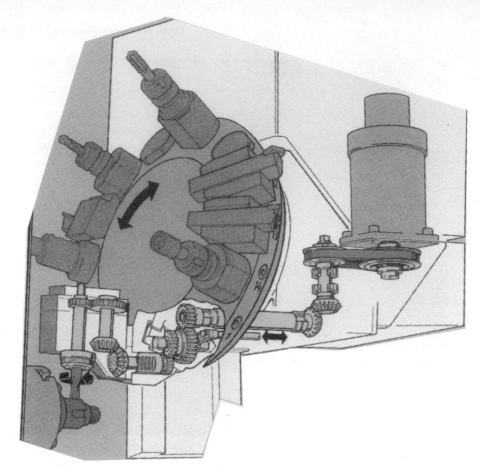

Figure 20.16 Live tooling on a CNC lathe.

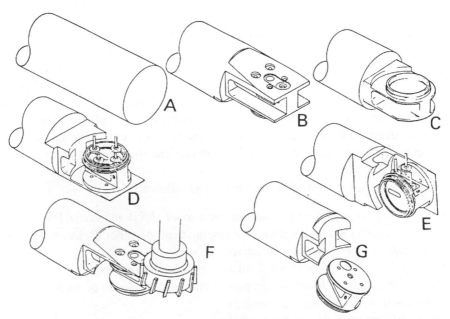

Figure 20.17 'One hit' manufacture of probe body.

6. Mill off excess metal on bottom of part to release it from billet (N.B. operation 3 milled chamfer around base of component to avoid leaving burrs when it is separated from the billet in this operation).
7. Complete finished part as shown.

Advantages

- Zero set-up time.
- Part location always known exactly.
- Part finished ready for assembly in fifteen minutes.
- Zero fixturing costs.
- Cheap standard form of raw material used.

Disadvantage

- Scrap billet material needed to hold part. The fitting of a bar feed could overcome this disadvantage.

20.9 The machining centre

The crank lever part shown in Figure 20.18 was required in small quantities, and had been made in small batches on horizontal milling machines using traditional fixturing techniques. Five operations were required, each requiring a fixture costing £1200 to design and make, with an elapsed time from part design to release for production of six months. The setting-up time was one hour for each operation. Variations in the casting made accurate location difficult, and the process was unreliable. The process is charted in Figure 20.19.

Manufacture of the part was moved to a machining centre, where all the machining could be carried out as one operation. The company had developed techniques for using low melting point alloy to hold components in place for machining, and this method was adopted for holding the lever casting. The casting

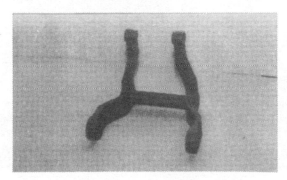

Figure 20.18 Crank lever.

Operational flexibility of crank lever

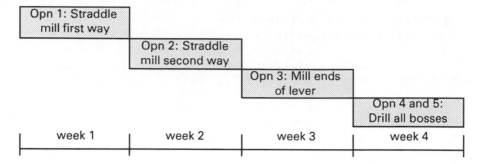

Figure 20.19 Original sequence of operations.

was positioned approximately and bolted down on a simple pillar, as shown in Figure 20.20. The machine then probed to determine the actual location, and proceeded to machine all faces. The new process timings are:

- Clamp on simple fixture using low melting point alloy – 1 minute.
- Probe front of job to check squareness – 50 seconds.
- Probe width of job to establish centre line – 50 seconds.
- Rotate 90°, probe end shape of casting – 50 seconds.
- Rotate 180°, probe second end of casting – 50 seconds.
- Machine all faces – 6 minutes.

The advantages are:

- Fixturing considerably simplified.
- One hit machining eliminates cumulative location problems of previous method.
- Lead time cut from four weeks to less than a day.

Figure 20.20 Revised set-up using low melt alloy.

20.10 Assembly operations

While the preceding sections have related to machining operations, the principles set out are equally applicable to assembly operations, even when carried out manually.

In the simplest of manual operations, it is possible to avoid operator idle time by having the workplace for the next task prepared, complete with all parts, materials and tools, so that the operator can move directly from the previous work station.

An assembly line can be treated like a machine. Tooling can be arranged to be flexible, programmable to carry out operations on a variety of parts, and programmable devices can be used for part positioning. Where retooling is necessary on line changeover, quick change devices should be fitted, and fixtures, etc., should be designed as replaceable units, with standardised clamping.

An example of this comes from a manufacturer of small electrical devices. The line changeover time from one product to another was of the order of half a day, and as a result product was scheduled in batches averaging three days. This resulted in a need for large stocks of finished product, and from time to time sudden customer demand would result in an out-of-stock situation which could not be rapidly remedied.

A study of the operation showed that certain stations within the line required little changeover, while others accounted for the majority of the delay. The line was therefore redesigned to allow these stations to be easily removed and replaced with spare units already set-up. This allowed the line changeover time to be reduced to under an hour, enabling the line to be changed during a lunch break, and allowed the batch quantities to be reduced to quantities produced in half a day. This resulted in substantial reduction in stocks of finished goods, and an improved response to sudden customer demand.

20.11 Rapid response manufacturing

While the above examples have stressed machine utilisation as the reason for reduced changeover time, it will have been obvious that these increased efficiencies have resulted in both stock savings and shortened lead times. The basic principles applied to the machines can also be applied to other aspects of manufacturing to provide a manufacturing operation capable of rapid response to customer demands (Figure 20.21). This requires changed practices and determined effort, together with substantial investment in modern equipment, but it is achievable, and results in improvement in both quality and customer satisfaction.

The requirements of such a system applied to a single part are:

1. Material supplies – Use stock bars/plate whenever possible *or* negotiate fast and reliable delivery of quality castings/parts.
2. Design service – Fast response to customer modification *plus* speedy and reliable CNC programming system.

Inputs   Outputs

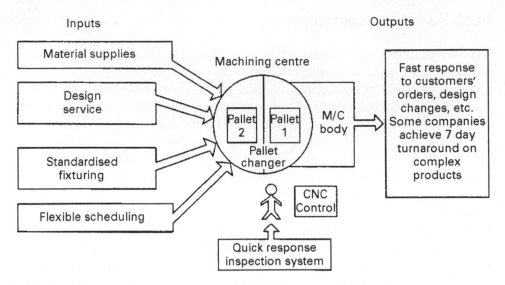

Figure 20.21 Rapid response manufacturing system.

3. Standardised fixturing – No jig and tool design or manufacturing delays.
4. Flexible scheduling – To cope with batch quantities of *one* part and to give lead times of *days* rather than *months*.
5. Quick response inspection system – Avoidance of inspection delays, perhaps by provision of on-line or on-machine quality control.

Bibliography

The following publications contain information likely to be of interest to readers of this book. Sources of the references made in the text are included in the list.

GENERAL

Boothroyd, G. *Fundamentals of Metal Machining and Machine Tools*. McGraw-Hill (1975).
Chapman, W. A. J. *Workshop Technology Part 3*. Third Edition. Edward Arnold (1975).
Crawford, R. J. *Plastics Engineering*. Peyanon Press.
Radford, J. D. and Richardson, D. B. *Production Engineering Technology*. Third Edition. Macmillan (1980).
Wright Baker H. (Editor) *Modern Workshop Technology Part 2*. Third Edition. Machine Tools and Manufacturing Processes. Macmillan (1967).

SPECIFICATION AND STANDARDISATION

The management of design for economic production. PD 6470: 1975. British Standards Institution (1975).
Handbook No. 18 Metric Standards for Engineering. British Standards Institution (1972).
Machinery's Screw Thread Book. Twentieth Edition. Machinery Publishing Co. (1972).
Woodward, D. W. (Editor) *Standards for Industry*. Heinemann (1965).

MACHINE TOOLS

Chironis. N. P. (Editor) *Machine Devices and Instrumentation*. McGraw-Hill (1966).
A Management Guide to Numerical Control Machine Tools. Institution of Production Engineers (1978).
Pustzai and Sava, *Introduction to Numerical Control*.
Schlesinger, G. *Testing Machine Tools*. Seventh Edition. Machinery Publishing Co. Ltd (1970).
Town. H. C. *Design and Construction of Machine Tools*. Iliffe (1971).
Tobias. S. A. *Vibrations of Vertical Milling Machines Under Test and Working Conditions*. Institution of Mechanical Engineers (173) (1959).

METAL CUTTING

A Guide to Flexible Manufacturing. I. Prod. E.
A Treatise on Milling and Milling Machines. Third Edition. Cincinnati Milling Machine Co. (1951).

Armarego, E. J. A. and Brown, R. H. *The Machining of Metals*. Prentice Hall (1969).
Backer, W. R. Marshall E. R. and Shaw, M. C. *The Size Effect in Metal Cutting*. Trans. ASME Vol. 74 (1952).
Black, P. H. *Theory of Metal Cutting*. McGraw-Hill (1961).
Gisbrook, H. Precision Grinding Research. *Journal I. Prod. E.* (May 1960).
Guest, J. J., *Proceedings I. Mech. E.* page 543 (Oct. 1915).
Harry, J. E. *Industrial Lasers and Their Applications*. McGraw-Hill.
King, R. I. and Hahn, R. S. *Modern Grinding Technology*.
Kronenberg, M. *Machining Science and Application*. Pergamon (1966).
Parnaby, J. The Design of Competitive Manufacturing Systems. *Int. J. Technology Management* (1986).
Parnaby J. and Donovan, J. R. *Education and Training in Manufacturing Systems*. I. Prod. E.
Ranky, P. *The Design of FMS* (1986).
Sandvik, *Modern Metal Cutting* (1994).
Sémon, G. *A Practical Guide to Electro-Discharge Machining*. Ateliers des Charmilles S. A. (Geneva).
Shaw, M. C. *Metal Cutting Principles*. Third Edition. MIT Press (1968).
Springborn, R. K. (Ed.) *Non-Traditional Machining Processes*. American Society of Tool & Manufacturing Engineers.
Wilson, John F. *Practice and Theory of Electro-Chemical Machining*. Wiley-Interscience.

METAL FORMING

Calladine, C. R. *Engineering Plasticity*. Pergamon (1969).
Crane, E. V. *Plastic Working in Presses*. Third Edition. Wiley (1944).
Feldmann, H. D. *Cold Forging of Steel*. Hutchinson (1961).
Grainger, J. A. *Presswork and Presses*. Second Edition. Machinery Publishing Co. (1952).
Makelt, H. *Mechanical Presses*. Edward Arnold (1968).
Rowe, G. W. *An Introduction to the Principles of Metalworking*. Edward Arnold (1965).
Schuler, L. *Metal Forming Handbook*. Fourth Edition. Louis Schuler-Göppingen Wuertt (1964).
Willis, J. *Deep Drawing*. Butterworth (1954).
Pugh, H. Ll. D. and Watkins, M. T. *Experimental Investigation of the Extrusion of Metals*. I. Prod. E. Brighton Conference (1961).
Tilsley, R. and Howard, F. *Cold Extrusion of Ferrous and Non-Ferrous Materials*. I. Prod. E. Brighton Conference (1960).

DIMENSIONAL AND QUALITY CONTROL

Handbook of Industrial Metrology. Society of Manufacturing Engineers. Prentice Hall (1967).
Adams, L. F. *Engineering Measurements and Instrumentation*. Hodder and Stoughton (1975).
Caplen, R. H. *A Practical Approach to Quality Control*. Third Edition. Business Books (1978).
Galyer, J. F. W. and Shotbolt, C. R. *Metrology for Engineers*. Cassell (1964).
Miller, L. *Engineering Dimensional Metrology*. Edward Arnold (1962).
Town, H. C. and Moore, H. *Inspection Machines, Measuring Systems and Instruments*. Batsford (1978).
Chree, Dr. Proceedings of the Physical Society of London. Vol. XVIII p. 35.
Evans, J. C. The Pneumatic Gauging Technique in its Application to Dimensional Measurement. *Journal I. Prod. E.* (1957).

DESIGN FOR MANUFACTURE

Corkett, Dooner, Meleka and Pym. *Design for Manufacture* (1991).
Niebel, B. W. and Draper, A. B. *Production Design and Process Engineering*. McGraw Hill.

JIT AND MODERN MANAGEMENT PHILOSOPHY

Black, J. T. *The Design of the Factory with a Future* (1991).
Goldratt, E. M. *The Goal*.
Goldratt, E. M. and Fox, R. E. *The Race*. North River Pressing (1986).

ROBOTICS

Groover, Weiss, et al. *Industrial Robotics Technology, Progamming and Application*, McGraw Hill.

MANUFACTURING MANAGEMENT

Bessant, J. and Blackwell, N.C.C. *Managing Advanced Manufacturing Technology*.
Chadwick, L. *The Essence of Management Accounting*, Prentice Hall.
Evans and West, *Applied Production and Operations Management*. Fifth Edition (1994).
Lockyer, Muhlemann and Oakland. *Production and Operations Management*. Pitman (1988).
Wild, R. *Production and Operations Management*. Third Edition. Holt, Reinhart & Winston (1984).

British Standards Specifications

The following standards, which relate to the subject matter of this book, may be purchased from the British Standards Institution at 2 Park Street, London, W1A 2BS. For certain standards, a shortened form of title is given.

(It is important lo consult the *current* standard, as revisions are made from time to time.)

Machine tools

BS 46 & 4235	*Keys and Keyways and Taper Pins*
BS 292	*Dimensions of Ball Bearings and Cylindrical Roller Bearings*
BS 426	*Lathe Centres*
BS 436	*Machine Cut Gears, Spur and Helical*
BS 721	*Machine Cut Gears, Worm Gearing*
BS 1089	*Workhead Spindles for Grinding Machines*
BS 1660	*Machine Tapers, Reduction Sleeves and Extension Sockets*
BS 1983	*Chucks for Machine Tools*
BS 2059	*Straight Sided Splines and Serrations*
BS 2485	*Tee Slots, Bolts, Nuts and Tenons*
BS 2771	*Electrical Equipment of Machine Tools*
BS 2917	*Graphical Symbols for Fluid Power Transmission Diagrams*
BS 3027	*Dimensions for Worm Gear Units*
BS 3134	*Dimensions of Tapered Roller Bearings*
BS 3550	*Involute Splines*
BS 3641	*Symbols for Machine Tools – Including NC symbols*
BS 3790	*Endless V-belt Drives*
BS 4185	*Machine Tool Components*
BS 4656	*The Accuracy of Machine Tools and Methods of Tooling*
BS 5063	*Lubricants for Machine Tools*

Cutting tools

BS 122	*Milling Cutters and Reamers*
BS 328	*Twist Drills, Combined Drills and Countersinks*
BS 949	*Screwing Taps*
BS 1296	*Single Point Cutting Tools*
BS 4193	*Dimensions for Throwaway Carbide Tips*

Grinding

BS 1089	*Workhead Spindles for Grinding Machines*
BS 2064	*Diamond Abrasive Wheels and Tools*
BS 4481	*Bonded Abrasive Products*

Metal forming and presswork

BS 18	*Methods for Tensile Testing of Metals*
BS 224	*Steel for Die Blocks for Drop Forging*
BS 240	*Brinell Hardness Test*
BS 427	*Vickers Hardness Test*
BS 860	*Table of Approximate Comparison of Hardness Scales*
BS 1639	*Methods for Bend Testing of Metals*
BS 3855	*Method for Modified Erichsen Cupping Test*
BS 4184	*Power Press Nomenclature*
DD 45	*Press Tool Die Sets*

Specification and standardisation

BS 308	*Engineering Drawing Practice*
BS 970	*Wrought Steels*
BS 1134	*Assessment of Surface Texture*
BS 1916	*Limits and Fits for Engineering*
BS 2045	*Preferred Numbers*
BS 4500	*ISO Limits and Fits*
BS 5775	*Letters, Symbols, Signs and Abbreviations*
See also PD 6470:	*The Management of Design for Economic Production.*

Measurement

BS 817	*Surface Plates and Tables*
BS 870	*External Micrometers*
BS 907	*Dial Gauges for Linear Measurement*
BS 958	*Precision Levels for Engineering Workshops*
BS 959	*Specification for Internal Micrometers*
BS 1054	*Engineers Comparators*
BS 1134	*Methods for the Assessment of Surface Texture*
BS 3064	*Sine Bars and Sine Tables*
BS 4311	*Slip (or Block) Gauges and Accessories*
BS 5204	*Specification for Straight Edges*
BS 5317	*Metric Length Bars and Their Accessories*

Gauges

BS 919	*Screw Gauges, Limits and Tolerances*
	Pt 1 Unified form; Pt. 2 *Whitworth and BA*; Pt. 3 *ISO Metric*
BS 969	*Plain Limit Gauges: Limits and Tolerances*
BS 1044	*Gauge Blanks, Plug, Ring, and Caliper Gauges*

Screw threads

BS 21	*Pipe Threads for Tubes and Fittings*
BS 84	*Parallel Screw Threads of Whitworth Form*
BS 93	*British Association (BA)* **Screw Threads**
BS 3643	*ISO Metric Screw Threads*
BS 5346	*ISO Metric Trapezoidal Screw Threads*

Application of statistical methods to quality control fraction-defective charts for quality control

BS 2564	*Control Chart Technique*
BS 6000	*Guide to the Use of BS6001*
BS 6001	*Sampling Procedures and Tables for Inspection by Attributes*

Tool design
BS 1098   *Jig Bushes*
BS 1935   *Spindle Noses and Adaptors, Multi-spindle Heads*
BS 5078   *Jig and Fixture Components*
DD 45   *Press Tool Die Sets*

Quality systems
BS 5750   *Quality Systems*

Answers to Exercises

Exercises – 3

4. 1435 kN   115 kJ
5. (a) 104
   (d) 15 tons

Presswork problems

7. (a) Pierce and blank
   (c) 70, 70.7, 119.3, 120
8. (b) 1414 kN,   4948 Nm
9. 135 kN
10. (a) 160 mm
    (b) Punch diameter = nom. size of hole
    (c) 86.5 mm assuming $D/d = 1.85$

Press mechanism problems

11. (a) 125 T

Exercises – 6

2. 210 mm; 8
5. 67 N; 59 N in plane of 50 kg load
10. (a) 0.0157 mm, tool below centre line
    (b) (i) 0.2667 mm;   (ii) 0.0801 mm; hole 37.84 mm diameter
11. Drivers/driven; 5.28, 3.68, 2.57, 1.80, 1.26, 0.88
15. No 2

Exercises – 9

1. (a) 1950 N/mm^2; 1672 W
   (b) 1100 N/mm^2; 2065 W
4. 4.17°55'
7. (a) 2273 N; 931 W; $k = 0.51$
   (b) (i) Force and power rise approx 50%; K unchanged
       (ii) Force and power fall slightly; K rises slightly
9. 60 rev/min
10. (b) 69%; 3900 N/mm^2; 450 mm^3/s
12. (a) 0.94
    (b) 25
    (c) 659 N/mm^2
13. (a) 1134 N at 22°
    (b) 26°
    (c) 760 N
    (d) 42°
14. (b) θ increases
    (c) 25°
15. 84; 2837 J/min

Exercises – 10

1. (a) $V_c T^n = C$
   (b) C falls as t is increased
   (c) 13.2 m/min
2. (a) 6.38×10^4 mm^3/min;
   (b) C decreases
   (c) C increases slightly
3. $V_c T^{0.14} = 84$; $T = 88$ min (high accuracy not possible from given data)
4. 56 rev/min
5. 2.2 min; 152
6. 190 rev/min
7. 42.4 m/min

Exercises – 11

2. 0.042 mm
3. 78.48 mm; 0.088 mm
6. Centre of cutter 61.5 mm from edge of work; 0.85 min
7. 181 N
8. 6 teeth

Exercises – 12

3. Traverse rate > 2.3 m/min
7. 0.000 05 mm, using Guest's method
10. $v_1 t_1 = v_2 t_2$ to maintain metal removal rate but $v_1^2 t_1 > v_2^2 t_2$ to lower force/grit;
 $v_2 = 10$ m/min, $t_2 = 0.06$ mm are possible values

Exercises – 13

2. (b) 2500 W

Exercises – 14

7. 0.3665 mm
8. 0.4243 mm

Exercises – 15

3. 0.1133 mm at $330°$
4. $+0.02$ mm; -0.03 mm
5. $h = 0.0012$ mm
7. $h = 9.788$ mm; error of $l = +0.028$ mm
8. 102.408 mm; $±0.089$ mm
9. (b) $47°18'$
11. 1.8 mm diameter 0.04 mm; $×50\ 000$

Exercises – 16

4. 99.820/99.733 mm diameter
5. T. P. centre line of slot from new datum 35.684 mm new posn. tol. for slot $±0.04$
 at MMC (alternatives possible)
7. $20°31'$; 0.0908 mm
8. 0.037 mm

Exercises – 17

3. (i) 47.502 mm;   (ii) $\sigma = 0.009$;   (iii) medium precision
4. Method 4: $\bar{X} = 2.34\ \mu$, $\sigma = 0.98\ \mu$
6. $\bar{x} = 84.599$ mm: $x = 0.011$ mm: $T > 8\sigma$ high precision
8. (i) RPI $= 2.84$;   (ii) mean, 55 mm $± 0.104/±0.067$; range, 0.389/0.319 mm
10. Producer's risk, 0.02; consumer's risk, 0.42
11. 0.983, 0.785, 0.446, 0.067
14. (a) 0.662
 (b) 290

16. (a) $p_a = 0.423$ 0.4271
       $p_r = 0.185$ 0.5729
   (b) $p_a = 0.809$ 0.8364
       $p_r = 0.019$ 0.1636
       $p_a = 0.238$ 0.3386
       $p_r = 0.311$ 0.7614
17. Hole, 25.028/25.000 mm diameter; shaft 24.9985/24.9815 mm diameter

Index

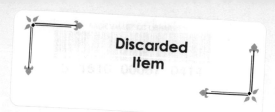